Passivity of Complex Dynamical Networks

Jin-Liang Wang · Huai-Ning Wu ·
Shun-Yan Ren

Passivity of Complex Dynamical Networks

Analysis, Control and Applications

Springer

Jin-Liang Wang
Tianjin Key Laboratory of Autonomous Intelligence Technology and Systems
School of Computer Science and Technology
Tiangong University
Tianjin, China

Huai-Ning Wu
Science and Technology on Aircraft Control Laboratory, School of Automation Science and Electrical Engineering
Beihang University
Beijing, China

Shun-Yan Ren
School of Mechanical Engineering
Tiangong University
Tianjin, China

ISBN 978-981-33-4286-6 ISBN 978-981-33-4287-3 (eBook)
https://doi.org/10.1007/978-981-33-4287-3

This Springer imprint is published by the registered company Springer Nature Singapore Pte Ltd.
The registered company address is: 152 Beach Road, #21-01/04 Gateway East, Singapore 189721, Singapore

Preface

Recently, the passivity problem for complex dynamical networks (CDNs) has received considerable attention due to its wide applications in various areas such as stability, complexity, signal processing, synchronization, and fuzzy control. But, the network models considered in the majority of existing studies have the same dimension of input and output vectors. Practically, the dimensions of input and output vectors in many real networks are different. Obviously, it is more interesting to study the passivity of CDNs with different dimensions of input and output. Moreover, the passivity is an advantageous tool to analyze the synchronization of CDNs. However, in many circumstances, CDNs need to realize synchronization in a finite time. Therefore, in order to better study the finite-time synchronization (FTS) of CDNs, it's essential to further investigate the finite-time passivity (FTP) for CDNs. On the other hand, in most existing works on the passivity of CDNs, there is a simplified assumption that the node state is only dependent on the time. In fact, such simplification does not match the peculiarities of some real networks such as food webs, reaction-diffusion neural networks (RDNNs), coupled reaction-diffusion neural networks (CRDNNs), chemical reactions, and so on. Evidently, it is also very challenging and interesting to study the passivity of partial differential CDNs, in which the input and output variables are varied with the time and space variables. Therefore, the investigation of the passivity about ordinary differential CDNs and partial differential CDNs has both practical and theoretical significance.

The aim of this book is to introduce recent research work on analysis, control and applications of passivity for CDNs. The book is organized as follows:

Chapter 1: The background of passivity for ordinary differential CDNs and partial differential CDNs is introduced, as well as the organization of this book, and some important definitions and useful lemmas are also provided in this chapter.

Chapter 2: This chapter investigates the input strict passivity and the output strict passivity for a generalized complex network with nonlinear, time-varying, non-symmetric, and delayed coupling. By constructing some suitable Lyapunov functionals, several sufficient conditions for ensuring the input strict passivity and the output strict passivity are derived for the CDN. Finally, two numerical examples are given to show the effectiveness of the obtained results.

Chapter 3: This chapter considers a multi-weighted complex network model with different dimensions of output and input vectors. First, we analyze the passivity of the proposed network model by employing some inequality techniques and Lyapunov functional method, and give a synchronization condition for the output strictly passive CDN with multi-weights (CDNMWs). Furthermore, by using pinning adaptive strategies to control a small portion of nodes or edges, some sufficient conditions for ensuring the passivity of CDNMWs are presented. In addition, when the CDNMWs under the designed pinning adaptive control strategies is the output strictly passive, two synchronization criteria are also established. Finally, two examples are given to show the validity of the obtained results.

Chapter 4: This chapter focuses on two kinds of CDNs with state and derivative coupling, respectively. Firstly, some important concepts about FTP, finite-time output strict passivity, and finite-time input strict passivity are introduced. By making use of state feedback controllers and adaptive state feedback controllers, several sufficient conditions are given to guarantee the FTP of these two network models. On the other hand, based on the obtained FTP results, some FTS criteria for the CDNs with state and derivative coupling are gained. Finally, two simulation examples are proposed to verify the availability of the derived results.

Chapter 5: This chapter studies the FTP problem for two classes of coupled neural networks (CNNs) with adaptive coupling weights. By selecting appropriate adaptive laws and controllers, several FTP conditions are given for CNNs with undirected and directed topologies. Furthermore, some FTS conditions are also established by employing the FTP of the CNNs. At last, two numeral examples are used to check the correctness of the obtained criteria.

Chapter 6: This chapter, respectively, studies the FTP of multi-weighted coupled neural networks (MWCNNs) with and without coupling delays by designing appropriate controllers. In addition, some sufficient conditions to guarantee the FTS of these MWCNNs are obtained under the condition that the MWCNNs are finite-time passive. Finally, two examples are also given to verify the proposed FTP and FTS criteria.

Chapter 7: This chapter is concerned with the passivity and stability problems of RDNNs, in which, the input and output variables are varied with the time and space variables. By utilizing the Lyapunov functional method combined with the inequality techniques, some sufficient conditions for ensuring the passivity and global exponential stability of RDNNs are derived. Furthermore, when the parameter uncertainties appear in RDNNs, several criteria for robust passivity and robust global exponential stability are also presented. Finally, a numerical example is provided to illustrate the effectiveness of the proposed criteria.

Chapter 8: In this chapter, we investigate the passivity problem of a class of delayed reaction-diffusion networks (DRDNs), in which the input and output variables are varied with the time and space variables. Firstly, we establish several criteria for the passivity, input strict passivity and output strict passivity by using the Lyapunov functional method combined with the inequality techniques, and also reveal the relationship between passivity and stability of the reaction-diffusion networks. Moreover, these results are applied to investigate the asymptotic stability and control design of a food web model with time-delay and reaction-diffusion terms. Numerical simulations are presented finally to demonstrate the effectiveness of the proposed results.

Chapter 9: This chapter presents a complex network model consisting of N uncertain RDNNs with multiple time-delays. We analyze the passivity and synchronization of the proposed network model, and derive several passivity and synchronization criteria based on some inequality techniques. In addition, by considering the difficulty in achieving passivity (synchronization) in such a network, an adaptive control scheme is also developed to ensure that the proposed network achieves passivity (synchronization). Finally, we design two numerical examples to verify the effectiveness of the derived passivity and synchronization criteria.

Tianjin, China — Jin-Liang Wang
Beijing, China — Huai-Ning Wu
Tianjin, China — Shun-Yan Ren
September 2020

Acknoweldgements This book was supported by the National Natural Science Foundation of China under Grant 61773285, the Tianjin Talent Development Special Support Program for Young Top-Notch Talent, the Natural Science Foundation of Tianjin, China, under Grant 19JCYBJC18700, and the Program for Innovative Research Team in University of Tianjin (No. TD13-5032). I'd like to begin by acknowledging my postgraduates Xiao Han, Jie Hou, Chen-Guang Liu, Rui Li, Han-Yu Wu, Rong-Guo Liang, Fang Wu, Min Cao, and Xin-Yu Du who have unselfishly given their valuable time in arranging these raw materials into something I'm proud of.

The original version of this book was revised: The belated corrections have been incorporated. The corrections to this book are available at https://doi.org/10.1007/978-981-33-4287-3_10

Contents

Symbols and Acronyms

$\mathbb{N}$	1, 2, 3, ...
$\mathbb{R}$	Field of real numbers
$\mathbb{R}^+$	Field of nonnegative real numbers
$\mathbb{R}^n$	n-dimensional Euclidean space
$\mathbb{R}^{m\times n}$	Space of all $m \times n$ real matrices
I_n	$n \times n$ real identity matrix
Ω	$\Omega = \{x = (x_1, x_2, \ldots, x_\sigma)^T \| |x_m| < l_m, m = 1, 2, \ldots, \sigma\}$ is an open bounded domain in $\mathbb{R}^\sigma$ with smooth boundary $\partial\Omega$
$\bar{\Omega}$	$\Omega \cup \partial\Omega$
mesΩ	The measure of Ω
$L^2(\Omega)$	The space of real functions on Ω which are L^2 for the Lebesgue measure
$\mathcal{G}$	$\mathcal{G} = \{\mathcal{D}, \mathcal{B}\}$ represents a directed graph, which is made up of a set of nodes $\mathcal{D} = \{1, 2, \ldots, N\}$ and a set of directed edges $\mathcal{B} \subseteq \mathcal{D} \times \mathcal{D}$
$P > 0$	Symmetric positive definite
$P \geqslant 0$	Symmetric positive semi-definite
$P < 0$	Symmetric negative definite
$P \leqslant 0$	Symmetric negative semi-definite
B^{T}	Transpose of matrix B
B^{-1}	Inverse of matrix B
$\otimes$	Kronecker product of two matrices
$\lambda_M(A)$	Maximum eigenvalue of matrix A
$\lambda_m(A)$	Minimum eigenvalue of matrix A
diag$(\cdots)$	Block-diagonal matrix
$\|\| \cdot \|\|$	Euclidean norm of a vector
$\|\|e(\cdot,t)\|\|_2$	$\|\|e(\cdot,t)\|\|_2 = (\int_\Omega \sum_{i=1}^n e_i^2(x,t)dx)^{\frac{1}{2}}, e(x,t) = (e_1(x,t), e_2(x,t), \ldots, e_n(x,t))^T \in \mathbb{R}^n, (x,t) \in \Omega \times \mathbb{R}$
$\|\|\Theta(\cdot,t)\|\|_\tau$	$\|\|\Theta(\cdot,t)\|\|_\tau = \sup_{-\tau\leqslant\theta\leqslant 0}\|\|\Theta(\cdot,t+\theta)\|\|_2, \Theta(x,t) = (\theta_1(x,t), \theta_2(x,t), \ldots, \theta_n(x,t))^T \in \mathbb{R}^n$
C	Set of continuous functions
$\| \cdot \|$	Absolute value
sup	Supremum

inf	Infimum
CDDNs	Complex delayed dynamical networks
CDNMWs	Complex dynamical networks with multi-weights
CDNs	Complex dynamical networks
CGNNs	Cohen-Grossberg neural networks
CNNs	Coupled neural networks
CRDNNs	Coupled reaction-diffusion neural networks
CURDNNs	Coupled uncertain reaction-diffusion neural networks
DNNs	Delayed neural networks
DRDNs	Delayed reaction-diffusion networks
FTISP	Finite-time input strict passivity
FTOSP	Finite-time output strict passivity
FTP	Finite-time passivity
FTS	Finite-time synchronization
IRDNNs	Interval RDNNs
MWCNNs	Multi-weighted coupled neural networks
NNs	Neural networks
RDNNs	Reaction-diffusion neural networks

Chapter 1
Introduction

1.1 Background

1.1.1 Passivity for Ordinary Differential CDNs

Today various complex networks can be seen everywhere and are becoming an important part of our daily life. Some of the most well-known examples include food webs, communication networks, social networks, power grids, cellular networks, World Wide Web, metabolic systems, disease transmission networks, etc. Therefore, the dynamical behavior of complex dynamical networks (CDNs) have been extensively studied by the researchers. Passivity describes the input and output characteristics of the systems, which can guarantee the internal stability of the systems to some extent, and has found successful applications in diverse areas [1–27] such as stability [23], complexity [24], signal processing [4], chaos control and synchronization [25, 26], and fuzzy control [27]. Evidently, it is very significant and interesting to study the passivity of the CDNs, and a wide variety of passivity criteria have been presented for various complex networks.

Unfortunately, in most existing works on the passivity of complex networks, they always assume that the node input has the same dimension as the output vector. Practically, the dimensions of input and output vectors in many real networks are different. Therefore, it is interesting to further study the passivity of CDNs, in which the input and output have different dimensions. But, there are very few works about the passivity of CDNs with different dimensions of input and output [28, 29]. On the other hand, passivity theory provides a powerful tool to analyze the synchronization of CDNs. But, in many practical applications, synchronization is realized over a finite time interval may be more reasonable. Therefore, it is more meaningful and valuable to study the finite-time synchronization (FTS) of CDNs [30–37]. Obviously, it is very important to define the finite-time passivity (FTP) for studying the FTS of

---

The original version of this book was revised: The belated corrections have been incorporated. The corrections to this book are available at https://doi.org/10.1007/978-981-33-4287-3_10

J.-L. Wang et al., *Passivity of Complex Dynamical Networks*,
https://doi.org/10.1007/978-981-33-4287-3_1

CDNs. Regretfully, very few results about FTP of CDNs have been reported [38–41]. Therefore, it is very interesting and important to further investigate the FTP of CDNs.

As we all known, most of existing results on the passivity are all based on the network models with single weight for not only the ordinary differential equation network models but also the partial differential equation network models. In fact, many real-world networks can be described by CDN models with multi-weights, where nodes are coupled by multiple coupling forms, such as transportation networks, complex biology networks, social networks, and so on. For instance, individuals or organizations can get in touch with each other by email, mobile phone, Facebook, etc. Considering that every way of communication has different influence, social networks should be modeled by multi-weighted complex networks. Furthermore, in the public traffic roads networks, taking every bus line as the network node, the edge means that two different bus lines have the same bus stops. In this case, there obviously exist different coupling weights in public traffic roads networks, such as coefficient of bus line length, passenger flow density, departing frequency, and so on. Consequently, it is of great importance to investigate the multi-weighted CDNs. However, very few researchers have studied the passivity and FTP problems of CDNs with multi-weights (CDNMWs).

1.1.2 Passivity for Partial Differential CDNs

In most existing works on the complex networks, there is a simplified assumption that the node state is only dependent on the time. In fact, such simplification does not match the peculiarities of some real networks. As a well-known example of complex networks, food webs attract increasing attention of researchers from different fields in recent years. A food web can be characterized by a model of complex network, in which a node represents a species. To our knowledge, species are usually inhomogeneously distributed in a bounded habitat and the different population densities of predators and preys may cause different population movements, thus it is important and interesting to investigate their spatial density in order to better protect and control their population. In such a case, the state variable of node will represent the spatial density of the species. Obviously, the spatial density of the species is seriously dependent on the time and space. In the modeling of chemical reactions, it is also essential to consider the diffusion effects. Practically, there are many reaction-diffusion phenomena in nature and discipline fields. Evidently, it is important to study the CDNs with spatially and temporally varying state variables [42–44]. Moreover, the input and output variables are also dependent on the time and space in many spatially and temporally CDNs. Therefore, it is important and interesting to study the passivity of partial differential CDNs, in which the input and output variables are varied with the time and space variables.

More recently, some authors have discussed the passivity for partial differential CDNs with spatially and temporally varying input and output variables [45, 46]. In [45], the passivity of a parabolic complex network model with time-varying delays and parametric uncertainties was analyzed. Wei et al. [46] investigated the passivity of impulsive coupled reaction-diffusion neural networks with and without time-varying delay, and several input strict passivity and output strict passivity conditions were derived for coupled reaction-diffusion neural networks (CRDNNs) by constructing suitable Lyapunov functionals and utilizing some inequality techniques. But, in [45, 46], the output and input dimensions are always assumed to be the same. To our knowledge, there are very few works on the passivity of partial differential CDNs with different dimensions of input and output. Moreover, many real-world networks such as public traffic roads networks, social networks, and so on, are supposed to be described by network models that have multiple coupling delays. Obviously, it is very meaningful to further investigate the passivity problem for partial differential CDNs with multiple delay couplings, but very few results on this topic have been reported.

1.2 Book Organization

As one of the most significant and interesting dynamical properties of the CDNs, passivity has received much attention. A wide variety of passivity criteria have been presented for various complex networks, but there are still some interesting and challenging problems deserving further investigation. Thus, the main aim of this book is to introduce some recent results on analysis and control of passivity for ordinary differential CDNs and partial differential CDNs, and this book can serve as a stepping stone to study the passivity of CDNs. The rest of this book is organized as follows:

Chapter 2 introduces a generalized complex network with non-linear, time-varying, non-symmetric, and delayed coupling, and considers the input strict passivity and output strict passivity of the proposed network model. By constructing suitable Lyapunov functionals, several input strict passivity and output strict passivity criteria are established for the complex network.

Chapter 3 focuses on the passivity problem for a CDNMWs with different dimensions of input and output vectors. By employing Lyapunov functional method and some inequality techniques, several passivity criteria are presented for such network model. Furthermore, the passivity of CDNMWs is studied by pinning a fraction of nodes or edges with adaptive strategies. In addition, some synchronization criteria for CDNMWs are also established by exploiting the output strict passivity.

Chapter 4 presents two complex network models, in which nodes are coupled by the states or derivative of the states. On one hand, the FTP, finite-time input strict passivity and finite-time output strict passivity of the presented network models are discussed by making use of adaptive state feedback controllers and state feedback controllers. Moreover, some FTS criteria for these network models are derived on the basis of the obtained FTP results.

Chapter 5 discusses the FTP for adaptive coupled neural networks with undirected and directed topologies. On one hand, several sufficient conditions for ensuring the FTP of directed and undirected coupled neural networks (CNNs) are derived by designing appropriate adaptive laws and controllers. On the other hand, the relationship between FTP and FTS of the adaptive coupled neural networks with undirected and directed topologies are also revealed.

Chapter 6 proposes two kinds of multi-weighted coupled neural networks (MWCNNs) with and without coupling delays. On one hand, the FTP of the proposed network models is analyzed, and some sufficient conditions are established. On the other hand, several FTS criteria for finite-time passive MWCNNs with and without coupling delays are presented.

Chapter 7 investigates the passivity and global exponential stability of reaction-diffusion neural networks (RDNNs) with Dirichlet boundary conditions, and several sufficient conditions are obtained by utilizing the Lyapunov functional method combined with the inequality techniques. Furthermore, when the parameter uncertainties appear in RDNNs, some robust passivity and robust global exponential stability conditions are also derived.

Chapter 8 gives some passivity, input strict passivity and output strict passivity conditions and reveal the relationship between passivity and stability for a class of delayed reaction-diffusion networks (DRDNs), in which the input and output variables are varied with the time and space variables. Moreover, these results are applied to investigate the asymptotic stability and control design of a food web model.

Chapter 9 proposes a CDN model consisting of N uncertain RDNNs with multiple time-delays. Based on some inequality techniques, several passivity and synchronization criteria for the proposed network model are established. Moreover, an adaptive control strategy is given to guarantee that the network can achieve passivity and synchronization.

1.3 Some Definitions

Definition 1.1 (*see* [47]) Function class QUAD(Δ, P): let $P = \text{diag}(p_1, p_2, \ldots, p_n)$ be a positive definite diagonal matrix and $\Delta = \text{diag}(\delta_1, \delta_2, \ldots, \delta_n)$ be a diagonal matrix. QUAD(Δ, P) denotes a class of continuous functions $f(x,t) : \mathbb{R}^n \times [0, +\infty) \to \mathbb{R}^n$ satisfying

$$(x-y)^T P\{[f(x,t) - f(y,t)] - \Delta(x-y)\} \leqslant -\eta(x-y)^T(x-y)$$

for some $\eta > 0$, all $x, y \in \mathbb{R}^n$ and $t > 0$.

It can be verified that many of the benchmark chaotic systems belong to "Function class QUAD" [48], such as the Lorenz system [49], the Chen system [50] and the Lü system [51].

Definition 1.2 (*see* [52]) Let $A = (a_{ij})_{m\times n} \in \mathbb{R}^{m\times n}$ and $B = (b_{ij})_{p\times q} \in \mathbb{R}^{p\times q}$. Then, the Kronecker product of A and B is defined as the matrix

$$A \otimes B = \begin{bmatrix} a_{11}B & a_{12}B & \cdots & a_{1n}B \\ a_{21}B & a_{22}B & \cdots & a_{2n}B \\ \vdots & \vdots & \cdots & \vdots \\ a_{m1}B & a_{m2}B & \cdots & a_{mn}B \end{bmatrix} \in \mathbb{R}^{mp\times nq}.$$

The Kronecker product has the following properties:

$$\begin{aligned} &(1)\ (A\otimes B)^T = A^T \otimes B^T; \\ &(2)\ (\alpha A)\otimes B = A\otimes(\alpha B); \\ &(3)\ (A+B)\otimes C = A\otimes C + B\otimes C; \\ &(4)\ (A\otimes B)(C\otimes D) = (AC)\otimes(BD), \end{aligned}$$

where $\alpha \in \mathbb{R}$, C, D are matrices with suitable dimensions.

1.4 Some Lemmas

Lemma 1.3 (see [53]) *Assume that an irreducible matrix* $A = (A_{ij})_{N\times N} \in \mathbb{R}^{N\times N}$ *satisfies* $\sum_{j=1}^{N} A_{ij} = 0$ *with* $A_{ij} \geqslant 0 (i \neq j)$. *Then, there exists a positive vector* $\kappa = (\kappa_1, \kappa_2, \ldots, \kappa_N)^T \in \mathbb{R}^N$ *such that*

(1) $A^T\kappa = 0$;
(2) $\tilde{A} = \Psi A + A^T \Psi$ *is symmetric and*

$$\sum_{j=1}^{N} \tilde{A}_{ij} = \sum_{j=1}^{N} \tilde{A}_{ji} = 0$$

for all $i = 1, 2, \ldots, N$, *where* $\Psi = \mathrm{diag}(\kappa_1, \kappa_2, \ldots, \kappa_N)$.

Lemma 1.4 *For any vectors* $x, y \in \mathbb{R}^n$ *and* $n \times n$ *square matrix* $Q > 0$, *the following matrix inequality holds:*

$$x^T y + y^T x \leqslant x^T Q x + y^T Q^{-1} y.$$

Lemma 1.5 (see [54]) *Let* $g(x)$ *be a real-valued function belonging to* $C^1(\Omega)$ *where* $g(x)|_{\partial\Omega} = 0$. *Then,*

$$\int_{\Omega} g^2(x)dx \leqslant l_m^2 \int_{\Omega} \left(\frac{\partial g(x)}{\partial x_m}\right)^2 dx, m = 1, 2, \ldots, \sigma.$$

Lemma 1.6 (see [55]) *Suppose that a nonnegative and continuous function $S(t)$ meets*

$$\dot{S}(t) \leqslant -\zeta S^{\iota}(t),\ t \geqslant 0,$$

where $0 < \iota < 1$ *and* $\mathbb{R} \ni \zeta > 0$. *Then, we derive*

$$S^{1-\iota}(t) \leqslant S^{1-\iota}(0) - \zeta(1-\iota)t,\ 0 \leqslant t \leqslant \nu,$$

and

$$S(t) = 0,\ t \geqslant \nu$$

with ν *given by*

$$\nu = \frac{S^{1-\iota}(0)}{\zeta(1-\iota)}.$$

Lemma 1.7 (see [56]) *The following inequality holds:*

$$(|\Phi_1| + |\Phi_2| + \cdots + |\Phi_n|)^{\kappa} \leqslant |\Phi_1|^{\kappa} + |\Phi_2|^{\kappa} + \cdots + |\Phi_n|^{\kappa}$$

for any $\Phi_i \in \mathbb{R}, 0 < \kappa \leqslant 1, i = 1, \ldots, n$.

References

1. López, N., & María, E. (2005). Several dissipativity and passivity implications in the linear discrete-time setting. *Mathematical Problems in Engineering*, *6*, 599–616.
2. López, E. N., & Fossas, E. (2004). Feedback passivity of nonlinear discrete-time systems with direct input-output link. *Automatica*, *40*(8), 1423–1428.
3. Niculescu, S., & Lozano, R. (2001). On the passivity of linear delay systems. *IEEE Transactions on Automatic Control*, *46*(3), 460–464.
4. Xie, L., Fu, M., & Li, H. (1998). Passivity analysis and passification for uncertain signal processing systems. *IEEE Transactions on Signal Processing*, *46*(9), 2394–2403.
5. Mahmoud, M., & Ismail, A. (2004). Passivity and passification of time-delay systems. *Journal of Mathematical Analysis and Applications*, *292*(1), 247–258.
6. Zhao, J., & Hill, D. (2008). Passivity and stability of switched systems: A multiple storage function method. *Systems and Control Letters*, *57*(2), 158–164.
7. Moylan, P. J., & Hill, D. J. (1980). Dissipative dynamical systems: Basic input-output and state properties. *Journal of the Franklin Institute*, *309*(5), 327–357.
8. Wang, J. L., Wu, H. N., & Yang, Z. C. (2011). Passivity analysis of impulsive complex networks. *International Journal of Automation and Computing*, *8*(4), 484–489.
9. Gao, H., Chen, T., & Chai, T. (2007). Passivity and passification for networked control systems. *SIAM Journal on Control and Optimization*, *46*(4), 1299–1322.

10. Chen, J. L., & Lee, L. (2001). Passivity approach to feedback connection stability for discrete-time descriptor systems. In *Proceedings of the 40th IEEE Conference on Decision and Control (Cat. No.01CH37228)* (Vol. 3, pp. 2865–2866).
11. Liu, X. (2007). Passivity analysis of uncertain fuzzy delayed systems. *Chaos, Solitons and Fractals*, *34*(3), 833–838.
12. Uang, H. J. (2005). On the dissipativity of nonlinear systems: Fuzzy control approach. *Fuzzy Sets and Systems*, *156*(2), 185–207.
13. Chang, W. J., Ku, C. C., Huang, P. H., & Chang, W. (2009). Fuzzy controller design for passive continuous-time affine T-S fuzzy models with relaxed stability conditions. *ISA Transactions*, *48*(3), 295–303.
14. Li, C., Zhang, H., & Liao, X. (2006). Passivity and passification of fuzzy systems with time delays. *Computers and Mathematics with Applications*, *52*(6), 1067–1078.
15. Lou, X., & Cui, B. (2007). Passivity analysis of integro-differential neural networks with time-varying delays. *Neurocomputing*, *70*(4), 1071–1078.
16. Park, J. (2007). Further results on passivity analysis of delayed cellular neural networks. *Chaos, Solitons and Fractals*, *34*(5), 1546–1551.
17. Chen, B., Li, H., Lin, C., & Zhou, Q. (2009). Passivity analysis for uncertain neural networks with discrete and distributed time-varying delays. *Physics Letters A*, *373*(14), 1242–1248.
18. Song, Q., Liang, J., & Wang, Z. (2009). Passivity analysis of discrete-time stochastic neural networks with time-varying delays. *Neurocomputing*, *72*(7–9), 1782–1788.
19. Song, Q., & Wang, Z. (2010). New results on passivity analysis of uncertain neural networks with time-varying delays. *International Journal of Computer Mathematics*, *87*(3), 668–678.
20. Li, C., & Liao, X. (2005). Passivity analysis of neural networks with time delay. *IEEE Transactions on Circuits and Systems II: Express Briefs*, *52*(8), 471–475.
21. Yao, J., Wang, H. O., Guan, Z. H., & Xu, W. (2009). Passive stability and synchronization of complex spatio-temporal switching networks with time delays. *Automatica*, *45*(7), 1721–1728.
22. Yao, J., Guan, Z. H., & Hill, D. J. (2009). Passivity-based control and synchronization of general complex dynamical networks. *Automatica*, *45*(9), 2107–2113.
23. Hill, D. J., & Moylan, P. J. (1977). Stability results for nonlinear feedback systems. *Automatica*, *13*(4), 377–382.
24. Chua, L. O. (1999). Passivity and complexity. *IEEE Transactions on Circuits and Systems I: Fundamental Theory and Applications*, *46*(1), 71–82.
25. Yu, W. (1999). Passive equivalence of chaos in lorenz system. *IEEE Transactions on Circuits and Systems I: Fundamental Theory and Applications*, *46*(7), 876–878.
26. Wu, C. W. (2001). Synchronization in arrays of coupled nonlinear systems: Passivity, circle criterion, and observer design. *IEEE Transactions on Circuits and Systems I: Fundamental Theory and Applications*, *48*(10), 1257–1261.
27. Calcev, G., Gorez, R., & Neyer, M. D. (1998). Passivity approach to fuzzy control systems. *Automatica*, *34*(3), 339–344.
28. Wang, J., Wu, H., Huang, T., Ren, S., & Wu, J. (2017). Passivity analysis of coupled reaction-diffusion neural networks with Dirichlet boundary conditions. *IEEE Transactions on Systems, Man, and Cybernetics: Systems*, *47*(8), 2148–2159.
29. Wang, J. L., Wu, H. N., Huang, T., Ren, S. Y., & Wu, J. (2017). Passivity of directed and undirected complex dynamical networks with adaptive coupling weights. *IEEE Transactions on Neural Networks and Learning Systems*, *28*(8), 1827–1839.
30. Qiu, S. H., Huang, Y. L., & Ren, S. Y. (2018). Finite-time synchronization of multi-weighted complex dynamical networks with and without coupling delay. *Neurocomputing*, *275*, 1250–1260.
31. Lu, X. D., Zhang, X. F., & Liu, Q. R. (2018). Finite-time synchronization of nonlinear complex dynamical networks on time scales via pinning impulsive control. *Neurocomputing*, *275*, 2104–2110.
32. Li, J. R., Jiang, H. J., Hu, C., & Yu, J. (2018). Analysis and discontinuous control for finite-time synchronization of delayed complex dynamical networks. *Chaos, Solitons and Fractals*, *114*, 291–305.

33. Li, H. L., Cao, J. D., Jiang, H. J., & Alsaedi, A. (2018). Graph theory-based finite-time synchronization of fractional-order complex dynamical networks. *Journal of the Franklin Institute, 355*(13), 5711–5789.
34. He, G., Fang, J. A., & Li, Z. (2017). Finite-time synchronization of cyclic switched complex networks under feedback control. *Journal of the Franklin Institute, 354*(9), 3780–3796.
35. Ren, H. W., Deng, F. Q., & Peng, Y. J. (2018). Finite time synchronization of Markovian jumping stochastic complex dynamical systems with mix delays via hybrid control strategy. *Neurocomputing, 272*, 683–693.
36. Mei, J., Jiang, M. H., Xu, W. M., & Wang, B. (2013). Finite-time synchronization control of complex dynamical networks with time delay. *Communications in Nonlinear Science and Numerical Simulation, 18*(9), 2462–2478.
37. Xu, Y. H., Zhou, W. N., Fang, J. A., Xie, C. R., & Tong, D. B. (2016). Finite-time synchronization of the complex dynamical network with non-derivative and derivative coupling. *Neurocomputing, 173*, 1356–1361.
38. Wang, J. L., Xu, M., Wu, H. N., & Huang, T. (2018). Finite-time passivity of coupled neural networks with multiple weights. *IEEE Transactions on Network Science and Engineering, 5*(3), 184–197.
39. Huang, Y. L., Chen, W. Z., & Wang, J. M. (2018). Finite-time passivity of delayed multi-weighted complex dynamical networks with different dimensional nodes. *Neurocomputing, 312*, 74–89.
40. Wang, J., Zhang, X., Wu, H., Huang, T., & Wang, Q. (2019). Finite-time passivity and synchronization of coupled reaction-diffusion neural networks with multiple weights. *IEEE Transactions on Cybernetics, 49*(9), 3385–3397.
41. Rajavel, S., Samidurai, R., Cao, J. D., Alsaedi, A., & Ahmad, B. (2017). Finite-time non-fragile passivity control for neural networks with time-varying delay. *Applied Mathematics and Computation, 297*, 145–158.
42. Balasubramaniam, P., & Vidhya, C. (2010). Global asymptotic stability of stochastic BAM neural networks with distributed delays and reaction-diffusion terms. *Journal of Computational and Applied Mathematics, 234*(12), 3458–3466.
43. Song, Q., & Wang, Z. (2009). Dynamical behaviors of fuzzy reaction-diffusion periodic cellular neural networks with variable coefficients and delays. *Applied Mathematical Modelling, 33*(9), 3533–3545.
44. Hu, C., Jiang, H., & Teng, Z. (2010). Impulsive control and synchronization for delayed neural networks with reaction-diffusion terms. *IEEE Transactions on Neural Networks, 21*(1), 67–81.
45. Wang, J.-L., & Wu, H.-N. (2012). Robust stability and robust passivity of parabolic complex networks with parametric uncertainties and time-varying delays. *Neurocomputing, 87*, 26–32.
46. Wei, P.-C., Wang, J.-L., Huang, Y.-L., Xu, B.-B., & Ren, S.-Y. (2015). Passivity analysis of impulsive coupled reaction-diffusion neural networks with and without time-varying delay. *Neurocomputing, 168*, 13–22.
47. Lu, W., & Chen, T. (2006). New approach to synchronization analysis of linearly coupled ordinary differential systems. *Physica D, 213*(2), 214–230.
48. Guo, W., Austin, F., & Chen, S. (2010). Global synchronization of nonlinearly coupled complex networks with non-delayed and delayed coupling. *Communications in Nonlinear Science and Numerical Simulation, 15*(6), 1631–1639.
49. Lorenz, E. N. (2004). *Deterministic Nonperiodic Flow*. New York, NY: Springer.
50. Chen, G., & Ueta, T. (1999). Yet another chaotic attractor. *International Journal of Bifurcation and Chaos, 9*(7), 1465–1466.
51. Lu, J., & Chen, G. (2002). A new chaotic attractor coined. *International Journal of Bifurcation and Chaos, 12*(3), 659–661.
52. Laub, A. (2005). Matrix Analysis for Scientists and Engineers. *Society for Industrial and Applied Mathematics*.
53. Yu, W., Chen, G., Lu, J., & Kurths, J. (2013). Synchronization via pinning control on general complex networks. *SIAM Journal on Control and Optimization, 51*(2), 1395–1416.

54. Lu, J. G. (2008). Global exponential stability and periodicity of reaction-diffusion delayed recurrent neural networks with Dirichlet boundary conditions. *Chaos, Solitons and Fractals*, *35*(1), 116–125.
55. Tang, Y. (1998). Terminal sliding mode control for rigid robots. *Automatica*, *34*(1), 51–56.
56. Huang, X., Lin, W., & Yang, B. (2005). Global finite-time stabilization of a class of uncertain nonlinear systems. *Automatica*, *41*(5), 881–888.

Chapter 2
Passivity Analysis of CDNs With Multiple Time-Varying Delays

2.1 Introduction

In the real world, complex networks can be seen everywhere, and have been viewed as a fundamental tool in understanding dynamical behavior and the response of real systems such as food webs, communication networks, social networks, power grids, cellular networks, World Wide Web, metabolic systems, disease transmission networks, and many others. The complex networks have been extensively studied by researchers in recent years, and many interesting results have been derived.

Many existing studies focused on investigating the geometry features [1–3], synchronization [4–15], control [16–22] and robust stability [23, 24] of complex networks. It should be pointed out that the above mentioned literature on complex networks are considered under some simplified assumptions. For instance, the coupling among the nodes of complex networks are linear [1, 6, 9–22], time invariant [1, 6, 9–12, 14–22], symmetric [6, 9, 13–15, 17] and non-delay [1, 6, 12, 14–16, 20, 21]. In fact, such simplification does not match the peculiarities of real networks in many circumstances. Firstly, the interplay of two different nodes in a network can not be described accurately by linear functions of their states because they are naturally nonlinear functions of states. Secondly, many real-world networks are more likely to have different coupling strengths for different connections, and coupling strengths are frequently varied with time. Moreover, the coupling topology is likely to directed and weighted in many real-world networks such as the food web, metabolic networks, World-Wide-Web, epidemic networks, document citation networks. In addition, it is well known that a signal or influence traveling through a complex network often is associated with time-delays due to the finite speeds of transmission and spreading as well as traffic congestions, this is very common in biological and physical networks, and absolute constant delay may be scarce and delays are frequently varied with time. Generally speaking, different nodes have different time-delay vectors and the elements of each node also have different time-varying delays. Therefore, a general model of CDN with nonlinear, time-varying, nonsymmetric and delayed coupling should be more accurate to describe the evolutionary process of the nodes.

J.-L. Wang et al., *Passivity of Complex Dynamical Networks*,
https://doi.org/10.1007/978-981-33-4287-3_2

Passivity [25–51] is an important concept of system theory and provides a nice tool for analyzing the stability of systems, and has found applications in diverse areas such as stability [47], complexity [48], signal processing [28], chaos control and synchronization [49, 50], and fuzzy control [51]. Therefore, the passivity theory has received lots of attention. In recent years, many researchers have studied the passivity of fuzzy systems [35–38] and NNs [39–44]. In [30], Zhao and Hill presented a concept of passivity for switched systems using multiple storage functions, and generalized Branicky's stability theorem of multiple Lyapunov functions by relaxing the non-increasing condition on values of Lyapunov-like functions. Song, Liang and Wang [42] investigated the passivity for a class of discrete-time stochastic neural networks with time-varying delays. They considered a generalized activation function, derived a delay-dependent passivity condition by constructing proper Lyapunov-Krasovskii functional and using stochastic analysis technique, and the case that parameter uncertainties was also considered [42]. To our knowledge, there are few works on the passivity of complex networks [45, 46], in which Yao et al. obtained some sufficient conditions on passivity properties for linear (or linearized) complex networks with and without coupling delays (constant delay). It is well known that the phenomena of time-delays are common in complex networks, and most of delays is notable. So it is crucial for us to take the delay into the consideration when we study complex networks. Moreover, absolute constant delay may be scarce and delays are frequently varied with time. Therefore, it is necessary to further study the passivity of complex networks with multiple time-varying delays. In [34], Chen and Lee developed input strict passivity to study both input-output and internal stability problems for feedback connections of discrete-time linear descriptor systems. By applying the input strict passivity and Lyapunov theory, Chang et al. [37] derived the relaxed stability conditions for continuous-time affine Takagi-Sugeno (T-S) fuzzy models. To the best of our knowledge, the input strict passivity of complex networks with time-varying delays has not yet been established. Therefore, it is interesting to study the input strict passivity of complex networks with multiple time-varying delays. As a natural extension of input strict passivity, we also introduce the output strict passivity in order to better study dynamical behavior of complex networks.

Motivated by the above discussions, we formulate a delayed dynamical network model with general topology. The objective of this chapter is to study the input strict passivity and output strict passivity of the proposed network model. Some sufficient conditions on input strict passivity and output strict passivity of CDN are obtained by Lyapunov functional method.

2.2 Network Model and Preliminaries

In this chapter, we consider a complex network consisting of N identical nodes with diffusive and delay coupling. Let $x_i(t) = (x_{i1}(t), x_{i2}(t), \ldots, x_{in}(t))^T \in \mathbb{R}^n$ denote the state vector of node i, $y_i(t) \in \mathbb{R}^n$ denote the output vector of node i, and $u_i(t) \in \mathbb{R}^n$ denote the input vector of node i, the mathematical model of the coupled network can be described as follows:

$$\begin{cases} \dot{x}_i(t) = f(x_i(t)) + \sum\limits_{j=1}^{N} G_{ij}(t)h(x_j(t)) + \sum\limits_{j=1}^{N} \hat{G}_{ij}(t)\hat{h}(\tilde{x}_j(t)) \\ \quad +B_i(t)u_i(t), \\ y_i(t) = C_i(t)x_i(t) + D_i(t)u_i(t), \end{cases} \tag{2.1}$$

where $\tilde{x}_i(t) = (x_{i1}(t - \tau_{i1}(t)), x_{i2}(t - \tau_{i2}(t)), \ldots, x_{in}(t - \tau_{in}(t)))^T$, $\tau_{il}(t)$ is the time-varying delay with $0 \leqslant \tau_{il}(t) \leqslant \tau_{il} \leqslant \tau, i = 1, 2, \ldots, N, l = 1, 2, \ldots, n$.

The function $f(\cdot) \in \mathbb{R}^n$ is continuously differentiable, $B_i(t), C_i(t)$ and $D_i(t)$ are known matrices with appropriate dimensions, $h(x_i(t)) \in \mathbb{R}^n$ and $\hat{h}(\tilde{x}_i(t)) = (\hat{h}_1(x_{i1}(t - \tau_{i1}(t))), \hat{h}_2(x_{i2}(t - \tau_{i2}(t))), \ldots, \hat{h}_n(x_{in}(t - \tau_{in}(t))))^T \in \mathbb{R}^n$ are vector-valued functions, describe the coupling relations between two nodes for non-delayed configuration and delayed one at time t, respectively, and $h(0) = \hat{h}(0) = 0$, $G(t) = (G_{ij}(t))_{N\times N}$ and $\hat{G}(t) = (\hat{G}_{ij}(t))_{N\times N}$ represent the topological structure of the complex network and coupling strength between nodes for non-delayed configuration and delayed one at time t, respectively, the diagonal elements of $G(t)$ and $\hat{G}(t)$ are defined as follows:

$$G_{ii}(t) = -\sum_{\substack{j=1\\ j\neq i}}^{N} G_{ij}(t), \quad \hat{G}_{ii}(t) = -\sum_{\substack{j=1\\ j\neq i}}^{N} \hat{G}_{ij}(t), \quad i = 1, 2, \ldots, N.$$

One should note that, in this model, the individual couplings between two connected nodes may be nonlinear, and the coupling configurations are not restricted to the symmetric and irreducible connections or the non-negative off-diagonal links.

Definition 2.1 (*see* [27, 45, 46]) The network (2.1) is called input strictly passive if there exist two constants β and $\gamma > 0$ such that

$$2\int_0^{t_p} y^T(s)u(s)ds \geqslant -\beta^2 + \gamma\int_0^{t_p} u^T(s)u(s)ds$$

for all $t_p \geqslant 0$.

Definition 2.2 (*see* [27, 45, 46]) The network (2.1) is called output strictly passive if there exist two constants β and $\gamma > 0$ such that

$$2\int_0^{t_p} y^T(s)u(s)ds \geqslant -\beta^2 + \gamma\int_0^{t_p} y^T(s)y(s)ds$$

for all $t_p \geqslant 0$.

By using Kronecker product, we can rewrite network (2.1) in a compact form as follows:

$$\begin{cases} \dot{x}(t) = F(x(t)) + (G(t) \otimes I_n)H(x(t)) + (\hat{G}(t) \otimes I_n)\hat{H}(\tilde{x}(t)) \\ \qquad\quad +B(t)u(t), \\ y(t) = C(t)x(t) + D(t)u(t), \end{cases} \tag{2.2}$$

where

$$\begin{aligned}
&x(t) = (x_1^T(t), x_2^T(t), \dots, x_N^T(t))^T \in \mathbb{R}^{nN}, \\
&y(t) = (y_1^T(t), y_2^T(t), \dots, y_N^T(t))^T \in \mathbb{R}^{nN}, \\
&F(x(t)) = (f^T(x_1(t)), f^T(x_2(t)), \dots, f^T(x_N(t)))^T \in \mathbb{R}^{nN}, \\
&\tilde{x}(t) = (\tilde{x}_1^T(t), \tilde{x}_2^T(t), \dots, \tilde{x}_N^T(t))^T \in \mathbb{R}^{nN}, \\
&\hat{H}(\tilde{x}(t)) = (\hat{h}^T(\tilde{x}_1(t)), \hat{h}^T(\tilde{x}_2(t)) \cdots, \hat{h}^T(\tilde{x}_N(t)))^T \in \mathbb{R}^{nN}, \\
&B(t) = \mathrm{diag}(B_1(t), B_2(t), \dots, B_N(t)), \\
&H(x(t)) = (h^T(x_1(t)), h^T(x_2(t)), \dots, h^T(x_N(t)))^T \in \mathbb{R}^{nN}, \\
&C(t) = \mathrm{diag}(C_1(t), C_2(t), \dots, C_N(t)), \\
&u(t) = (u_1^T(t), u_2^T(t), \dots, u_N^T(t))^T \in \mathbb{R}^{nN}, \\
&D(t) = \mathrm{diag}(D_1(t), D_2(t), \dots, D_N(t)).
\end{aligned}$$

2.3 Main Results

In [52, 53], and many others, authors make the assumption that the function $f(\cdot)$ is in the QUAD function class. In this chapter, we make similar assumptions:

(A1) There exist positive definite diagonal matrix $P = \mathrm{diag}(p_1, p_2, \dots, p_n)$ and a diagonal matrix $\Delta = \mathrm{diag}(\delta_1, \delta_2, \dots, \delta_n)$ such that f satisfies the following inequality:

$$x^T P(f(x) - \Delta x) \leqslant -\eta x^T x$$

for some $\eta > 0$ and all $x \in \mathbb{R}^n$.

In order to obtain our main results, another assumption is introduced.

(A2) Suppose that there exist two constants $L_1, L_2 > 0$ such that

$$\|h(z)\| \leqslant L_1\|z\|, \quad \|\hat{h}(z)\| \leqslant L_2\|z\|$$

hold for any $z \in \mathbb{R}^n$.

For the convenience, we denote

$$\hat{P} = \mathrm{diag}(P, P, \dots, P), \quad \hat{\Delta} = \mathrm{diag}(\Delta, \Delta, \dots, \Delta).$$

2.3.1 Input Strict Passivity of CDNs

Theorem 2.3 *Let (A1) and (A2) hold, and* $\dot{\tau}_{il}(t) \leqslant \sigma < 1, i = 1, 2, \ldots, N, l = 1, 2, \ldots, n$. *Then, the network (2.1) is input strictly passive if there exist a matrix* $Z = \text{diag}(Z_1, Z_2, \ldots, Z_N)$, $Z_i = \text{diag}(z_{i1}, z_{i2}, \ldots, z_{in})$, $z_{il} > 0$, *and two positive constants* ς, γ *such that*

$$\frac{(\hat{P}B(t) - C^T(t))(B^T(t)\hat{P} - C(t))}{\varsigma} + \hat{P}(G(t) \otimes I_n)(G(t) \otimes I_n)^T \hat{P} + 2\hat{P}\hat{\Delta}$$
$$+(-2\eta + L_1^2 + aL_2^2)I_{nN} + \frac{\hat{P}(\hat{G}(t) \otimes I_n)Z^{-1}(\hat{G}(t) \otimes I_n)^T \hat{P}}{1 - \sigma} \leqslant 0, \tag{2.3}$$

$$D(t) + D^T(t) - (\varsigma + \gamma)I_{nN} \geqslant 0, \tag{2.4}$$

where $a = \max\{z_{il}, i = 1, 2, 3, \ldots, N, \ l = 1, 2, \ldots, n\}$.

Proof Construct the Lyapunov functional for system (2.2) as follows:

$$V(x(t)) = x^T(t)\hat{P}x(t) + \sum_{i=1}^{N}\sum_{l=1}^{n} \int_{t-\tau_{il}(t)}^{t} z_{il}\hat{h}_l^2(x_{il}(\alpha))d\alpha, \tag{2.5}$$

where $\hat{h}(x_i(\alpha)) = (\hat{h}_1(x_{i1}(\alpha)), \hat{h}_2(x_{i2}(\alpha)), \ldots, \hat{h}_n(x_{in}(\alpha)))^T, i = 1, 2, \ldots, N$. The derivative of $V(x(t))$ satisfies

$$\begin{aligned}
\dot{V}(x(t)) &= 2x^T(t)\hat{P}\dot{x}(t) + \hat{H}^T(x(t))Z\hat{H}(x(t)) \\
&\quad - \sum_{i=1}^{N}\sum_{l=1}^{n}(1 - \dot{\tau}_{il}(t))z_{il}\hat{h}_l^2(x_{il}(t - \tau_{il}(t))) \\
&\leqslant 2x^T(t)\hat{P}F(x(t)) + 2x^T(t)\hat{P}(G(t) \otimes I_n)H(x(t)) \\
&\quad + 2x^T(t)\hat{P}(\hat{G}(t) \otimes I_n)\hat{H}(\tilde{x}(t)) + 2x^T(t)\hat{P}B(t)u(t) \\
&\quad + \hat{H}^T(x(t))Z\hat{H}(x(t)) - (1 - \sigma)\hat{H}^T(\tilde{x}(t))Z\hat{H}(\tilde{x}(t)),
\end{aligned}$$

where $\hat{H}(x(t)) = (\hat{h}^T(x_1(t)), \hat{h}^T(x_2(t)), \ldots, \hat{h}^T(x_N(t)))^T$. Then, we can get

$$\begin{aligned}
&\dot{V}(x(t)) - 2y^T(t)u(t) + \gamma u^T(t)u(t) \\
\leqslant\ & 2x^T(t)\hat{P}(G(t) \otimes I_n)H(x(t)) + 2x^T(t)\hat{P}(\hat{G}(t) \otimes I_n)\hat{H}(\tilde{x}(t)) \\
&+2x^T(t)\hat{P}F(x(t)) + 2x^T(t)\hat{P}B(t)u(t) + \hat{H}^T(x(t))Z\hat{H}(x(t)) \\
&-(1 - \sigma)\hat{H}^T(\tilde{x}(t))Z\hat{H}(\tilde{x}(t)) - 2x^T(t)C^T(t)u(t) \\
&-u^T(t)(D(t) + D^T(t))u(t) + \gamma u^T(t)u(t).
\end{aligned}$$

Applying Lemma 1.4, we can easily obtain

$$\begin{aligned}
&2x^T(t)\hat{P}(G(t)\otimes I_n)H(x(t)) \\
\leqslant &H^T(x(t))H(x(t)) + x^T(t)\hat{P}(G(t)\otimes I_n)(G(t)\otimes I_n)^T\hat{P}x(t),
\end{aligned} \tag{2.6}$$

$$\begin{aligned}
&2x^T(t)(\hat{P}B(t) - C^T(t))u(t) \\
\leqslant &\frac{x^T(t)(\hat{P}B(t) - C^T(t))(B^T(t)\hat{P} - C(t))x(t)}{\varsigma} + \varsigma u^T(t)u(t),
\end{aligned} \tag{2.7}$$

$$\begin{aligned}
&2x^T(t)\hat{P}(\hat{G}(t)\otimes I_n)\hat{H}(\tilde{x}(t)) \\
\leqslant &(1-\sigma)\hat{H}^T(\tilde{x}(t))Z\hat{H}(\tilde{x}(t)) \\
&+ \frac{x^T(t)\hat{P}(\hat{G}(t)\otimes I_n)Z^{-1}(\hat{G}(t)\otimes I_n)^T\hat{P}x(t)}{1-\sigma}.
\end{aligned} \tag{2.8}$$

We have from inequalities (2.4), (2.6), (2.7) and (2.8) that

$$\begin{aligned}
&\dot{V}(x(t)) - 2y^T(t)u(t) + \gamma u^T(t)u(t) \\
\leqslant &2x^T(t)\hat{P}F(x(t)) + H^T(x(t))H(x(t)) + \hat{H}^T(x(t))Z\hat{H}(x(t)) \\
&+ x^T(t)\hat{P}(G(t)\otimes I_n)(G(t)\otimes I_n)^T\hat{P}x(t) \\
&+ \frac{x^T(t)(\hat{P}B(t) - C^T(t))(B^T(t)\hat{P} - C(t))x(t)}{\varsigma} \\
&+ \frac{x^T(t)\hat{P}(\hat{G}(t)\otimes I_n)Z^{-1}(\hat{G}(t)\otimes I_n)^T\hat{P}x(t)}{1-\sigma}.
\end{aligned}$$

According to (A1) and (A2), we can obtain

$$\begin{aligned}
x^T(t)\hat{P}F(x(t)) &= \sum_{i=1}^{N} x_i^T(t)Pf(x_i(t)) \\
&\leqslant \sum_{i=1}^{N}(-\eta x_i^T(t)x_i(t) + x_i^T(t)P\Delta x_i(t)) \\
&= x^T(t)(-\eta I_{nN} + \hat{P}\hat{\Delta})x(t),
\end{aligned} \tag{2.9}$$

$$\begin{aligned}
H^T(x(t))H(x(t)) &= \sum_{i=1}^{N} h^T(x_i(t))h(x_i(t)) \\
&\leqslant \sum_{i=1}^{N} L_1^2 x_i^T(t))x_i(t) \\
&= L_1^2 x^T(t)x(t),
\end{aligned} \tag{2.10}$$

$$\begin{aligned}\hat{H}^T(x(t))Z\hat{H}(x(t)) &= \sum_{i=1}^{N}\hat{h}^T(x_i(t))Z_i\hat{h}(x_i(t))\\ &\leqslant \sum_{i=1}^{N} aL_2^2 x_i^T(t))x_i(t)\\ &= aL_2^2 x^T(t)x(t).\end{aligned} \tag{2.11}$$

Using (2.3), (2.9), (2.10) and (2.11), we have

$$\begin{aligned}&\dot{V}(x(t)) - 2y^T(t)u(t) + \gamma u^T(t)u(t)\\ &\leqslant x^T(t)\Bigg[(-2\eta + L_1^2 + aL_2^2)I_{nN} + \hat{P}(G(t)\otimes I_n)(G(t)\otimes I_n)^T\hat{P}\\ &\quad + \frac{(\hat{P}B(t) - C^T(t))(B^T(t)\hat{P} - C(t))}{\varsigma} + 2\hat{P}\hat{\Delta}\\ &\quad + \frac{\hat{P}(\hat{G}(t)\otimes I_n)Z^{-1}(\hat{G}(t)\otimes I_n)^T\hat{P}}{1-\sigma}\Bigg]x(t)\\ &\leqslant 0.\end{aligned} \tag{2.12}$$

By integrating Eq. (2.12) with respect to t over the time period 0 to t_p, we get

$$2\int_0^{t_p} y^T(t)u(t)dt \geqslant V(x(t_p)) - V(x(0)) + \gamma\int_0^{t_p} u^T(t)u(t)dt.$$

From the construction of $V(x(t))$, we have $V(x(0)) \geqslant 0$ and $V(x(t_p) \geqslant 0$. Thus,

$$2\int_0^{t_p} y^T(t)u(t)dt \geqslant -V(x(0)) + \gamma\int_0^{t_p} u^T(t)u(t)dt$$

for all $t_p \geqslant 0$. The proof is completed.

Theorem 2.4 *Let (A1) and (A2) hold, and $\dot{\tau}_{il}(t) \leqslant \sigma < 1, i = 1, 2, \ldots, N, l = 1, 2, \ldots, n$. Then, the network (2.1) is input strictly passive if there exist two positive constants ξ and γ with $\xi > \gamma$ such that*

$$\begin{aligned}&2\hat{P}\hat{\Delta} + \frac{L_2^2\hat{P}(\hat{G}(t)\otimes I_n)(\hat{G}(t)\otimes I_n)^T\hat{P}}{1-\sigma} + \frac{(\hat{P}B(t) - C^T(t))(B^T(t)\hat{P} - C(t))}{\xi - \gamma}\\ &+ (-2\eta + 1 + L_1^2)I_{nN} + \hat{P}(G(t)\otimes I_n)(G(t)\otimes I_n)^T\hat{P} \leqslant 0,\end{aligned} \tag{2.13}$$

$$D(t) + D^T(t) - \xi I_{nN} \geqslant 0. \tag{2.14}$$

Proof Define the following Lyapunov functional for system (2.2):

$$V(x(t)) = x^T(t)\hat{P}x(t) + \sum_{i=1}^{N}\sum_{l=1}^{n}\int_{t-\tau_{il}(t)}^{t} x_{il}^2(\alpha)d\alpha. \tag{2.15}$$

The derivative of $V(x(t))$ satisfies

$$\begin{aligned}\dot{V}(x(t)) \leqslant & 2x^T(t)\hat{P}\dot{x}(t) + x^T(t)x(t) - (1-\sigma)\tilde{x}^T(t)\tilde{x}(t) \\ = & 2x^T(t)\hat{P}F(x(t)) + 2x^T(t)\hat{P}(G(t)\otimes I_n)H(x(t)) \\ & + 2x^T(t)\hat{P}(\hat{G}(t)\otimes I_n)\hat{H}(\tilde{x}(t)) + 2x^T(t)\hat{P}B(t)u(t) \\ & + x^T(t)x(t) - (1-\sigma)\tilde{x}^T(t)\tilde{x}(t).\end{aligned}$$

Then, we can get

$$\begin{aligned}&\dot{V}(x(t)) - 2y^T(t)u(t) + \gamma u^T(t)u(t) \\ \leqslant & 2x^T(t)\hat{P}F(x(t)) + x^T(t)x(t) + 2x^T(t)\hat{P}(G(t)\otimes I_n)H(x(t)) \\ & + 2x^T(t)\hat{P}(\hat{G}(t)\otimes I_n)\hat{H}(\tilde{x}(t)) + 2x^T(t)\hat{P}B(t)u(t) - 2x^T(t)C^T(t)u(t) \\ & - (1-\sigma)\tilde{x}^T(t)\tilde{x}(t) - u^T(t)(D(t)+D^T(t))u(t) + \gamma u^T(t)u(t) \\ \leqslant & 2x^T(t)\hat{P}F(x(t)) + x^T(t)x(t) + 2x^T(t)\hat{P}(G(t)\otimes I_n)H(x(t)) \\ & + 2x^T(t)\hat{P}(\hat{G}(t)\otimes I_n)\hat{H}(\tilde{x}(t)) + 2x^T(t)\hat{P}B(t)u(t) \\ & - (1-\sigma)\tilde{x}^T(t)\tilde{x}(t) - 2x^T(t)C^T(t)u(t) - (\xi-\gamma)u^T(t)u(t).\end{aligned}$$

Applying Lemma 1.4, we can easily obtain

$$\begin{aligned}& 2x^T(t)(\hat{P}B(t) - C^T(t))u(t) \\ \leqslant & \frac{x^T(t)(\hat{P}B(t) - C^T(t))(B^T(t)\hat{P} - C(t))x(t)}{\xi-\gamma} \\ & + (\xi-\gamma)u^T(t)u(t),\end{aligned} \tag{2.16}$$

$$\begin{aligned}& 2x^T(t)\hat{P}(\hat{G}(t)\otimes I_n)\hat{H}(\tilde{x}(t)) \\ \leqslant & \frac{(1-\sigma)\hat{H}^T(\tilde{x}(t))\hat{H}(\tilde{x}(t))}{L_2^2} \\ & + \frac{L_2^2 x^T(t)\hat{P}(\hat{G}(t)\otimes I_n)(\hat{G}(t)\otimes I_n)^T\hat{P}x(t)}{1-\sigma}.\end{aligned} \tag{2.17}$$

It follows from (2.6), (2.16) and (2.17) that

$$
\begin{aligned}
&\dot{V}(x(t)) - 2y^T(t)u(t) + \gamma u^T(t)u(t) \\
\leqslant & 2x^T(t)\hat{P}F(x(t)) + x^T(t)x(t) - (1-\sigma)\tilde{x}^T(t)\tilde{x}(t) \\
&+ x^T(t)\hat{P}(G(t)\otimes I_n)(G(t)\otimes I_n)^T\hat{P}x(t) + H^T(x(t))H(x(t)) \\
&+ \frac{x^T(t)(\hat{P}B(t) - C^T(t))(B^T(t)\hat{P} - C(t))x(t)}{\xi - \gamma} \\
&+ \frac{(1-\sigma)\hat{H}^T(\tilde{x}(t))\hat{H}(\tilde{x}(t))}{L_2^2} \\
&+ \frac{L_2^2 x^T(t)\hat{P}(\hat{G}(t)\otimes I_n)(\hat{G}(t)\otimes I_n)^T\hat{P}x(t)}{1-\sigma}.
\end{aligned}
$$

According to (A2), we have

$$
\begin{aligned}
\hat{H}^T(\tilde{x}(t))\hat{H}(\tilde{x}(t)) = & \sum_{i=1}^{N}\hat{h}^T(\tilde{x}_i(t))\hat{h}(\tilde{x}_i(t)) \\
\leqslant & \sum_{i=1}^{N} L_2^2\tilde{x}_i^T(t)\tilde{x}_i(t) \\
= & L_2^2\tilde{x}^T(t)\tilde{x}(t).
\end{aligned} \tag{2.18}
$$

It follows from inequalities (2.9), (2.10), (2.13) and (2.18) that

$$
\begin{aligned}
&\dot{V}(x(t)) - 2y^T(t)u(t) + \gamma u^T(t)u(t) \\
\leqslant & x^T(t)\Bigg[(-2\eta + 1 + L_1^2)I_{nN} + 2\hat{P}\hat{\Delta} + \frac{L_2^2\hat{P}(\hat{G}(t)\otimes I_n)(\hat{G}(t)\otimes I_n)^T\hat{P}}{1-\sigma} \\
&+ \frac{(\hat{P}B(t) - C^T(t))(B^T(t)\hat{P} - C(t))}{\xi - \gamma} + \hat{P}(G(t)\otimes I_n)(G(t)\otimes I_n)^T\hat{P}\Bigg]x(t) \\
\leqslant & 0.
\end{aligned} \tag{2.19}
$$

Similarly, we can conclude that the network (2.1) is input strictly passive. The proof is completed.

Remark 2.5 The conditions in Theorems 2.3 and 2.4 do not restrict the network configurations to be symmetric, irreducible and non-negative off-diagonal. Also, Theorems 2.3 and 2.4 don't require the restrictive assumption that the derivative of the time-varying delays is non-negative in some previous results (see, e.g., [11]). Therefore, our criteria may be more general and verifiable.

2.3.2 Output Strict Passivity of CDNs

Theorem 2.6 *Let (A1) and (A2) hold, and* $\dot{\tau}_{il}(t) \leqslant \sigma < 1, i = 1, 2, \ldots, N, l = 1, 2, \ldots, n$. *Then the network (2.1) is output strictly passive if there exist a matrix* $Z = \text{diag}(Z_1, Z_2, \ldots, Z_N)$, $Z_i = \text{diag}(z_{i1}, z_{i2}, \ldots, z_{in})$, $z_{il} > 0$, *and two positive constants* ς, γ *such that*

$$\begin{aligned}
&(-2\eta + L_1^2 + aL_2^2)I_{nN} + 2\hat{P}\hat{\Delta} + \hat{P}(G(t) \otimes I_n)(G(t) \otimes I_n)^T \hat{P} \\
&+ \frac{W(t)W^T(t)}{\varsigma} + \frac{\hat{P}(\hat{G}(t) \otimes I_n)Z^{-1}(\hat{G}(t) \otimes I_n)^T \hat{P}}{1 - \sigma} \\
&+ \gamma C^T(t)C(t) \leqslant 0, \qquad (2.20) \\
&D(t) + D^T(t) - \gamma D^T(t)D(t) - \varsigma I_{nN} \geqslant 0, \qquad (2.21)
\end{aligned}$$

where $W(t) = \hat{P}B(t) - C^T(t) + \gamma C^T(t)D(t)$, $a = \max\{z_{il},\ i = 1, 2, 3, \ldots, N, l = 1, 2, \ldots, n\}$.

Proof Construct the same Lyapunov function as (2.5) for system (2.2). Then,we can obtain

$$\begin{aligned}
\dot{V}(x(t)) \leqslant &2x^T(t)\hat{P}F(x(t)) + 2x^T(t)\hat{P}(G(t) \otimes I_n)H(x(t)) \\
&+ 2x^T(t)\hat{P}(\hat{G}(t) \otimes I_n)\hat{H}(\tilde{x}(t)) + 2x^T(t)\hat{P}B(t)u(t) \\
&+ \hat{H}^T(x(t))Z\hat{H}(x(t)) - (1 - \sigma)\hat{H}^T(\tilde{x}(t))Z\hat{H}(\tilde{x}(t)).
\end{aligned}$$

Then, we can get

$$\begin{aligned}
&\dot{V}(x(t)) - 2y^T(t)u(t) + \gamma y^T(t)y(t) \\
\leqslant &2x^T(t)\hat{P}F(x(t)) + 2x^T(t)\hat{P}(G(t) \otimes I_n)H(x(t)) + 2x^T(t)\hat{P}(\hat{G}(t) \otimes I_n)\hat{H}(\tilde{x}(t)) \\
&+ 2x^T(t)\hat{P}B(t)u(t) + \hat{H}^T(x(t))Z\hat{H}(x(t)) - (1 - \sigma)\hat{H}^T(\tilde{x}(t))Z\hat{H}(\tilde{x}(t)) \\
&- 2x^T(t)C^T(t)u(t) - u^T(t)(D(t) + D^T(t))u(t) + \gamma(C(t)x(t) \\
&+ D(t)u(t))^T(C(t)x(t) + D(t)u(t)) \\
\leqslant &2x^T(t)\hat{P}F(x(t)) + 2x^T(t)\hat{P}(G(t) \otimes I_n)H(x(t)) + 2x^T(t)\hat{P}(\hat{G}(t) \otimes I_n)\hat{H}(\tilde{x}(t)) \\
&+ 2x^T(t)(\hat{P}B(t) - C^T(t) + \gamma C^T(t)D(t))u(t) + \hat{H}^T(x(t))Z\hat{H}(x(t)) \\
&- (1 - \sigma)\hat{H}^T(\tilde{x}(t))Z\hat{H}(\tilde{x}(t)) - u^T(t)(D(t) + D^T(t) - \gamma D^T(t)D(t))u(t) \\
&+ \gamma x^T(t)C^T(t)C(t)x(t).
\end{aligned}$$

Applying Lemma 1.4, we have

$$2x^T(t)W(t)u(t) \leqslant \frac{x^T(t)W(t)W^T(t)x(t)}{\varsigma} + \varsigma u^T(t)u(t). \qquad (2.22)$$

Using (2.6), (2.8) and (2.22), we can easily obtain

$$
\begin{aligned}
&\dot{V}(x(t)) - 2y^T(t)u(t) + \gamma y^T(t)y(t) \\
\leqslant &2x^T(t)\hat{P}F(x(t)) + H^T(x(t))H(x(t)) + x^T(t)\hat{P}(G(t)\otimes I_n)(G(t)\otimes I_n)^T\hat{P}x(t) \\
&+ \frac{x^T(t)W(t)W^T(t)x(t)}{\varsigma} + \hat{H}^T(x(t))Z\hat{H}(x(t)) + \gamma x^T(t)C^T(t)C(t)x(t) \\
&+ \frac{x^T(t)\hat{P}(\hat{G}(t)\otimes I_n)Z^{-1}(\hat{G}(t)\otimes I_n)^T\hat{P}x(t)}{1-\sigma} - u^T(t)(D(t) \\
&+ D^T(t) - \gamma D^T(t)D(t) - \varsigma I_{nN})u(t).
\end{aligned}
$$

It follows from inequalities (2.9), (2.10), (2.11), (2.20) and (2.21) that

$$
\begin{aligned}
&\dot{V}(x(t)) - 2y^T(t)u(t) + \gamma y^T(t)y(t) \\
\leqslant &x^T(t)\Bigg[(-2\eta + L_1^2 + aL_2^2)I_{nN} + 2\hat{P}\hat{\Delta} + \hat{P}(G(t)\otimes I_n)(G(t)\otimes I_n)^T\hat{P} \\
&+ \frac{W(t)W^T(t)}{\varsigma} + \frac{\hat{P}(\hat{G}(t)\otimes I_n)Z^{-1}(\hat{G}(t)\otimes I_n)^T\hat{P}}{1-\sigma} + \gamma C^T(t)C(t)\Bigg]x(t) \\
\leqslant &0.
\end{aligned}
\tag{2.23}
$$

By integrating Eq. (2.23) with respect to t over the time period 0 to t_p, we get

$$
2\int_0^{t_p} y^T(t)u(t)dt \geqslant V(x(t_p)) - V(x(0)) + \gamma\int_0^{t_p} y^T(t)y(t)dt.
$$

From the construction of $V(x(t))$, we have $V(x(0)) \geqslant 0$ and $V(x(t_p) \geqslant 0$. Thus,

$$
2\int_0^{t_p} y^T(t)u(t)dt \geqslant -V(x(0)) + \gamma\int_0^{t_p} y^T(t)y(t)dt
$$

for all $t_p \geqslant 0$. The proof is completed.

Theorem 2.7 *Let (A1) and (A2) hold, and* $\dot{\tau}_{il}(t) \leqslant \sigma < 1,\ i = 1, 2, \ldots, N,\ l = 1, 2, \ldots, n$. *Then, the network (2.1) is output strictly passive if there exist two positive constants* ξ, γ *such that*

$$
\begin{aligned}
&(-2\eta+1+L_1^2)I_{nN}+\gamma C^T(t)C(t)+\frac{L_2^2\hat{P}(\hat{G}(t)\otimes I_n)(\hat{G}(t)\otimes I_n)^T\hat{P}}{1-\sigma} \\
&+\frac{W(t)W^T(t)}{\xi}+\hat{P}(G(t)\otimes I_n)(G(t)\otimes I_n)^T\hat{P}+2\hat{P}\hat{\Delta}\leqslant 0, \qquad (2.24)\\
&D(t)+D^T(t)-\gamma D^T(t)D(t)-\xi I_{nN}\geqslant 0, \qquad (2.25)
\end{aligned}
$$

where $W(t)=\hat{P}B(t)-C^T(t)+\gamma C^T(t)D(t)$.

Proof Construct the same Lyapunov functional as (2.15) for system (2.2). Then, we can obtain

$$
\begin{aligned}
\dot{V}(x(t))\leqslant &2x^T(t)\hat{P}F(x(t))+2x^T(t)\hat{P}(G(t)\otimes I_n)H(x(t))\\
&+2x^T(t)\hat{P}(\hat{G}(t)\otimes I_n)\hat{H}(\tilde{x}(t))+2x^T(t)\hat{P}B(t)u(t)\\
&+x^T(t)x(t)-(1-\sigma)\tilde{x}^T(t)\tilde{x}(t).
\end{aligned}
$$

Then, we can get

$$
\begin{aligned}
&\dot{V}(x(t))-2y^T(t)u(t)+\gamma y^T(t)y(t)\\
\leqslant &2x^T(t)\hat{P}F(x(t))+x^T(t)x(t)+2x^T(t)\hat{P}(G(t)\otimes I_n)H(x(t))\\
&+2x^T(t)\hat{P}(\hat{G}(t)\otimes I_n)\hat{H}(\tilde{x}(t))+2x^T(t)\hat{P}B(t)u(t)\\
&-(1-\sigma)\tilde{x}^T(t)\tilde{x}(t)-2x^T(t)C^T(t)u(t)-u^T(t)(D(t)+D^T(t))u(t)\\
&+\gamma(C(t)x(t)+D(t)u(t))^T(C(t)x(t)+D(t)u(t))\\
\leqslant &2x^T(t)\hat{P}F(x(t))+x^T(t)(I_{nN}+\gamma C^T(t)C(t))x(t)+2x^T(t)\hat{P}(G(t)\otimes I_n)H(x(t))\\
&+2x^T(t)\hat{P}(\hat{G}(t)\otimes I_n)\hat{H}(\tilde{x}(t))+2x^T(t)(\hat{P}B(t)-C^T(t)\\
&+\gamma C^T(t)D(t))u(t)-(1-\sigma)\tilde{x}^T(t)\tilde{x}(t)-u^T(t)(D(t)+D^T(t)\\
&-\gamma D^T(t)D(t))u(t).
\end{aligned}
$$

Applying Lemma 1.4, we have

$$
2x^T(t)W(t)u(t)\leqslant\frac{x^T(t)W(t)W^T(t)x(t)}{\xi}+\xi u^T(t)u(t). \qquad (2.26)
$$

Using (2.6), (2.17), (2.25) and (2.26), we can get

$$
\begin{aligned}
&\dot{V}(x(t)) - 2y^T(t)u(t) + \gamma y^T(t)y(t) \\
\leqslant& 2x^T(t)\hat{P}F(x(t)) + x^T(t)(I_{nN} + \gamma C^T(t)C(t))x(t) + H^T(x(t))H(x(t)) \\
&+ x^T(t)\hat{P}(G(t)\otimes I_n)(G(t)\otimes I_n)^T\hat{P}x(t) + \frac{x^T(t)W(t)W^T(t)x(t)}{\xi} \\
&+ \frac{(1-\sigma)\hat{H}^T(\tilde{x}(t))\hat{H}(\tilde{x}(t))}{L_2^2} + \frac{L_2^2 x^T(t)\hat{P}(\hat{G}(t)\otimes I_n)(\hat{G}(t)\otimes I_n)^T\hat{P}x(t)}{1-\sigma} \\
&- (1-\sigma)\tilde{x}^T(t)\tilde{x}(t) - u^T(t)(D(t) + D^T(t) - \gamma D^T(t)D(t) - \xi I_{nN})u(t) \\
\leqslant& 2x^T(t)\hat{P}F(x(t)) + x^T(t)(I_{nN} + \gamma C^T(t)C(t))x(t) + H^T(x(t))H(x(t)) \\
&+ x^T(t)\hat{P}(G(t)\otimes I_n)(G(t)\otimes I_n)^T\hat{P}x(t) + \frac{x^T(t)W(t)W^T(t)x(t)}{\xi} \\
&+ \frac{(1-\sigma)\hat{H}^T(\tilde{x}(t))\hat{H}(\tilde{x}(t))}{L_2^2} + \frac{L_2^2 x^T(t)\hat{P}(\hat{G}(t)\otimes I_n)(\hat{G}(t)\otimes I_n)^T\hat{P}x(t)}{1-\sigma} \\
&- (1-\sigma)\tilde{x}^T(t)\tilde{x}(t).
\end{aligned}
$$

It follows from inequalities (2.9), (2.10), (2.18) and (2.24) that

$$
\begin{aligned}
&\dot{V}(x(t)) - 2y^T(t)u(t) + \gamma y^T(t)y(t) \\
\leqslant& x^T(t)[(-2\eta + 1 + L_1^2)I_{nN} + \gamma C^T(t)C(t) + \frac{L_2^2\hat{P}(\hat{G}(t)\otimes I_n)(\hat{G}(t)\otimes I_n)^T\hat{P}}{1-\sigma} \\
&+ \frac{W(t)W^T(t)}{\xi} + \hat{P}(G(t)\otimes I_n)(G(t)\otimes I_n)^T\hat{P} + 2\hat{P}\hat{\Delta}]x(t) \\
\leqslant& 0.
\end{aligned} \tag{2.27}
$$

Similarly, we can conclude that the network (2.1) is output strictly passive. The proof is completed.

Remark 2.8 The conditions in Theorems 2.6 and 2.7 do not restrict the network configurations to be symmetric, irreducible and non-negative off-diagonal. Also, Theorems 2.6 and 2.7 don't require the restrictive assumption that the derivative of the time-varying delays is non-negative in some previous results (see, e.g., [11]). Therefore, our criteria may be more general and verifiable.

2.4 Examples

In this section, we give two examples and their simulation to show the effectiveness of the proposed theoretical results.

Example 2.9 Consider a complex network, in which each node is a 3-dimensional nonlinear system described by

$$\begin{pmatrix} \dot{x}_1 \\ \dot{x}_2 \\ \dot{x}_3 \end{pmatrix} = \begin{pmatrix} 10(x_2 - x_1) \\ 2x_1 - x_1x_3 - 10x_2 \\ x_1x_2 - \frac{8}{3}x_3 \end{pmatrix}.$$

Take

$$G(t) = \begin{pmatrix} -0.2 & 0.1 & 0.1 \\ 0.2 & 0.3 & -0.5 \\ 0.1 & 0 & -0.1 \end{pmatrix}, \ \hat{G}(t) = \begin{pmatrix} -1.2 & 0.4 & 0.8 \\ -0.2 & 0.1 & 0.1 \\ 0.2 & 0.3 & -0.5 \end{pmatrix},$$

$$B_i(t) = C_i(t) = \begin{pmatrix} 0.7 & 0 & 0 \\ 0 & 0.6 & 0 \\ 0 & 0 & 0.5 \end{pmatrix}, \quad D_i(t) = \begin{pmatrix} 0.8 & 0 & 0 \\ 0 & 0.8 & 0 \\ 0 & 0 & 0.8 \end{pmatrix}$$

and

$$h(x_i(t)) = (\sin(x_{i1}(t))/5, \sin(x_{i2}(t))/5, \sin(x_{i3}(t))/5)^T,$$
$$\hat{h}(x_i(t)) = (\sin(x_{i1}(t))/6, \sin(x_{i2}(t))/6, \sin(x_{i3}(t))/6)^T, \ i = 1, 2, 3.$$

It is obvious that the couplings are not restricted to linear, symmetric and the non-negative off-diagonal, and we can take $\eta = \frac{8}{3}$, $P = I_3$, $\Delta = \mathrm{diag}(0, 0, 0)$, $L_1 = \frac{1}{5}$, and $L_2 = \frac{1}{6}$.

In the following, we analyze the input strict passivity of complex network with different time-varying delays.

Setting $\tau_{il}(t) = 1 - \frac{1}{2i+l}e^{-t}$, then $0 \leqslant \tau_{il}(t) \leqslant \tau_{il} = \tau = 1, \dot{\tau}_{il}(t) = \frac{1}{2i+l}e^{-t} \leqslant \frac{1}{3} < 1$, for $t \geqslant 0$, $i, l = 1, 2, 3$.

We can find positive-definite matrix $Z = I_9$ satisfying (2.3) and (2.4) with $\varsigma = 0.5, \gamma = 1$. According to Theorem 2.3, we know that complex network (2.1) with above given parameters is input strictly passive. Set $u_i(t) = (\sin((i+1)\pi t), \sin((i+2)\pi t), \sin((i+3)\pi t))^T$. The simulation results are shown in Figs. 2.1 and 2.2.

Example 2.10 Consider a CDN, in which each node is a 3-dimensional nonlinear system described by

$$\begin{pmatrix} \dot{x}_1 \\ \dot{x}_2 \\ \dot{x}_3 \end{pmatrix} = \begin{pmatrix} 8(x_2 - x_1) \\ 2x_1 - x_1x_3 - 10x_2 \\ x_1x_2 - \frac{5}{3}x_3 \end{pmatrix}.$$

Take

$$G(t) = \begin{pmatrix} -0.4 & 0.2 & 0.2 \\ 0.5 & 0.1 & -0.6 \\ 0.1 & 0.2 & -0.3 \end{pmatrix}, \ \hat{G}(t) = \begin{pmatrix} -1.1 & 0.4 & 0.7 \\ -0.3 & 0.2 & 0.1 \\ 0.2 & 0.5 & -0.7 \end{pmatrix},$$

$$B_i(t) = C_i(t) = \begin{pmatrix} 0.7 & 0 & 0 \\ 0 & 0.7 & 0 \\ 0 & 0 & 0.7 \end{pmatrix}, \quad D_i(t) = \begin{pmatrix} 0.9 & 0 & 0 \\ 0 & 0.9 & 0 \\ 0 & 0 & 0.9 \end{pmatrix}$$

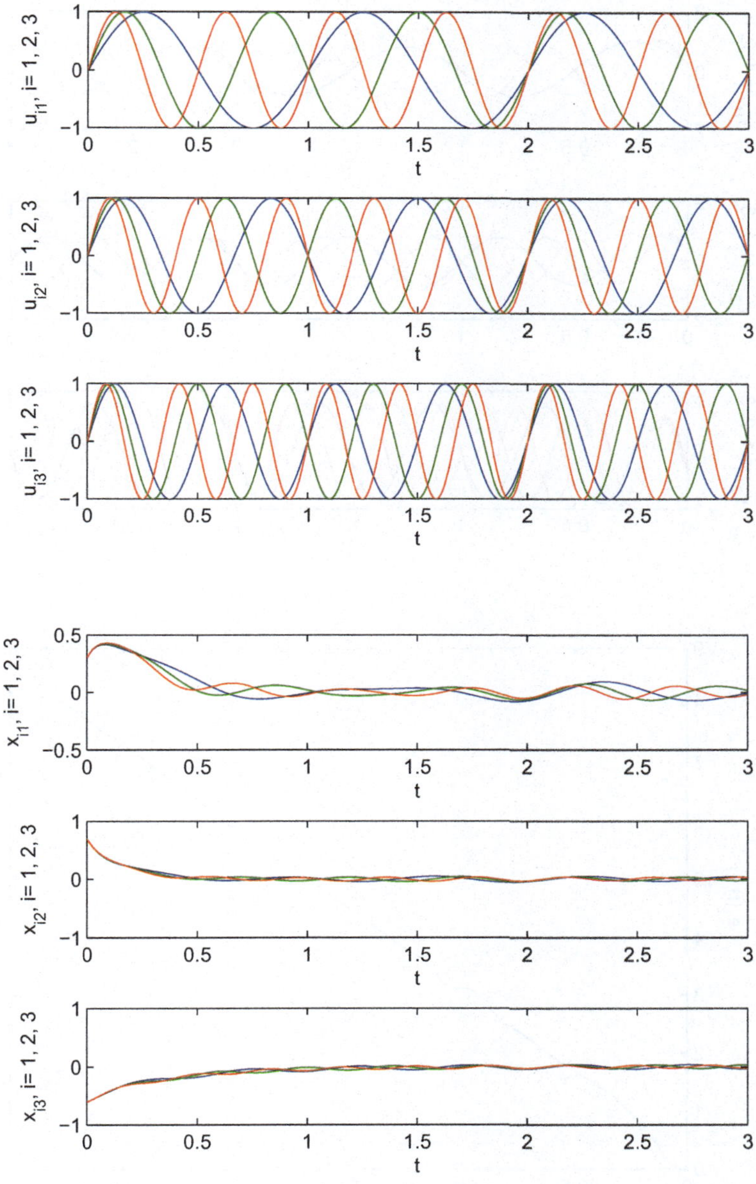

Fig. 2.1 The change process of input variables $u_{il}(i, l = 1, 2, 3)$, state variables $x_{il}(i, l = 1, 2, 3)$.

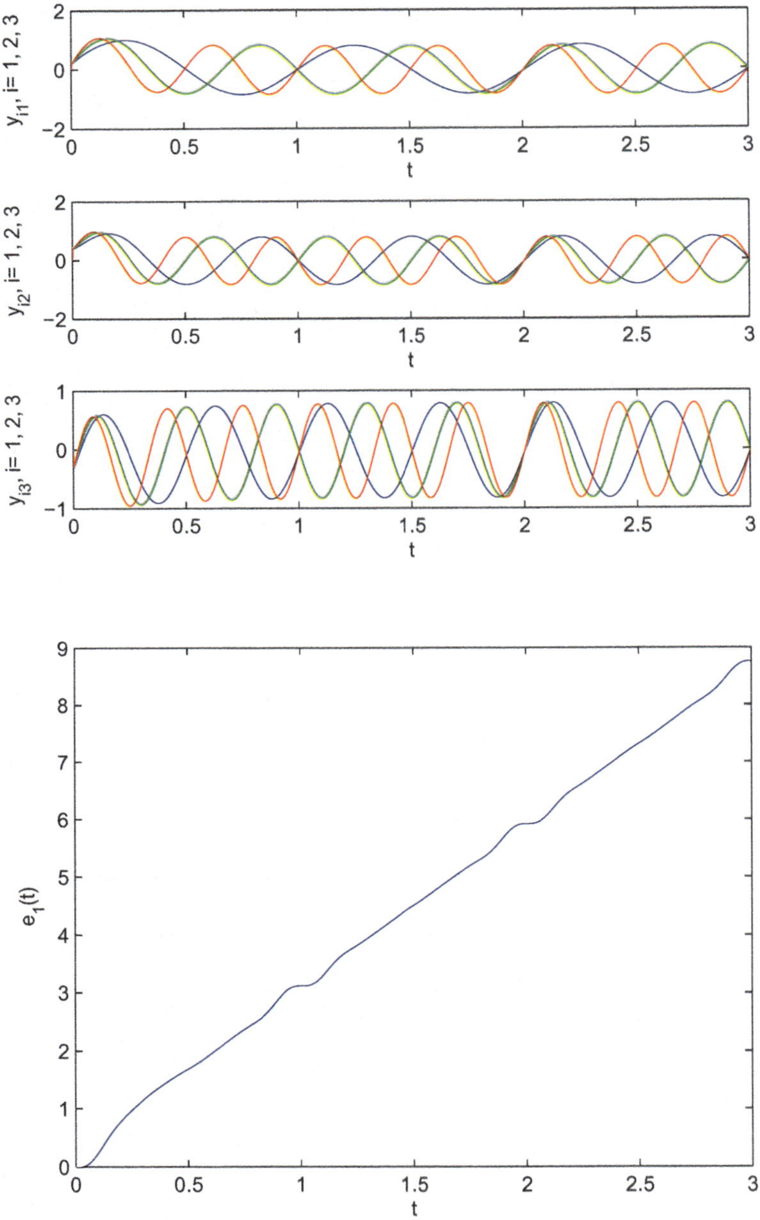

Fig. 2.2 The change process of output variables $y_{il}(i, l = 1, 2, 3)$ and $e_1(t)$, where $e_1(t) = 2\int_0^t y^T(s)u(s)ds - \int_0^t u^T(s)u(s)ds$

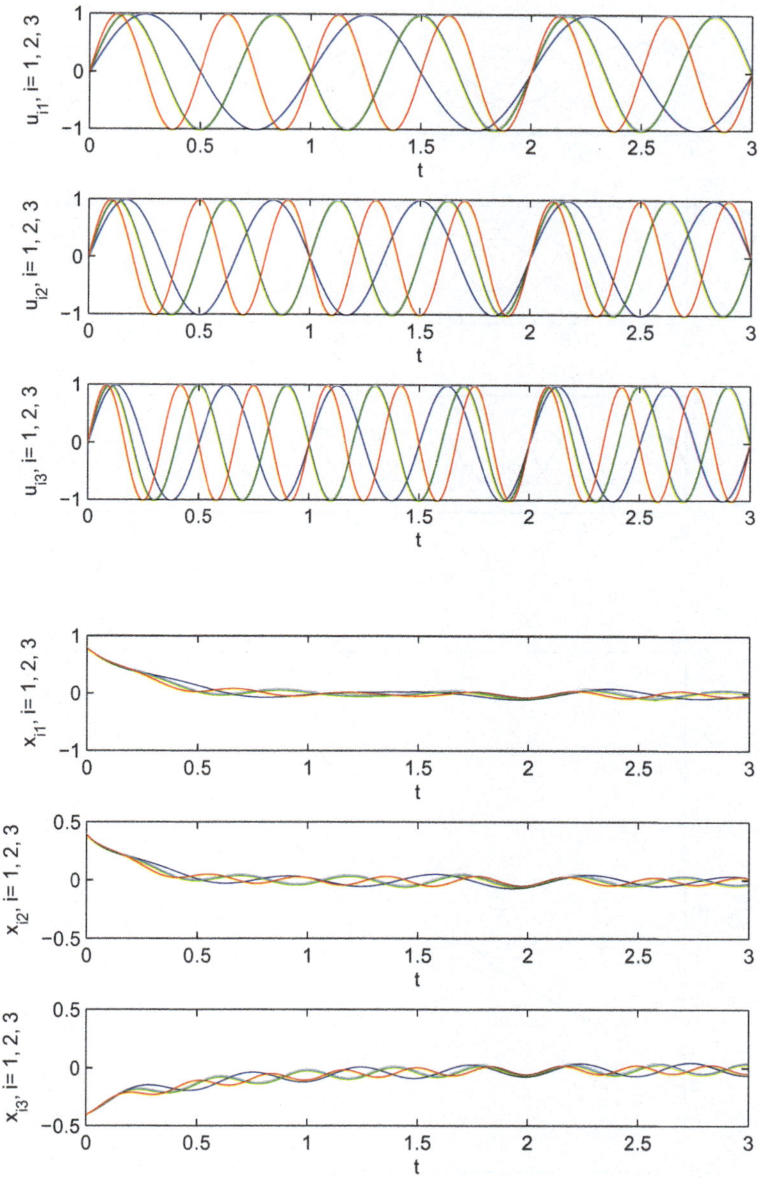

Fig. 2.3 The change process of input variables $u_{il}(i, l = 1, 2, 3)$, state variables $x_{il}(i, l = 1, 2, 3)$

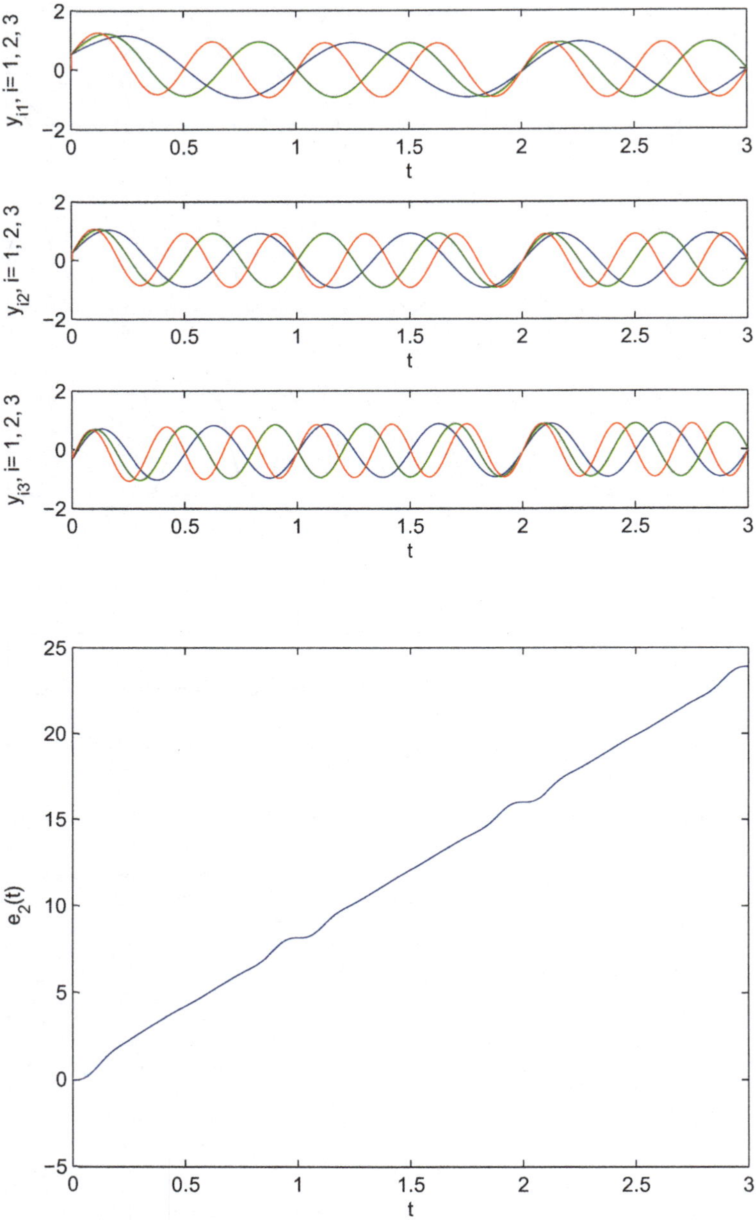

Fig. 2.4 The change process of output variables $y_{il}(i, l = 1, 2, 3)$ and $e_2(t)$, where $e_2(t) = 2\int_0^t y^T(s)u(s)ds - 0.1\int_0^t y^T(s)y(s)ds$

and

$$h(x_i(t)) = (\sin(x_{i1}(t))/4, \sin(x_{i2}(t))/4, \sin(x_{i3}(t))/4)^T,$$
$$\hat{h}(x_i(t)) = (\sin(x_{i1}(t))/5, \sin(x_{i2}(t))/5, \sin(x_{i3}(t))/5)^T, \ i = 1, 2, 3.$$

It is obvious that the couplings are not restricted to linear, symmetric and the non-negative off-diagonal, and we can take $\eta = \frac{5}{3}$, $P = I_3$, $\Delta = \text{diag}(0, 0, 0)$, $L_1 = \frac{1}{4}$, and $L_2 = \frac{1}{5}$.

In the following, we analyze the output strict passivity of complex network with different time-varying delays.

Setting $\tau_{il}(t) = 1 - \frac{1}{2i+2l}e^{-t}$, then $0 \leqslant \tau_{il}(t) \leqslant \tau_{il} = \tau = 1$, $\dot{\tau}_{il}(t) = \frac{1}{2i+2l}e^{-t} \leqslant \frac{1}{4} < 1$, for $t \geqslant 0$, $i, l = 1, 2, 3$.

We can find positive constants $\gamma = 0.1$ and $\xi = 1.5$ satisfying (2.24) and (2.25). According to Theorem 2.7, we know that complex network (2.1) with above given parameters is output strictly passive. Set $u_i(t) = (\sin((i+1)\pi t), \sin((i+2)\pi t), \sin((i+3)\pi t))^T$. The simulation results are shown in Figs. 2.3 and 2.4.

2.5 Conclusion

A generalized complex network with nonlinear, time-varying, nonsymmetric and delayed coupling has been introduced. We have considered the input strict passivity and output strict passivity of the proposed network model. Several input strict passivity and output strict passivity criteria have been established by constructing suitable Lyapunov functionals. Illustrative simulations have been provided to verify the correctness and effectiveness of the theoretical results. In future works, we will study the passivity of complex networks with impulsive effects and parameter uncertainties.

References

1. Zhou, J., & Lu, J. A. (2007). Topology identification of weighted complex dynamical networks. *Physica A, 386*(1), 481–491.
2. Hong, 2., Kim, B. J., Choi, M. Y., & Park, H. (2004). Factors that predict better synchronizability on complex networks. *Physical Review. E, 69*(6), 067105.
3. Nishikawa, T., Motter, A., Lai, Y. C., & Hoppensteadt, F. (2003). Heterogeneity in oscillator networks: Are smaller worlds easier to synchronize? *Physical Review Letters, 91*(1), 014101.

4. Wang, J. L., & Wu, H. N. (2012). Local and global exponential output synchronization of complex delayed dynamical networks. *Nonlinear Dynamics*, *67*(1), 497–504.
5. Wang, J. L., Yang, Z. C., Huang, T., & Xiao, M. (2011). Local and global exponential synchronization of complex delayed dynamical networks with general topology. *Discrete and Continuous Dynamical Systems. Series B 16*(1), 393–408.
6. Fan, J., & Wang, X. (2005). On synchronization in scale-free dynamical networks. *Physica A*, *349*(3–4), 443–451.
7. Lu, W., Atay, F., & Jost, J. (2007). Synchronization of discrete-time dynamical networks with time-varying couplings. *SIAM Journal on Mathematical Analysis*, *39*(4), 1231–1259.
8. Yu, W., Cao, J., & Lu, J. (2008). Global synchronization of linearly hybrid coupled networks with time-varying delay. *SIAM Journal Applied Dynamical Systems*, *7*(1), 108–133.
9. Li, C., & Chen, G. (2004). Synchronization in general complex dynamical networks with coupling delays. *Physica A*, *343*, 263–278.
10. Wang, Q., Duan, Z., Chen, G., & Feng, Z. (2008). Synchronization in a class of weighted complex networks with coupling delays. *Physica A*, *387*(22), 5616–5622.
11. Wu, X. (2008). Synchronization-based topology identification of weighted general complex dynamical networks with time-varying coupling delay. *Physica A*, *387*(4), 997–1008.
12. Cai, S., Zhou, J., Xiang, L., & Liu, Z. (2008). Robust impulsive synchronization of complex delayed dynamical networks. *Physics Letters A*, *372*(30), 4990–4995.
13. Chen, M. (2010). Synchronization in complex dynamical networks with random sensor delay. *IEEE Transactions on Circuits and Systems II: Express Briefs*, *57*(1), 46–50.
14. Li, X., & Chen, G. (2003). Synchronization and desynchronization of complex dynamical networks: An engineering viewpoint. *IEEE Transactions on Circuits and Systems I: Fundamental Theory and Applications*, *50*(11), 1381–1390.
15. Li, Z., & Chen, G. (2006). Global synchronization and asymptotic stability of complex dynamicalnetworks. *IEEE Transactions on Circuits and Systems II: Express Briefs*, *53*(1), 28–33.
16. Lu, J., & Chen, G. (2005). A time-varying complex dynamical network model and its controlled synchronization criteria. *IEEE Transactions on Automatic Control*, *50*(6), 841–846.
17. Xiang, L., Chen, Z., Liu, Z., Chen, F., & Yuan, Z. (2008). Pinning control of complex dynamical networks with heterogeneous delays. *Computers and Mathematics with Applications*, *56*(5), 1423–1433.
18. Xiang, L., Liu, Z., Chen, Z., & Yuan, Z. (2008). Pinning weighted complex networks with heterogeneous delays by a small number of feedback controllers. *Science in China Series F: Information Sciences*, *51*(5), 511–523.
19. Zhang, Q., Lu, J., Lu, J., & Tse, C. K. (2008). Adaptive feedback synchronization of a general complex dynamical network with delayed nodes. *IEEE Transactions on Circuits and Systems II: Express Briefs*, *55*(2), 183–187.
20. Zhou, J., Lu, J.-A., Lü, J. (2008). Pinning adaptive synchronization of a general complex dynamical network. *Automatica 44*(4), 996–1003.
21. Chen, T., Liu, X., & Lu, W. (2007). Pinning complex networks by a single controller. *IEEE Transactions on Circuits and Systems I: Regular Papers*, *54*(6), 1317–1326.
22. Li, Z., Feng, G., & Hill, D. (2006). Controlling complex dynamical networks with coupling delays to a desired orbit. *Physics Letters A*, *370*(5–6), 533–535.
23. Lestas, I. C., & Vinnicombe, G. (2007). Scalable robust stability for nonsymmetric heterogeneous networks. *Automatica*, *43*(4), 714–723.
24. Wang, Y. W., Yang, M., Wang, H. O., & Guan, Z. H. (2009). Robust stabilization of complex switched networks with parametric uncertainties and delays via impulsive control. *IEEE Transactions on Circuits and Systems I: Regular Papers*, *56*(9), 2100–2108.
25. López, N., & María, E. (2005). Several dissipativity and passivity implications in the linear discrete-time setting. *Mathematical Problems in Engineering*, *6*, 599–616.
26. López, E. N., & Fossas, E. (2004). Feedback passivity of nonlinear discrete-time systems with direct input-output link. *Automatica*, *40*(8), 1423–1428.
27. Niculescu, S., & Lozano, R. (2001). On the passivity of linear delay systems. *IEEE Transactions on Automatic Control*, *46*(3), 460–464.

28. Xie, L., Fu, M., & Li, H. (1998). Passivity analysis and passification for uncertain signal processing systems. *IEEE Transactions on Signal Processing*, *46*(9), 2394–2403.
29. Mahmoud, M., & Ismail, A. (2004). Passivity and passification of time-delay systems. *Journal of Mathematical Analysis and Applications*, *292*(1), 247–258.
30. Zhao, J., & Hill, D. (2008). Passivity and stability of switched systems: A multiple storage function method. *Systems and Control Letters*, *57*(2), 158–164.
31. Moylan, P. J., & Hill, D. J. (1980). Dissipative dynamical systems: Basic input-output and state properties. *Journal of the Franklin Institute*, *309*(5), 327–357.
32. Wang, J. L., Wu, H. N., & Yang, Z. C. (2011). Passivity analysis of impulsive complex networks. *International Journal of Automation and Computing*, *8*(4), 484–489.
33. Gao, H., Chen, T., & Chai, T. (2007). Passivity and passification for networked control systems. *SIAM Journal on Control and Optimization*, *46*(4), 1299–1322.
34. Chen, J. L., & Lee, L. (2001). Passivity approach to feedback connection stability for discrete-time descriptor systems. In *Proceedings of the 40th IEEE Conference on Decision and Control (Cat. No.01CH37228)* (Vol. 3, pp. 2865–2866).
35. Liu, X. (2007). Passivity analysis of uncertain fuzzy delayed systems. *Chaos, Solitons and Fractals*, *34*(3), 833–838.
36. Uang, H. J. (2005). On the dissipativity of nonlinear systems: Fuzzy control approach. *Fuzzy Sets and Systems*, *156*(2), 185–207.
37. Chang, W. J., Ku, C. C., Huang, P. H., & Chang, W. (2009). Fuzzy controller design for passive continuous-time affine T-S fuzzy models with relaxed stability conditions. *ISA Transactions*, *48*(3), 295–303.
38. Li, C., Zhang, H., & Liao, X. (2006). Passivity and passification of fuzzy systems with time delays. *Computers and Mathematics with Applications*, *52*(6), 1067–1078.
39. Lou, X., & Cui, B. (2007). Passivity analysis of integro-differential neural networks with time-varying delays. *Neurocomputing*, *70*(4), 1071–1078.
40. Park, J. (2007). Further results on passivity analysis of delayed cellular neural networks. *Chaos, Solitons and Fractals*, *34*(5), 1546–1551.
41. Chen, B., Li, H., Lin, C., & Zhou, Q. (2009). Passivity analysis for uncertain neural networks with discrete and distributed time-varying delays. *Physics Letters A*, *373*(14), 1242–1248.
42. Song, Q., Liang, J., & Wang, Z. (2009). Passivity analysis of discrete-time stochastic neural networks with time-varying delays. *Neurocomputing*, *72*(7–9), 1782–1788.
43. Song, Q., & Wang, Z. (2010). New results on passivity analysis of uncertain neural networks with time-varying delays. *International Journal of Computer Mathematics*, *87*(3), 668–678.
44. Li, C., & Liao, X. (2005). Passivity analysis of neural networks with time delay. *IEEE Transactions on Circuits and Systems II: Express Briefs*, *52*(8), 471–475.
45. Yao, J., Wang, H. O., Guan, Z. H., & Xu, W. (2009). Passive stability and synchronization of complex spatio-temporal switching networks with time delays. *Automatica*, *45*(7), 1721–1728.
46. Yao, J., Guan, Z. H., & Hill, D. J. (2009). Passivity-based control and synchronization of general complex dynamical networks. *Automatica*, *45*(9), 2107–2113.
47. Hill, D. J., & Moylan, P. J. (1977). Stability results for nonlinear feedback systems. *Automatica*, *13*(4), 377–382.
48. Chua, L. O. (1999). Passivity and complexity. *IEEE Transactions on Circuits and Systems I: Fundamental Theory and Applications*, *46*(1), 71–82.
49. Yu, W. (1999). Passive equivalence of chaos in lorenz system. *IEEE Transactions on Circuits and Systems I: Fundamental Theory and Applications*, *46*(7), 876–878.
50. Wu, C. W. (2001). Synchronization in arrays of coupled nonlinear systems: Passivity, circle criterion, and observer design. *IEEE Transactions on Circuits and Systems I: Fundamental Theory and Applications*, *48*(10), 1257–1261.
51. Calcev, G., Gorez, R., & Neyer, M. D. (1998). Passivity approach to fuzzy control systems. *Automatica*, *34*(3), 339–344.
52. Lu, W., & Chen, T. (2006). New approach to synchronization analysis of linearly coupled ordinary differential systems. *Physica D*, *213*(2), 214–230.

53. Guo, W., Austin, F., & Chen, Sh. (2010). Global synchronization of nonlinearly coupled complex networks with non-delayed and delayed coupling. *Communications in Nonlinear Science and Numerical Simulation, 15*(6), 1631–1639.

Chapter 3
Passivity Analysis and Pinning Control of Multi-weighted CDNs

3.1 Introduction

Recently, many researchers have paid much attention to the dynamical behaviors of complex networks, especially the synchronization of CDNs [1–10]. In [7], Cheng and Cao investigated the globally exponential synchronization for a CDN with discrete time-delays by means of linear matrix inequality technique and Lyapunov functional method. Xu et al. [8] discussed finite-time adaptive synchronization of CDNs by designing the bounded delay feedback controller. In [9], a sufficient condition was presented to ensure the exponential synchronization of the CDNs based on the Lyapunov functional.

In most of existing works, the synchronization problem of CDNs was transformed into the stability problem of error systems. Considering that stability problem can be dealt with by exploiting the passivity property of system in many situations, many researchers have widely studied the passivity problem of CDNs [11–26]. In [11], Yao et al. respectively discussed CDNs with and without coupling delays, and several criteria for passivity were obtained by applying Lyapunov functional. Wang et al. [12] considered the passivity for a generalized CDN with multiple time-varying delays. In [14], the authors introduced a complex switching network, and addressed passivity property of such network model. Su and Shen [20] studied the passivity of delayed CDNs by exploiting sampled-data control strategy. However, the network models considered in [11–26] have the same dimensions of input and output vectors. Practically, the dimensions of input and output vectors in many real networks are different. But, there are very few works about the passivity of CDNs with different dimensions of input and output [27, 28]. Wang et al. [27] analyzed the passivity of two CRDNNs with different dimensions of input and output. In [28], the authors respectively discussed the passivity of directed and undirected CRDNNs with different dimensions of input and output by designing suitable adaptive strategies.

On the other hand, in these existing works on the passivity of CDNs with different dimensions of input and output [27, 28], the network models with single weight were discussed. In practice, many real systems should be modeled by complex networks

J.-L. Wang et al., *Passivity of Complex Dynamical Networks*,
https://doi.org/10.1007/978-981-33-4287-3_3

with multi-weights, such as transportation networks, social networks, and so on. For instance, in social network, individuals or organizations can get in touch with others by mobile phone, Facebook, MSN, and so on. Considering that every way of communication has different weight, thus social networks should be modeled by multi-weighted complex networks. Furthermore, in the public traffic roads networks, taking every bus line as the network node, the edge means that two different bus lines have the same bus stops. In this case, there obviously exist different coupling weights in public traffic roads networks, such as coefficient of bus line length, passenger flow density, departing frequency, and so on [29]. However, there are rarely works about dynamical behaviors of CDNMWs [29–31]. By changing transfers coefficient, congestion degrees and passenger flow density between different bus lines, An et al. [29] investigated the global synchronization problem for a multi-weighted complex network. In [30], the authors derived a criterion about global synchronization of the public traffic roads networks with multi-weights based on the Lyapunov stability theory. Wang and Wu [31] took into consideration synchronization and $\mathcal{H}_\infty$ synchronization for CRDNNs with hybrid couplings.

Furthermore, since complex networks usually are not passive by themselves, there is a need to adopt some control strategies for ensuring passivity of CDNs. In recent years, in order to realize the desired dynamical behaviors in complex networks, some researchers have designed a great deal of pinning control strategies, including nodes-based pinning control strategy [32–35] and edges-based pinning control strategy [36, 37]. Yang et al. [32] considered the synchronization of CRDNNs with time-varying delay by adding impulsive controller to a small portion of nodes. In [33], the authors studied the synchronization of nonlinear dynamical networks with multiple stochastic disturbances using different pinning schemes. Tang et al. [34] investigated distributed robust pinning synchronization of complex networks with parameter uncertainties and stochastic coupling. Wang et al. [36] discussed the synchronization problem of two kinds of CRDNNs using edges-based adaptive strategy. In [37], Yu et al. presented a distributed adaptive control scheme for synchronization in CDNs. Obviously, it is also very important to study the passivity of CDNMWs by utilizing nodes-based and edges-based pinning control strategies. To the best of our knowledge, the nodes-based and edges-based pinning passivity problems of CDNMWs with different dimensions of input and output have not yet been investigated.

The aim of this chapter is to study the passivity of CDNMWs with different dimensions of input and output. First, several passivity criteria for CDNMWs are established by using some inequality techniques, and a synchronization condition for the output strictly passive CDNMWs is presented. Second, we derive several sufficient conditions for ensuring passivity of CDNMWs by using nodes-based pinning control strategy, and present a synchronization criterion for the output strictly passive CDNMWs. Third, we study the passivity of CDNMWs by exploiting edges-based pinning control method, and establish a synchronization condition for the output strictly passive CDNMWs.

3.2 Preliminaries

3.2.1 Network Model

The CDNMWs to be investigated in this chapter is given by

$$\begin{aligned}\dot{w}_i(t)=&g(w_i(t))+a_1\sum_{j=1}^{N}A^1_{ij}H_1w_j(t)+a_2\sum_{j=1}^{N}A^2_{ij}H_2w_j(t)\\&+\cdots+a_q\sum_{j=1}^{N}A^q_{ij}H_qw_j(t)+Bu_i(t),\end{aligned}\tag{3.1}$$

where $i=1,2,\ldots,N$, $u_i(t)\in\mathbb{R}^m$ denotes the input vector; $w_i(t)=(w_{i1}(t), w_{i2}(t),\ldots,w_{in}(t))^T\in\mathbb{R}^n$ is the state vector of the ith node; $B\in\mathbb{R}^{n\times m}$ is a known matrix; $g:\mathbb{R}^n\to\mathbb{R}^n$ is a continuously differentiable vector function; $0<a_s\in\mathbb{R}$ $(s=1,2,\ldots,q)$ denotes the coupling strength of the sth coupling form; $H_s\in\mathbb{R}^{n\times n}>0$ $(s=1,2,\ldots,q)$ is the inner coupling matrix for the sth coupling form; $A^s=(A^s_{ij})_{N\times N}\in\mathbb{R}^{N\times N}$ $(s=1,2,\ldots,q)$ is the coupling configuration matrix representing coupling weights in the sth coupling form, where $\mathbb{R}\ni A^s_{ij}$ is defined as follows: if there exists a connection between node i and node j, then $\mathbb{R}\ni A^s_{ij}=A^s_{ji}>0$; otherwise, $\mathbb{R}\ni A^s_{ij}=A^s_{ji}=0$ $(i\neq j)$; and the diagonal elements of matrix A^s are defined by

$$A^s_{ii}=-\sum_{\substack{j=1\\ j\neq i}}^{N}A^s_{ij},\ i=1,2,\ldots,N.$$

Remark 3.1 Practically, many real networks can be described by multi-weighted CDN models, that is, nodes in networks are coupled by multiple coupling forms, such as social networks, transportation networks, and so on. However, very few researchers have considered the passivity about CDNMWs. Hence, it is very significant to further investigate the passivity for CDNMWs.

Remark 3.2 Throughout this chapter, we always suppose that network (3.1) is connected, and it has the same topology structure for different coupling forms. In other words, $A^s_{ij}=0\,(s\in\{1,2,\ldots,q\})$ if and only if $A^{\varsigma}_{ij}=0$ for $\varsigma=1,2,\ldots,s-1,s+1,\ldots,q$.

To obtain our results, we introduce an useful assumption as follows:

(A1). If there exist a matrix $0<P=\mathrm{diag}(p_1,p_2,\ldots,p_n)\in\mathbb{R}^{n\times n}$ and a diagonal matrix $\Delta=\mathrm{diag}(\delta_1,\delta_2,\ldots,\delta_n)\in\mathbb{R}^{n\times n}$ such that $g(\cdot)$ satisfies the following inequality:

$$\begin{aligned}&(\eta_1-\eta_2)^T P[g(\eta_1)-g(\eta_2)-\Delta(\eta_1-\eta_2)]\\ &\leqslant -\beta(\eta_1-\eta_2)^T(\eta_1-\eta_2),\end{aligned}$$

for some $0<\beta\in\mathbb{R}$ and any $\eta_1,\eta_2\in\mathbb{R}^n$.

3.2.2 *Definitions*

Definition 3.3 (*see* [38]) A system with supply rate $\vartheta(u,y)$ is dissipative if there is a nonnegative function $S:[0,+\infty)\to[0,+\infty)$, called the storage function, such that

$$\int_{t_1}^{t_2}\vartheta(u(t),y(t))dt\geqslant S(t_2)-S(t_1)$$

for any $t_1,t_2\in[0,+\infty)$ and $t_2\geqslant t_1$, where $u(t)\in\mathbb{R}^p$, $y(t)\in\mathbb{R}^q$ are the input and output of the system.

Definition 3.4 (*see* [22]) If a system is dissipative with respect to

$$\omega(u(t),y(t))=y^T(t)Pu(t),$$

where $P\in\mathbb{R}^{q\times p}$ is a constant matrix, then this system is passive.

Definition 3.5 (*see* [22]) If a system is dissipative with respect to

$$\omega(u(t),y(t))=y^T(t)Pu(t)-u^T(t)Q_1u(t)-y^T(t)Q_2y(t),$$

where $Q_1\in\mathbb{R}^{p\times p}\geqslant 0$, $Q_2\in\mathbb{R}^{q\times q}\geqslant 0$, $\lambda_m(Q_1)+\lambda_m(Q_2)>0$ and $P\in\mathbb{R}^{q\times p}$, then this system is strictly passive.

The system is input strictly passive if $Q_1>0$ and output strictly passive if $Q_2>0$.

3.3 Passivity of CDNMWs

In this section, we investigate the passivity of CDNMWs. By utilizing Lyapunov functional and some inequality technique, several criteria for passivity of CDNMWs are obtained. In addition, a synchronization condition is also established for CDNMWs under the condition that network is output strictly passive.

3.3.1 Passivity Analysis

Suppose that $\hat{w} = (\hat{w}_1, \hat{w}_2, \ldots, \hat{w}_n)^T \in \mathbb{R}^n$ is an equilibrium point of an isolated node of the network (3.1). Then, one gets

$$\dot{\hat{w}} = g(\hat{w}) = 0.$$

Letting $z_i(t) = w_i(t) - \hat{w}$, we have

$$\dot{z}_i(t) = g(w_i(t)) - g(\hat{w}) + \sum_{s=1}^{q}\sum_{j=1}^{N} a_s A_{ij}^s H_s z_j(t) + Bu_i(t), \tag{3.2}$$

where $i = 1, 2, \ldots, N$.

The output vector $y_i(t) \in \mathbb{R}^d$ of the network (3.2) is defined as follows:

$$y_i(t) = F_1 z_i(t) + F_2 u_i(t),$$

where $F_1 \in \mathbb{R}^{d\times n}$ and $F_2 \in \mathbb{R}^{d\times m}$ are known matrices.

Theorem 3.6 *If there exists a matrix $M \in \mathbb{R}^{dN\times mN}$ such that*

$$\begin{pmatrix} W_1 & E_1 \\ E_1^T & -(I_N \otimes F_2^T)M - M^T(I_N \otimes F_2) \end{pmatrix} \leqslant 0, \tag{3.3}$$

where $W_1 = 2I_N \otimes (P\Delta - \beta I_n) + \sum_{s=1}^{q} a_s A^s \otimes (PH_s + H_s P)$, $E_1 = I_N \otimes (PB) - (I_N \otimes F_1^T)M$, then the network (3.2) is passive.

Proof Define the following Lyapunov functional for the network (3.2):

$$V_1(t) = \sum_{i=1}^{N} z_i^T(t) P z_i(t). \tag{3.4}$$

Then, one obtains

$$\begin{aligned} \dot{V}_1(t) = 2\sum_{i=1}^{N} z_i^T(t) P\Big[& g(w_i(t)) - g(\hat{w}) + Bu_i(t) \\ & + \sum_{s=1}^{q}\sum_{j=1}^{N} a_s A_{ij}^s H_s z_j(t)\Big]. \end{aligned} \tag{3.5}$$

According to (**A1**), we have

$$\begin{aligned} & z_i^T(t)P(g(w_i(t)) - g(\hat{w})) \\ = & (w_i(t) - \hat{w})^T P(g(w_i(t)) - g(\hat{w})) \\ \leqslant & z_i^T(t)(P\Delta - \beta I_n)z_i(t). \end{aligned} \tag{3.6}$$

By (3.5) and (3.6), one gets

$$\begin{aligned} \dot{V}_1(t) \leqslant & 2\sum_{i=1}^N z_i^T(t)(P\Delta - \beta I_n)z_i(t) + 2\sum_{i=1}^N z_i^T(t)PBu_i(t) \\ & +2\sum_{s=1}^q\sum_{i=1}^N\sum_{j=1}^N a_s A_{ij}^s z_i^T(t)PH_s z_j(t) \\ = & 2z^T(t)[I_N \otimes (P\Delta - \beta I_n)]z(t) \\ & +2\sum_{s=1}^q a_s z^T(t)[A^s \otimes (PH_s)]z(t) \\ & +2z^T(t)[I_N \otimes (PB)]u(t), \end{aligned} \tag{3.7}$$

where $z(t) = (z_1^T(t), z_2^T(t), \ldots, z_N^T(t))^T$, $u(t) = (u_1^T(t), u_2^T(t), \ldots, u_N^T(t))^T$.
From (3.7), one has

$$\begin{aligned} & \dot{V}_1(t) - 2y^T(t)Mu(t) \\ \leqslant & 2z^T(t)[I_N \otimes (P\Delta - \beta I_n)]z(t) + 2\sum_{s=1}^q a_s z^T(t)[A^s \otimes (PH_s)]z(t) \\ & +2z^T(t)[I_N \otimes (PB)]u(t) - 2[z^T(t)(I_N \otimes F_1^T) + u^T(t)(I_N \otimes F_2^T)]Mu(t) \\ = & \xi^T(t)\begin{pmatrix} W_1 & E_1 \\ E_1^T & W_2 \end{pmatrix}\xi(t), \end{aligned}$$

where $W_2 = -(I_N \otimes F_2^T)M - M^T(I_N \otimes F_2)$, $y(t) = (y_1^T(t), y_2^T(t), \ldots, y_N^T(t))^T$, $\xi(t) = (z^T(t), u^T(t))^T$.
By (3.3), one gets

$$2y^T(t)Mu(t) \geqslant \dot{V}_1(t). \tag{3.8}$$

From (3.8), one obtains

$$2\int_{t_1}^{t_2} y^T(s)Mu(s)ds \geqslant V_1(t_2) - V_1(t_1)$$

for any $t_1, t_2 \in [0, +\infty)$ and $t_2 \geqslant t_1$. Namely,

$$\int_{t_1}^{t_2} y^T(s)Mu(s)ds \geqslant \hat{S}(t_2) - \hat{S}(t_1),$$

where $\hat{S}(t) = \frac{V_1(t)}{2}$.

Theorem 3.7 *If there exist matrices $M \in \mathbb{R}^{dN\times mN}$ and $0 < Q_1 \in \mathbb{R}^{mN\times mN}$ such that*

$$\begin{pmatrix} W_1 & E_1 \\ E_1^T & W_3 \end{pmatrix} \leqslant 0, \tag{3.9}$$

where $W_1=2I_N \otimes (P\Delta - \beta I_n) + \sum_{s=1}^{q} a_s A^s \otimes (PH_s + H_sP)$, $E_1=I_N \otimes (PB) - (I_N \otimes F_1^T)M$, $W_3 = Q_1 - (I_N \otimes F_2^T)M - M^T(I_N \otimes F_2)$, then the network (3.2) *is input strictly passive.*

Proof By (3.7), we can obtain

$$\begin{aligned}
&\dot{V}_1(t) - 2y^T(t)Mu(t) + u^T(t)Q_1u(t)\\
&\leqslant 2z^T(t)[I_N \otimes (P\Delta - \beta I_n)]z(t)\\
&\quad +2z^T(t)[I_N \otimes (PB)]u(t) + u^T(t)Q_1u(t)\\
&\quad +2\sum_{s=1}^{q} a_s z^T(t)[A^s \otimes (PH_s)]z(t)\\
&\quad -2[z^T(t)(I_N \otimes F_1^T) + u^T(t)(I_N \otimes F_2^T)]Mu(t)\\
&= \xi^T(t)\begin{pmatrix} W_1 & E_1 \\ E_1^T & W_3 \end{pmatrix}\xi(t).
\end{aligned}$$

Therefore,

$$2y^T(t)Mu(t) - u^T(t)Q_1u(t) \geqslant \dot{V}_1(t). \tag{3.10}$$

From (3.10), we have

$$2\int_{t_1}^{t_2} y^T(s)Mu(s)ds - \int_{t_1}^{t_2} u^T(s)Q_1u(s)ds \geqslant V_1(t_2) - V_1(t_1).$$

for any $t_1, t_2 \in [0, +\infty)$ and $t_2 \geqslant t_1$. That is to say,

$$\int_{t_1}^{t_2} \left(y^T(s)Mu(s) - u^T(s)\frac{Q_1}{2}u(s)\right) ds \geqslant \hat{S}(t_2) - \hat{S}(t_1).$$

Theorem 3.8 *If there exist matrices $M \in \mathbb{R}^{dN \times mN}$ and $0 < Q_2 \in \mathbb{R}^{dN \times dN}$ such that*

$$\begin{pmatrix} W_4 & E_2 \\ E_2^T & W_5 \end{pmatrix} \leqslant 0, \tag{3.11}$$

where $W_4 = 2I_N \otimes (P\Delta - \beta I_n) + \sum_{s=1}^{q} a_s A^s \otimes (PH_s + H_s P) + (I_N \otimes F_1^T)Q_2 (I_N \otimes F_1)$, $E_2 = I_N \otimes (PB) - (I_N \otimes F_1^T)M + (I_N \otimes F_1^T)Q_2(I_N \otimes F_2)$, $W_5 = -(I_N \otimes F_2^T)M - M^T(I_N \otimes F_2) + (I_N \otimes F_2^T)Q_2(I_N \otimes F_2)$, *then the network* (3.2) *is output strictly passive.*

Proof By (3.7), we obtain

$$\begin{aligned}
&\dot{V}_1(t) - 2y^T(t)Mu(t) + y^T(t)Q_2 y(t) \\
\leqslant\ & 2z^T(t)[I_N \otimes (P\Delta - \beta I_n)]z(t) + 2z^T(t)[I_N \otimes (PB)]u(t) \\
&+[z^T(t)(I_N \otimes F_1^T) + u^T(t)(I_N \otimes F_2^T)]Q_2[(I_N \otimes F_1)z(t) + (I_N \otimes F_2)u(t)] \\
&+2\sum_{s=1}^{q} a_s z^T(t)[A^s \otimes (PH_s)]z(t) - 2[z^T(t)(I_N \otimes F_1^T) \\
&+u^T(t)(I_N \otimes F_2^T)]Mu(t) \\
=\ & \xi^T(t)\begin{pmatrix} W_4 & E_2 \\ E_2^T & W_5 \end{pmatrix}\xi(t).
\end{aligned}$$

Therefore,

$$2y^T(t)Mu(t) - y^T(t)Q_2 y(t) \geqslant \dot{V}_1(t). \tag{3.12}$$

From (3.12), we have

$$2\int_{t_1}^{t_2} y^T(s)Mu(s)ds - \int_{t_1}^{t_2} y^T(s)Q_2 y(s)ds \geqslant V_1(t_2) - V_1(t_1),$$

for any $t_1, t_2 \in [0, +\infty)$ and $t_2 \geqslant t_1$. That is to say,

$$\int_{t_1}^{t_2} \left(y^T(s)Mu(s) - y^T(s)\frac{Q_2}{2}y(s)\right) ds \geqslant \hat{S}(t_2) - \hat{S}(t_1).$$

Remark 3.9 In the past decades, many researchers have analyzed the passivity of CDNs [11–26]. But, the network models considered in [11–26] have the same dimensions of input and output vectors. On the other hand, in most of existing works on the passivity of CDNs, the network models with single weight were discussed. To our knowledge, this is the first paper to analyze the passivity of CDNMWs with different dimensions of input and output vectors.

3.3.2 Synchronization Criteria

Definition 3.10 The CDNMWs (3.1) is synchronized if

$$\lim_{t \to +\infty} \|w_i(t) - w_j(t)\| = 0$$

for all $i, j = 1, 2, \ldots, N$ under the condition that $u_i(t) = 0, i = 1, 2, \ldots, N$.

Theorem 3.11 *Suppose that $S : [0, +\infty) \to [0, +\infty)$ is continuously differentiable and satisfies the following condition:*

$$\upsilon_1(\|z(t)\|) \leqslant S(t) \leqslant \upsilon_2(\|z(t)\|), \tag{3.13}$$

where $\upsilon_1, \upsilon_2 : [0, +\infty) \to [0, +\infty)$ are continuous and strictly monotonically increasing functions, $\upsilon_1(s)$ and $\upsilon_2(s)$ are positive for $s > 0$ with $\upsilon_1(0) = \upsilon_2(0) = 0$, then the network (3.2) *is asymptotically stable if it is output strictly passive with respect to storage function $S(t)$ and matrix $F_1 \in \mathbb{R}^{n \times n}$ is nonsingular.*

Proof If the network (3.2) is output strictly passive with respect to storage function $S(t)$, then there exist matrices $M \in \mathbb{R}^{nN \times mN}$ and $0 < Q_2 \in \mathbb{R}^{nN \times nN}$ such that

$$S(t+\gamma) - S(t) \leqslant \int_t^{t+\gamma} y^T(s) M u(s) ds - \int_t^{t+\gamma} y^T(s) Q_2 y(s) ds \tag{3.14}$$

for any $t \in [0, +\infty)$ and $\mathbb{R} \ni \gamma > 0$. Then, we can derive from (3.14) that

$$\begin{aligned} \frac{S(t+\gamma) - S(t)}{\gamma} \leqslant & \frac{\int_t^{t+\gamma} y^T(s) M u(s) ds}{\gamma} \\ & - \frac{\int_t^{t+\gamma} y^T(s) Q_2 y(s) ds}{\gamma}. \end{aligned} \tag{3.15}$$

Letting $\gamma \to 0$ in (3.15), one obtains

$$\dot{S}(t) \leqslant y^T(t) M u(t) - y^T(t) Q_2 y(t). \tag{3.16}$$

Taking $u(t) = 0$, we can get

$$\begin{aligned} \dot{S}(t) &\leqslant -z^T(t)(I_N \otimes F_1^T) Q_2 (I_N \otimes F_1) z(t) \\ &\leqslant -\lambda_m\Big((I_N \otimes F_1^T) Q_2 (I_N \otimes F_1)\Big)\|z(t)\|^2. \end{aligned} \tag{3.17}$$

From (3.13) and (3.17), we can get the network (3.2) is asymptotically stable.

According to Theorems 3.8 and 3.11, we can derive a conclusion as follows.

Corollary 3.12 *If there exist matrices $M \in \mathbb{R}^{nN\times mN}$, $F_2 \in \mathbb{R}^{n\times m}$, $0 < Q_2 \in \mathbb{R}^{nN\times nN}$ and a nonsingular matrix $F_1 \in \mathbb{R}^{n\times n}$ such that*

$$\begin{pmatrix} W_4 & E_2 \\ E_2^T & W_5 \end{pmatrix} \leqslant 0, \tag{3.18}$$

where $W_4 = 2I_N \otimes (P\Delta - \beta I_n) + \sum_{s=1}^{q} a_s A^s \otimes (PH_s + H_s P) + (I_N \otimes F_1^T)Q_2 (I_N \otimes F_1)$, $E_2 = I_N \otimes (PB) - (I_N \otimes F_1^T)M + (I_N \otimes F_1^T)Q_2(I_N \otimes F_2)$, $W_5 = -(I_N \otimes F_2^T)M - M^T(I_N \otimes F_2) + (I_N \otimes F_2^T)Q_2(I_N \otimes F_2)$, *then the network* (3.1) *is synchronized.*

Remark 3.13 In this section, by constructing appropriate Lyapunov functional, some sufficient conditions for ensuring passivity, input strict passivity and output strict passivity of the CDNMWs (3.2) are established (see Theorems 3.6, 3.7, and 3.8). In addition, we also study the synchronization of CDNMWs (3.1) by utilizing the output strict passivity.

3.4 Nodes-Based Pinning Passivity of CDNMWs

The network (3.1) under the nodes-based pinning adaptive controller is described as follows:

$$\begin{aligned}
\dot{w}_i(t) &= g(w_i(t)) + \sum_{s=1}^{q}\sum_{j=1}^{N} a_s A_{ij}^s H_s w_j(t) + Bu_i(t) \\
&\quad + v_i(t), \qquad i = 1, 2, \dots, l, \\
\dot{w}_i(t) &= g(w_i(t)) + \sum_{s=1}^{q}\sum_{j=1}^{N} a_s A_{ij}^s H_s w_j(t) + Bu_i(t), \\
&\qquad i = l+1, l+2, \dots, N,
\end{aligned} \tag{3.19}$$

where

$$\begin{aligned}
v_i(t) &= -\sum_{s=1}^{q} a_s k_i^s(t) H_s \left(w_i(t) - \frac{1}{N}\sum_{k=1}^{N} w_k(t) \right), \\
\dot{k}_i^s(t) &= \beta_i^s \left(w_i(t) - \frac{1}{N}\sum_{k=1}^{N} w_k(t) \right)^T \frac{PH_s + H_s P}{2} \left(w_i(t) \right. \\
&\quad \left. - \frac{1}{N}\sum_{k=1}^{N} w_k(t) \right),
\end{aligned} \tag{3.20}$$

in which $1 \leqslant l < N$, $PH_s + H_s P > 0$, $\mathbb{R} \ni \beta_i^s > 0$, $\mathbb{R} \ni k_i^s(0) > 0$.

Define $w^*(t) = \frac{1}{N}\sum_{k=1}^{N} w_k(t)$. Then, one gets

$$\begin{aligned}\dot{w}^*(t) &= \frac{1}{N}\sum_{k=1}^{N} g(w_k(t)) + \frac{1}{N}\sum_{k=1}^{N} Bu_k(t) + \frac{1}{N}\sum_{k=1}^{l} v_k(t)\\ &\quad + \frac{1}{N}\sum_{s=1}^{q}\sum_{j=1}^{N} a_s\left(\sum_{k=1}^{N} A_{kj}^s\right) H_s w_j(t)\\ &= \frac{1}{N}\sum_{k=1}^{N} g(w_k(t)) + \frac{1}{N}\sum_{k=1}^{N} Bu_k(t) + \frac{1}{N}\sum_{k=1}^{l} v_k(t).\end{aligned}$$

Letting $\hat{z}_i(t) = w_i(t) - w^*(t)$, we have

$$\begin{aligned}\dot{\hat{z}}_i(t) &= g(w_i(t)) - \frac{1}{N}\sum_{k=1}^{N} g(w_k(t)) + Bu_i(t) + \sum_{s=1}^{q}\sum_{j=1}^{N} a_s A_{ij}^s H_s \hat{z}_j(t)\\ &\quad - \frac{1}{N}\sum_{k=1}^{l} v_k(t) - \frac{1}{N}\sum_{k=1}^{N} Bu_k(t) - \sum_{s=1}^{q} a_s k_i^s(t) H_s \hat{z}_i(t),\\ \dot{k}_i^s(t) &= \beta_i^s \hat{z}_i^T(t)\frac{PH_s + H_s P}{2}\hat{z}_i(t),\ i = 1, 2, \ldots, l,\end{aligned} \tag{3.21}$$

where $k_i^s(t) \equiv 0$ for $i = l+1, l+2, \ldots, N$.

The output vector $y_i(t) \in \mathbb{R}^d$ of network (3.21) is defined as follows:

$$y_i(t) = F_1\hat{z}_i(t) + F_2u_i(t),$$

where $F_1 \in \mathbb{R}^{d\times n}$ and $F_2 \in \mathbb{R}^{d\times m}$ are known matrices.

3.4.1 Passivity Analysis

Theorem 3.14 *If there exist matrices $\hat{M} \in \mathbb{R}^{dN\times mN}$ and $\hat{k}^s = \mathrm{diag}(\hat{k}_1^s, \hat{k}_2^s, \ldots, \hat{k}_l^s, 0, \ldots, 0) \in \mathbb{R}^{N\times N}$ such that*

$$\begin{pmatrix} \hat{W}_1 & \hat{E}_1 \\ \hat{E}_1^T & \hat{W}_2 \end{pmatrix} \leqslant 0, \tag{3.22}$$

where $\hat{W}_1 = 2I_N \otimes (P\Delta - \beta I_n) + \sum_{s=1}^{q} a_s(A^s - \hat{k}^s) \otimes (PH_s + H_sP)$, $\hat{E}_1 = I_N \otimes (PB) - (I_N \otimes F_1^T)\hat{M}$, $\hat{W}_2 = -(I_N \otimes F_2^T)\hat{M} - \hat{M}^T(I_N \otimes F_2)$, $\hat{k}_i^s > 0$ for $i = 1, 2, \ldots, l$, then the network (3.21) is passive.

Proof Define a Lyapunov functional for the network (3.21) as follows:

$$V_2(t) = \sum_{i=1}^{N} \hat{z}_i^T(t) P \hat{z}_i(t) + \sum_{s=1}^{q} \sum_{i=1}^{l} \frac{a_s}{\beta_i^s} (k_i^s(t) - \hat{k}_i^s)^2. \tag{3.23}$$

Then,

$$\begin{aligned}
\dot{V}_2(t) &= 2\sum_{i=1}^{N} \hat{z}_i^T(t) P \dot{\hat{z}}_i(t) + 2\sum_{s=1}^{q} \sum_{i=1}^{l} \frac{a_s}{\beta_i^s} (k_i^s(t) - \hat{k}_i^s)\dot{k}_i^s(t) \\
&= 2\sum_{i=1}^{N} \hat{z}_i^T(t) P \Bigg[g(w_i(t)) - g(w^*(t)) + g(w^*(t)) - \frac{1}{N}\sum_{k=1}^{N} g(w_k(t)) \\
&\quad + \sum_{s=1}^{q} \sum_{j=1}^{N} a_s A_{ij}^s H_s \hat{z}_j(t) + B u_i(t) - \frac{1}{N}\sum_{k=1}^{N} B u_k(t) - \frac{1}{N}\sum_{k=1}^{l} v_k(t) \\
&\quad - \sum_{s=1}^{q} a_s k_i^s(t) H_s \hat{z}_i(t) \Bigg] + \sum_{s=1}^{q} \sum_{i=1}^{N} a_s (k_i^s(t) - \hat{k}_i^s) \hat{z}_i^T(t)(P H_s \\
&\quad + H_s P)\hat{z}_i(t),
\end{aligned}$$

where $\hat{k}_i^s = 0$ for $i = l+1, l+2, \ldots, N$.

Since

$$\begin{aligned}
\sum_{i=1}^{N} \hat{z}_i(t) &= \sum_{i=1}^{N} \left(w_i(t) - \frac{1}{N}\sum_{k=1}^{N} w_k(t) \right) \\
&= \sum_{i=1}^{N} w_i(t) - \sum_{k=1}^{N} w_k(t) \\
&= 0,
\end{aligned}$$

we obtain

$$\sum_{i=1}^{N} \hat{z}_i^T(t) P \left(g(w^*(t)) - \frac{1}{N}\sum_{k=1}^{N} g(w_k(t)) \right) = 0, \tag{3.24}$$

$$\frac{1}{N}\sum_{i=1}^{N} \hat{z}_i^T(t) P \sum_{k=1}^{N} B u_k(t) = 0, \tag{3.25}$$

$$\frac{1}{N}\sum_{i=1}^{N} \hat{z}_i^T(t) P \sum_{k=1}^{l} v_k(t) = 0. \tag{3.26}$$

In view of (3.24)–(3.26), one has

$$\begin{aligned}
\dot{V}_2(t) &\leqslant 2\sum_{i=1}^{N}\hat{z}_i^T(t)(P\Delta-\beta I_n)\hat{z}_i(t)+2\sum_{s=1}^{q}\sum_{i=1}^{N}\sum_{j=1}^{N}a_sA_{ij}^s\hat{z}_i^T(t)PH_s\hat{z}_j(t)\\
&\quad+2\sum_{i=1}^{N}\hat{z}_i^T(t)PBu_i(t)-\sum_{s=1}^{q}\sum_{i=1}^{N}a_s\hat{k}_i^s\hat{z}_i^T(t)(PH_s+H_sP)\hat{z}_i(t)\\
&=\hat{z}^T(t)\Big\{\sum_{s=1}^{q}a_s(A^s-\hat{k}^s)\otimes(PH_s+H_sP)\\
&\quad+2I_N\otimes(P\Delta-\beta I_n)\Big\}\hat{z}(t)+2\hat{z}^T(t)[I_N\otimes(PB)]u(t),
\end{aligned}\tag{3.27}$$

where $\hat{z}(t)=(\hat{z}_1^T(t),\hat{z}_2^T(t),\ldots,\hat{z}_N^T(t))^T$, $u(t)=(u_1^T(t),u_2^T(t),\ldots,u_N^T(t))^T$.

Therefore,

$$\begin{aligned}
&\dot{V}_2(t)-2y^T(t)\hat{M}u(t)\\
&\leqslant\hat{z}^T(t)\Big\{\sum_{s=1}^{q}a_s(A^s-\hat{k}^s)\otimes(PH_s+H_sP)\\
&\quad+2I_N\otimes(P\Delta-\beta I_n)\Big\}\hat{z}(t)\\
&\quad+2\hat{z}^T(t)[I_N\otimes(PB)]u(t)-2[\hat{z}^T(t)(I_N\otimes F_1^T)\\
&\quad+u^T(t)(I_N\otimes F_2^T)]\hat{M}u(t)\\
&=\hat{\xi}^T(t)\begin{pmatrix}\hat{W}_1 & \hat{E}_1\\ \hat{E}_1^T & \hat{W}_2\end{pmatrix}\hat{\xi}(t),
\end{aligned}\tag{3.28}$$

where $\hat{\xi}(t)=(\hat{z}^T(t),\ u^T(t))^T$, $y(t)=(y_1^T(t),y_2^T(t),\ldots,y_N^T(t))^T$.

By (3.22) and (3.28), one gets

$$2y^T(t)\hat{M}u(t)\geqslant\dot{V}_2(t).\tag{3.29}$$

From (3.29), we have

$$2\int_{t_1}^{t_2}y^T(s)\hat{M}u(s)ds\geqslant V_2(t_2)-V_2(t_1)$$

for any $t_1,t_2\in[0,+\infty)$ and $t_2\geqslant t_1$. Namely,

$$\int_{t_1}^{t_2}y^T(s)\hat{M}u(s)ds\geqslant\tilde{S}(t_2)-\tilde{S}(t_1),$$

where $\tilde{S}(t)=\frac{V_2(t)}{2}$.

Theorem 3.15 *If there exist matrices* $\hat{M} \in \mathbb{R}^{dN \times mN}$, $0 < \hat{Q}_1 \in \mathbb{R}^{mN \times mN}$ *and* $\hat{k}^s = \mathrm{diag}(\hat{k}_1^s, \hat{k}_2^s, \dots, \hat{k}_l^s, 0, \dots, 0) \in \mathbb{R}^{N \times N}$ *such that*

$$\begin{pmatrix} \hat{W}_1 & \hat{E}_1 \\ \hat{E}_1^T & \hat{W}_3 \end{pmatrix} \leqslant 0, \tag{3.30}$$

where $\hat{W}_1 = 2I_N \otimes (P\Delta - \beta I_n) + \sum_{s=1}^{q} a_s(A^s - \hat{k}^s) \otimes (PH_s + H_sP)$, $\hat{E}_1 = I_N \otimes (PB) - (I_N \otimes F_1^T)\hat{M}$, $\hat{W}_3 = \hat{Q}_1 - (I_N \otimes F_2^T)\hat{M} - \hat{M}^T(I_N \otimes F_2)$, $\hat{k}_i^s > 0$ *for* $i = 1, 2, \dots, l$, *then the network* (3.21) *is input strictly passive.*

Proof By (3.27), one has

$$\begin{aligned}
&\dot{V}_2(t) - 2y^T(t)\hat{M}u(t) + u^T(t)\hat{Q}_1u(t) \\
&\leqslant \hat{z}^T(t)\Big\{\sum_{s=1}^{q} a_s(A^s - \hat{k}^s) \otimes (PH_s + H_sP) \\
&\quad +2I_N \otimes (P\Delta - \beta I_n)\Big\}\hat{z}(t) \\
&\quad +2\hat{z}^T(t)[I_N \otimes (PB)]u(t) - 2[\hat{z}^T(t)(I_N \otimes F_1^T) \\
&\quad +u^T(t)(I_N \otimes F_2^T)]\hat{M}u(t) + u^T(t)\hat{Q}_1u(t) \\
&= \hat{\xi}^T(t)\begin{pmatrix} \hat{W}_1 & \hat{E}_1 \\ \hat{E}_1^T & \hat{W}_3 \end{pmatrix}\hat{\xi}(t).
\end{aligned}$$

Therefore,

$$2y^T(t)\hat{M}u(t) - u^T(t)\hat{Q}_1u(t) \geqslant \dot{V}_2(t). \tag{3.31}$$

From (3.31), we have

$$2\int_{t_1}^{t_2} y^T(s)\hat{M}u(s)ds - \int_{t_1}^{t_2} u^T(s)\hat{Q}_1u(s)ds \geqslant V_2(t_2) - V_2(t_1)$$

for any $t_1, t_2 \in [0, +\infty)$ and $t_2 \geqslant t_1$. That is to say,

$$\int_{t_1}^{t_2} \left(y^T(s)\hat{M}u(s) - u^T(s)\frac{\hat{Q}_1}{2}u(s)\right) ds \geqslant \tilde{S}(t_2) - \tilde{S}(t_1).$$

Theorem 3.16 *If there exist matrices* $\hat{M} \in \mathbb{R}^{dN \times mN}$, $0 < \hat{Q}_2 \in \mathbb{R}^{dN \times dN}$ *and* $\hat{k}^s = \mathrm{diag}(\hat{k}_1^s, \hat{k}_2^s, \dots, \hat{k}_l^s, 0, \dots, 0) \in \mathbb{R}^{N \times N}$ *such that*

$$\begin{pmatrix} \hat{W}_4 & \hat{E}_2 \\ \hat{E}_2^T & \hat{W}_5 \end{pmatrix} \leqslant 0, \tag{3.32}$$

where $\hat{W}_4 = 2I_N \otimes (P\Delta - \beta I_n) + \sum_{s=1}^{q} a_s(A^s - \hat{k}^s) \otimes (PH_s + H_s P) + (I_N \otimes F_1^T)\hat{Q}_2(I_N \otimes F_1)$, $\hat{E}_2 = I_N \otimes (PB) - (I_N \otimes F_1^T)\hat{M} + (I_N \otimes F_1^T)\hat{Q}_2(I_N \otimes F_2)$, $\hat{W}_5 = -(I_N \otimes F_2^T)\hat{M} - \hat{M}^T(I_N \otimes F_2) + (I_N \otimes F_2^T)\hat{Q}_2(I_N \otimes F_2)$, $\hat{k}_i^s > 0$ *for* $i = 1, 2, \ldots, l$, *then the network (3.21) is output strictly passive.*

Proof By (3.27), one gets

$$
\begin{aligned}
&\dot{V}_2(t) - 2y^T(t)\hat{M}u(t) + y^T(t)\hat{Q}_2 y(t) \\
&\leqslant \hat{z}^T(t)\Big\{\sum_{s=1}^{q} a_s(A^s - \hat{k}^s) \otimes (PH_s + H_s P) \\
&\quad + 2I_N \otimes (P\Delta - \beta I_n)\Big\}\hat{z}(t) \\
&\quad + 2\hat{z}^T(t)[I_N \otimes (PB)]u(t) - 2[\hat{z}^T(t)(I_N \otimes F_1^T) \\
&\quad + u^T(t)(I_N \otimes F_2^T)]\hat{M}u(t) + [\hat{z}^T(t)(I_N \otimes F_1^T) \\
&\quad + u^T(t)(I_N \otimes F_2^T)]\hat{Q}_2[(I_N \otimes F_1)\hat{z}(t) + (I_N \otimes F_2)u(t)] \\
&= \hat{\xi}^T(t)\begin{pmatrix} \hat{W}_4 & \hat{E}_2 \\ \hat{E}_2^T & \hat{W}_5 \end{pmatrix}\hat{\xi}(t).
\end{aligned}
$$

Therefore,

$$
2y^T(t)\hat{M}u(t) - y^T(t)\hat{Q}_2 y(t) \geqslant \dot{V}_2(t). \tag{3.33}
$$

From (3.33), we have

$$
2\int_{t_1}^{t_2} y^T(s)\hat{M}u(s)ds - \int_{t_1}^{t_2} y^T(s)\hat{Q}_2 y(s)ds \geqslant V_2(t_2) - V_2(t_1)
$$

for any $t_1, t_2 \in [0, +\infty)$ and $t_2 \geqslant t_1$. Namely,

$$
\int_{t_1}^{t_2}\left(y^T(s)\hat{M}u(s) - y^T(s)\frac{\hat{Q}_2}{2}y(s)\right)ds \geqslant \tilde{S}(t_2) - \tilde{S}(t_1).
$$

3.4.2 Synchronization Criteria

Theorem 3.17 *The CDNMWs* (3.1) *under the nodes-based pinning adaptive controller* (3.20) *is synchronized if the network* (3.21) *is output strictly passive with respect to storage function* $\tilde{S}(t) = \frac{V_2(t)}{2}$ *and matrix* $F_1 \in \mathbb{R}^{n\times n}$ *is nonsingular.*

Proof If the network (3.21) is output strictly passive with respect to storage function $\tilde{S}(t)$, then there exist matrices $\hat{M} \in \mathbb{R}^{nN \times mN}$ and $0 < \hat{Q}_2 \in \mathbb{R}^{nN \times nN}$ such that

$$\tilde{S}(t+\gamma) - \tilde{S}(t) \leqslant \int_t^{t+\gamma} y^T(s)\hat{M}u(s)ds - \int_t^{t+\gamma} y^T(s)\hat{Q}_2 y(s)ds \tag{3.34}$$

for any $t \in [0, +\infty)$ and $\mathbb{R} \ni \gamma > 0$. Then, we can get

$$\frac{\tilde{S}(t+\gamma) - \tilde{S}(t)}{\gamma} \leqslant \frac{\int_t^{t+\gamma} y^T(s)\hat{M}u(s)ds}{\gamma} - \frac{\int_t^{t+\gamma} y^T(s)\hat{Q}_2 y(s)ds}{\gamma}. \tag{3.35}$$

Letting $\gamma \to 0$ in (3.35), one obtains

$$\dot{\tilde{S}}(t) \leqslant y^T(t)\hat{M}u(t) - y^T(t)\hat{Q}_2 y(t).$$

Taking $u(t) = 0$, we can get

$$\begin{aligned}\dot{\tilde{S}}(t) &\leqslant -\hat{z}^T(t)(I_N \otimes F_1^T)\hat{Q}_2(I_N \otimes F_1)\hat{z}(t) \\ &\leqslant -\lambda_m\Big((I_N \otimes F_1^T)\hat{Q}_2(I_N \otimes F_1)\Big)\|\hat{z}(t)\|^2.\end{aligned} \tag{3.36}$$

According to (3.36), $\tilde{S}(t)$ is non-increasing and bounded. Consequently, both the error vector $\hat{z}(t)$ and the adaptive feedback gains $k_i^s(t)$ are bounded. Since $k_i^s(t)$ is monotonically increasing (see (3.20)), we have

$$\lim_{t\to+\infty} k_i^s(t) = \bar{k}_i^s > 0.$$

Thus, from the definition of $\tilde{S}(t)$, we can conclude that $\lim_{t\to+\infty} \sum_{i=1}^N \hat{z}_i^T(t)P\hat{z}_i(t)$ exists and is a non-negative real number. Next, we shall prove that $\lim_{t\to+\infty} \sum_{i=1}^N \hat{z}_i^T(t)$ $P\hat{z}_i(t) = 0$. If this assumption does not hold, we obtain

$$\lim_{t\to+\infty} \sum_{i=1}^N \hat{z}_i^T(t)P\hat{z}_i(t) = \omega \in \mathbb{R} > 0.$$

Then, we can easily find a $L \in \mathbb{R} > 0$ satisfying $\sum_{i=1}^{N} \hat{z}_i^T(t)P\hat{z}_i(t) > \frac{\omega}{2}$ for $t \geqslant L$. Therefore,

$$\|\hat{z}(t)\|^2 > \frac{\omega}{2\lambda_M(P)}, \quad t \geqslant L. \tag{3.37}$$

From (3.36) and (3.37), one obtains

$$\dot{\tilde{S}}(t) < -\frac{\omega\hat{\gamma}}{2\lambda_M(P)}, \quad t \geqslant L. \tag{3.38}$$

where $\hat{\gamma} = \lambda_m\Big((I_N \otimes F_1^T)\hat{Q}_2(I_N \otimes F_1)\Big)$.

By (3.38), we can derive

$$\begin{aligned} -\tilde{S}(L) \leqslant \tilde{S}(+\infty) - \tilde{S}(L) &= \int_L^{+\infty} \dot{\tilde{S}}(t)dt \\ &< -\int_L^{+\infty} \frac{\omega\hat{\gamma}}{2\lambda_M(P)}dt \\ &= -\infty. \end{aligned}$$

This is in contradiction with the assumption, so we have

$$\lim_{t\to+\infty} \sum_{i=1}^{N} \hat{z}_i^T(t)P\hat{z}_i(t) = 0.$$

Then, we can easily get $\lim_{t\to+\infty} \|\hat{z}(t)\| = 0$. Therefore, the network (3.1) is synchronized under the nodes-based pinning adaptive controller (3.20).

From the proof of Theorems 3.16 and 3.17, we can get a conclusion as follows.

Corollary 3.18 *If there exist matrices $\hat{M} \in \mathbb{R}^{nN\times mN}$, $0 < \hat{Q}_2 \in \mathbb{R}^{nN\times nN}$, $\hat{k}^s = \text{diag}(\hat{k}_1^s, \hat{k}_2^s, \ldots, \hat{k}_l^s, 0, \ldots, 0) \in \mathbb{R}^{N\times N}$, $F_2 \in \mathbb{R}^{n\times m}$, and nonsingular matrix $F_1 \in \mathbb{R}^{n\times n}$ such that*

$$\begin{pmatrix} \hat{W}_4 & \hat{E}_2 \\ \hat{E}_2^T & \hat{W}_5 \end{pmatrix} \leqslant 0, \tag{3.39}$$

where $\hat{W}_4 = 2I_N \otimes (P\Delta - \beta I_n) + \sum_{s=1}^{q} a_s(A^s - \hat{k}^s) \otimes (PH_s + H_sP) + (I_N \otimes F_1^T)\hat{Q}_2(I_N \otimes F_1)$, $\hat{E}_2 = I_N \otimes (PB) - (I_N \otimes F_1^T)\hat{M} + (I_N \otimes F_1^T)\hat{Q}_2(I_N \otimes F_2)$, $\hat{W}_5 = -(I_N \otimes F_2^T)\hat{M} - \hat{M}^T(I_N \otimes F_2) + (I_N \otimes F_2^T)\hat{Q}_2(I_N \otimes F_2)$, $\hat{k}_i^s > 0$ for $i = 1, 2, \ldots, l$, then the network (3.1) *is synchronized under the nodes-based pinning adaptive controller* (3.20).

Remark 3.19 In the last few years, a great deal of authors have analyzed the passivity of CDNs. But, very few researchers have considered the passivity control problem for CDNs [39, 40] , especially the pinning passivity of CDNMWs has not yet been discussed. In order to ensure passivity, input strict passivity and output strict passivity of the CDNMWs (3.1) (see Theorems 3.14, 3.15, and 3.16), a nodes-based pinning adaptive controller is developed. In addition, a synchronization condition is also presented for CDNMWs (3.1) by utilizing output strict passivity and nodes-based pinning adaptive controller (3.20).

3.5 Edges-Based Pinning Passivity of CDNMWs

Suppose that $\hat{\mathcal{B}}$ is a subset of undirected edges $\mathcal{B} \subset \{1, 2, \ldots, N\} \times \{1, 2, \ldots, N\}$, and network (3.1) is connected through the undirected edges $\hat{\mathcal{B}}$.

The network (3.1) with pinning adaptive coupling weights is represented by

$$\dot{w}_i(t) = g(w_i(t)) + \sum_{s=1}^{q}\sum_{j=1}^{N} a_s A^s_{ij}(t) H_s w_j(t) + Bu_i(t), \tag{3.40}$$

in which

$$\begin{aligned} \dot{A}^s_{ij}(t) &= \chi^s_{ij}(w_i(t) - w_j(t))^T (PH_s + H_s P)(w_i(t) \\ &\quad - w_j(t)), \qquad \text{if } \ (i, j) \in \hat{\mathcal{B}}, \\ \dot{A}^s_{ij}(t) &= 0, \qquad\qquad\quad \text{if } \ (i, j) \in \mathcal{B} - \hat{\mathcal{B}}, \end{aligned} \tag{3.41}$$

where $i = 1, 2, \ldots, N$, $PH_s + H_s P > 0, \mathbb{R} \ni \chi^s_{ij} = \chi^s_{ji} > 0$.

Define $w^*(t) = \frac{1}{N}\sum_{k=1}^{N} w_k(t)$. Then,

$$\dot{w}^*(t) = \frac{1}{N}\sum_{k=1}^{N} g(w_k(t)) + \frac{1}{N}\sum_{k=1}^{N} Bu_k(t).$$

Letting $\tilde{z}_i(t) = w_i(t) - w^*(t)$, one has

$$\begin{aligned} \dot{\tilde{z}}_i(t) &= g(w_i(t)) - \frac{1}{N}\sum_{k=1}^{N} g(w_k(t)) - \frac{1}{N}\sum_{k=1}^{N} Bu_k(t) \\ &\quad + Bu_i(t) + \sum_{s=1}^{q}\sum_{j=1}^{N} a_s A^s_{ij}(t) H_s \tilde{z}_j(t). \end{aligned} \tag{3.42}$$

The output vector $y_i(t) \in \mathbb{R}^d$ of the network (3.42) is defined as follows:

$$y_i(t) = F_1\tilde{z}_i(t) + F_2u_i(t),$$

where $F_1 \in \mathbb{R}^{d\times n}$ and $F_2 \in \mathbb{R}^{d\times m}$ are known matrices.

3.5.1 Passivity Analysis

Denote

$$F_{ij}^s = \begin{cases} A_{ij}^s(0), & \text{if } (i,j) \in \mathcal{B} - \hat{\mathcal{B}}, \\ -\sum_{\substack{j=1 \\ j\neq i}}^N F_{ij}^s, & \text{if } i = j, \\ 0, & \text{otherwise.} \end{cases}$$

$$\Phi_{ij}^s(t) = \begin{cases} A_{ij}^s(t), & \text{if } (i,j) \in \hat{\mathcal{B}}, \\ -\sum_{\substack{j=1 \\ j\neq i}}^N \Phi_{ij}^s(t), & \text{if } i = j, \\ 0, & \text{otherwise.} \end{cases}$$

Theorem 3.20 *If there exists a matrix* $\tilde{M} \in \mathbb{R}^{dN\times mN}$ *such that*

$$(I_N \otimes F_2^T)\tilde{M} + \tilde{M}^T(I_N \otimes F_2) > 0, \tag{3.43}$$

then the network (3.42) *is passive under the adaptive law* (3.41).

Proof Define the following Lyapunov functional for the network (3.42):

$$V_3(t) = \sum_{i=1}^N \tilde{z}_i^T(t)P\tilde{z}_i(t) + \sum_{s=1}^q \sum_{i=1}^N \sum_{(i,j)\in\hat{\mathcal{B}}} \frac{a_s}{4\chi_{ij}^s}(A_{ij}^s(t) - \hat{A}_{ij}^s)^2, \tag{3.44}$$

where $\hat{A}_{ij}^s = \hat{A}_{ji}^s (i \neq j)$ are nonnegative constants, and $\hat{A}_{ij}^s = 0 (i \neq j)$ if and only if $\dot{A}_{ij}^s(t) = 0$.

Then, we can obtain

$$\begin{aligned}\dot{V}_3(t) &= 2\sum_{i=1}^N \tilde{z}_i^T(t)P\dot{\tilde{z}}_i(t) + \frac{1}{2}\sum_{s=1}^q \sum_{i=1}^N \sum_{(i,j)\in\hat{\mathcal{B}}} \frac{a_s}{\chi_{ij}^s}(A_{ij}^s(t) - \hat{A}_{ij}^s)\dot{A}_{ij}^s(t) \\ &= 2\sum_{i=1}^N \tilde{z}_i^T(t)P\big[g(w_i(t)) - g(w^*(t)) + g(w^*(t))\end{aligned}$$

$$
\begin{aligned}
&-\frac{1}{N}\sum_{k=1}^{N} g(w_k(t)) + \sum_{s=1}^{q}\sum_{j=1}^{N} a_s A^s_{ij}(t) H_s \tilde{z}_j(t) \\
&+Bu_i(t) - \frac{1}{N}\sum_{k=1}^{N} Bu_k(t)\big] \\
&+\frac{1}{2}\sum_{s=1}^{q}\sum_{i=1}^{N}\sum_{(i,j)\in\hat{\mathcal{B}}} a_s (A^s_{ij}(t) - \hat{A}^s_{ij})(\tilde{z}_i(t) \\
&-\tilde{z}_j(t))^T (PH_s + H_s P)(\tilde{z}_i(t) - \tilde{z}_j(t)).
\end{aligned}
$$

Define $\hat{A}^s = (\hat{A}^s_{ij})_{N\times N}$, where $\hat{A}^s_{ii} = -\sum_{\substack{j=1\\ j\neq i}}^{N} \hat{A}^s_{ij}$, $i = 1, 2, \ldots, N$. Obviously,

$$
\begin{aligned}
&\sum_{i=1}^{N}\sum_{(i,j)\in\hat{\mathcal{B}}} (A^s_{ij}(t) - \hat{A}^s_{ij})(\tilde{z}_i(t) - \tilde{z}_j(t))^T (PH_s \\
&+H_s P)(\tilde{z}_i(t) - \tilde{z}_j(t)) \\
=&\sum_{i=1}^{N}\sum_{\substack{j=1\\ j\neq i}}^{N}(\Phi^s_{ij}(t) - \hat{A}^s_{ij})(\tilde{z}_i(t) - \tilde{z}_j(t))^T (PH_s \\
&+H_s P)(\tilde{z}_i(t) - \tilde{z}_j(t)) \\
=&\sum_{i=1}^{N}\sum_{\substack{j=1\\ j\neq i}}^{N}(\Phi^s_{ij}(t) - \hat{A}^s_{ij})\big[\tilde{z}_i^T(t)(PH_s + H_s P)\tilde{z}_i(t) \\
&-2\tilde{z}_i^T(t)(PH_s + H_s P)\tilde{z}_j(t) + \tilde{z}_j^T(t)(PH_s + H_s P)\tilde{z}_j(t)\big] \\
=&-2\sum_{i=1}^{N}\sum_{\substack{j=1\\ j\neq i}}^{N}(\Phi^s_{ij}(t) - \hat{A}^s_{ij})\tilde{z}_i^T(t)(PH_s + H_s P)\tilde{z}_j(t) \\
&+\sum_{i=1}^{N}\left(\sum_{\substack{j=1\\ j\neq i}}^{N}\Phi^s_{ij}(t) - \sum_{\substack{j=1\\ j\neq i}}^{N}\hat{A}^s_{ij}\right)\tilde{z}_i^T(t)(PH_s + H_s P)\tilde{z}_i(t) \\
&+\sum_{j=1}^{N}\left(\sum_{\substack{i=1\\ i\neq j}}^{N}\Phi^s_{ij}(t) - \sum_{\substack{i=1\\ i\neq j}}^{N}\hat{A}^s_{ij}\right)\tilde{z}_j^T(t)(PH_s + H_s P)\tilde{z}_j(t) \\
=&-2\sum_{i=1}^{N}\sum_{\substack{j=1\\ j\neq i}}^{N}(\Phi^s_{ij}(t) - \hat{A}^s_{ij})\tilde{z}_i^T(t)(PH_s + H_s P)\tilde{z}_j(t) \\
&-\sum_{i=1}^{N}(\Phi^s_{ii}(t) - \hat{A}^s_{ii})\tilde{z}_i^T(t)(PH_s + H_s P)\tilde{z}_i(t)
\end{aligned}
$$

$$-\sum_{j=1}^{N}(\Phi_{jj}^{s}(t)-\hat{A}_{jj}^{s})\tilde{z}_{j}^{T}(t)(PH_{s}+H_{s}P)\tilde{z}_{j}(t)$$
$$=-2\sum_{i=1}^{N}\sum_{j=1}^{N}(\Phi_{ij}^{s}(t)-\hat{A}_{ij}^{s})\tilde{z}_{i}^{T}(t)(PH_{s}+H_{s}P)\tilde{z}_{j}(t). \tag{3.45}$$

Since $A_{ij}^{s}(t)=\Phi_{ij}^{s}(t)+F_{ij}^{s}$, we can get from (3.45) that

$$\begin{aligned}
\dot{V}_{3}(t) \leqslant\ & 2\sum_{i=1}^{N}\tilde{z}_{i}^{T}(t)(P\Delta-\beta I_{n})\tilde{z}_{i}(t)\\
&+\sum_{s=1}^{q}\sum_{i=1}^{N}\sum_{j=1}^{N}a_{s}\hat{A}_{ij}^{s}\tilde{z}_{i}^{T}(t)(PH_{s}+H_{s}P)\tilde{z}_{j}(t)\\
&+\sum_{s=1}^{q}\sum_{i=1}^{N}\sum_{j=1}^{N}a_{s}F_{ij}^{s}\tilde{z}_{i}^{T}(t)(PH_{s}+H_{s}P)\tilde{z}_{j}(t)\\
&+2\sum_{i=1}^{N}\tilde{z}_{i}^{T}(t)PBu_{i}(t)\\
=\ & \tilde{z}^{T}(t)\Big\{2I_{N}\otimes(P\Delta-\beta I_{n})+\sum_{s=1}^{q}a_{s}(F^{s}\\
&+\hat{A}^{s})\otimes(PH_{s}+H_{s}P)\Big\}\tilde{z}(t)\\
&+2\tilde{z}^{T}(t)[I_{N}\otimes(PB)]u(t),
\end{aligned}$$

where $\tilde{z}(t)=(\tilde{z}_{1}^{T}(t),\tilde{z}_{2}^{T}(t),\ldots,\tilde{z}_{N}^{T}(t))^{T}$, $F^{s}=(F_{ij}^{s})_{N\times N}$, $u(t)=(u_{1}^{T}(t),u_{2}^{T}(t),\ldots,u_{N}^{T}(t))^{T}$.

Therefore,

$$\begin{aligned}
&\dot{V}_{3}(t)-2y^{T}(t)\tilde{M}u(t)\\
&\leqslant \tilde{z}^{T}(t)\Big\{2I_{N}\otimes(P\Delta-\beta I_{n})\\
&\quad+\sum_{s=1}^{q}a_{s}(F^{s}+\hat{A}^{s})\otimes(PH_{s}+H_{s}P)\Big\}\tilde{z}(t)\\
&\quad+2\tilde{z}^{T}(t)[I_{N}\otimes(PB)]u(t)-2[\tilde{z}^{T}(t)(I_{N}\otimes F_{1}^{T})\\
&\quad+u^{T}(t)(I_{N}\otimes F_{2}^{T})]\tilde{M}u(t),
\end{aligned} \tag{3.46}$$

where $y(t)=(y_{1}^{T}(t),y_{2}^{T}(t),\ldots,y_{N}^{T}(t))^{T}$.

Obviously,

$$
\begin{aligned}
&2\tilde{z}^T(t)[I_N \otimes (PB) - (I_N \otimes F_1^T)\tilde{M}]u(t) \\
&\leqslant \tilde{z}^T(t)[I_N \otimes (PB) - (I_N \otimes F_1^T)\tilde{M}][(I_N \otimes F_2^T)\tilde{M} \\
&\quad +\tilde{M}^T(I_N \otimes F_2)]^{-1}[I_N \otimes (B^TP) - \tilde{M}^T(I_N \otimes F_1)]\tilde{z}(t) \\
&\quad +u^T(t)[(I_N \otimes F_2^T)\tilde{M} + \tilde{M}^T(I_N \otimes F_2)]u(t).
\end{aligned} \tag{3.47}
$$

By (3.46) and (3.47), we have

$$
\begin{aligned}
&\dot{V}_3(t) - 2y^T(t)\tilde{M}u(t) \\
&\leqslant \tilde{z}^T(t)\Big\{2I_N \otimes (P\Delta - \beta I_n) + \sum_{s=1}^{q} a_s F^s \otimes (PH_s + H_sP) \\
&\quad + \sum_{\substack{s=1 \\ s\neq k}}^{q} a_s \hat{A}^s \otimes (PH_s + H_sP) + a_k\hat{A}^k \otimes (PH_k + H_kP) + [I_N \otimes (PB) \\
&\quad -(I_N \otimes F_1^T)\tilde{M}][(I_N \otimes F_2^T)\tilde{M} + \tilde{M}^T(I_N \otimes F_2)]^{-1}[I_N \otimes (B^TP) \\
&\quad -\tilde{M}^T(I_N \otimes F_1)]\Big\}\tilde{z}(t).
\end{aligned} \tag{3.48}
$$

According to the definition of $\hat{A}^k$, there obviously exists a unitary matrix $\Phi = (\phi_1, \phi_2, \ldots, \phi_N) \in \mathbb{R}^{N\times N}$ such that $\Phi^T\hat{A}^k\Phi = \Lambda$ with $\Lambda = \mathrm{diag}(\lambda_1, \lambda_2, \ldots, \lambda_N)$. $\lambda_i, i = 1, 2, \ldots, N$, are the eigenvalues of $\hat{A}^k$ and $0 = \lambda_1 > \lambda_2 \geqslant \lambda_3 \geqslant \cdots \geqslant \lambda_N$. Let $e(t) = (e_1^T(t), e_2^T(t), \ldots, e_N^T(t))^T = (\Phi^T \otimes I_n)\tilde{z}(t)$. Since $\phi_1 = \frac{1}{\sqrt{N}}(1, 1, \ldots, 1)$, one has $e_1(t) = (\phi_1^T \otimes I_n)\tilde{z}(t) = 0$. Then, one obtains

$$
\begin{aligned}
&\dot{V}_3(t) - 2y^T(t)\tilde{M}u(t) \\
&\leqslant \tilde{z}^T(t)\Big\{2I_N \otimes (P\Delta - \beta I_n) + \sum_{s=1}^{q} a_s F^s \otimes (PH_s \\
&\quad +H_sP) + \sum_{\substack{s=1 \\ s\neq k}}^{q} a_s \hat{A}^s \otimes (PH_s + H_sP) \\
&\quad +[I_N \otimes (PB) - (I_N \otimes F_1^T)\tilde{M}][(I_N \otimes F_2^T)\tilde{M} \\
&\quad +\tilde{M}^T(I_N \otimes F_2)]^{-1}[I_N \otimes (B^TP) - \tilde{M}^T(I_N \otimes F_1)] \\
&\quad +a_k(\Phi \otimes I_n)[\Lambda \otimes (PH_k + H_kP)](\Phi^T \otimes I_n)\Big\}\tilde{z}(t) \\
&\leqslant \tilde{z}^T(t)\Big\{2I_N \otimes (P\Delta - \beta I_n) + \sum_{s=1}^{q} a_s F^s \otimes (PH_s \\
&\quad +H_sP) + [I_N \otimes (PB) - (I_N \otimes F_1^T)\tilde{M}][(I_N \otimes F_2^T)\tilde{M} \\
&\quad +\tilde{M}^T(I_N \otimes F_2)]^{-1}[I_N \otimes (B^TP) - \tilde{M}^T(I_N \otimes F_1)]\Big\}\tilde{z}(t) \\
&\quad +a_ke^T(t)[\Lambda \otimes (PH_k + H_kP)]e(t).
\end{aligned}
$$

Because $PH_k + H_kP > 0$, we have

$$e^T(t)[\Lambda \otimes (PH_k + H_kP)]e(t) \\ \leqslant \lambda_2 e^T(t)[I_N \otimes (PH_k + H_kP)]e(t).$$

Therefore,

$$\begin{aligned}
&\dot{V}_3(t) - 2y^T(t)\tilde{M}u(t) \\
&\leqslant \tilde{z}^T(t)\Big\{2I_N \otimes (P\Delta - \beta I_n) + \sum_{s=1}^{q} a_s F^s \otimes (PH_s \\
&+H_sP) + [I_N \otimes (PB) - (I_N \otimes F_1^T)\tilde{M}][(I_N \otimes F_2^T)\tilde{M} \\
&+\tilde{M}^T(I_N \otimes F_2)]^{-1}[I_N \otimes (B^TP) - \tilde{M}^T(I_N \otimes F_1)]\Big\}\tilde{z}(t) \\
&+a_k\lambda_2 e^T(t)[I_N \otimes (PH_k + H_kP)]e(t) \\
&= \tilde{z}^T(t)\Big\{2I_N \otimes (P\Delta - \beta I_n) + \sum_{s=1}^{q} a_s F^s \otimes (PH_s \\
&+H_sP) + [I_N \otimes (PB) - (I_N \otimes F_1^T)\tilde{M}][(I_N \otimes F_2^T)\tilde{M} \\
&+\tilde{M}^T(I_N \otimes F_2)]^{-1}[I_N \otimes (B^TP) - \tilde{M}^T(I_N \otimes F_1)] \\
&+a_k\lambda_2[I_N \otimes (PH_k + H_kP)]\Big\}\tilde{z}(t).
\end{aligned}$$

Select sufficiently large $\hat{A}_{ij}^k$ such that

$$2\lambda_M\big(P\Delta - \beta I_n\big) + \lambda_M(\Psi) + \lambda_M(\Upsilon) + a_k\lambda_2\theta \leqslant 0,$$

where $\Psi = \sum_{s=1}^{q} a_s F^s \otimes (PH_s + H_sP)$, $\Upsilon = [I_N \otimes (PB) - (I_N \otimes F_1^T)\tilde{M}][(I_N \otimes F_2^T)\tilde{M} + \tilde{M}^T(I_N \otimes F_2)]^{-1}[I_N \otimes (B^TP) - \tilde{M}^T(I_N \otimes F_1)]$, $\theta = \lambda_m(PH_k + H_kP)$, one has

$$2y^T(t)\tilde{M}u(t) \geqslant \dot{V}_3(t).$$

Therefore,

$$\int_{t_1}^{t_2} y^T(s)\tilde{M}u(s)ds \geqslant \bar{S}(t_2) - \bar{S}(t_1)$$

for any $t_1, t_2 \in [0, +\infty)$ and $t_2 \geqslant t_1$, where $\bar{S}(t) = \frac{V_3(t)}{2}$.

Similar to the proof of Theorem 3.20, we can get the following conclusions.

Theorem 3.21 *If there exist matrices* $\tilde{M} \in \mathbb{R}^{dN\times mN}$ *and* $0 < \tilde{Q}_1 \in \mathbb{R}^{mN\times mN}$ *such that*

$$-\tilde{Q}_1 + (I_N \otimes F_2^T)\tilde{M} + \tilde{M}^T(I_N \otimes F_2) > 0, \tag{3.49}$$

then the network (3.42) *is input strictly passive under the adaptive law* (3.41).

Theorem 3.22 *If there exist matrices* $\tilde{M} \in \mathbb{R}^{dN\times mN}$ *and* $0 < \tilde{Q}_2 \in \mathbb{R}^{dN\times dN}$ *such that*

$$(I_N \otimes F_2^T)\tilde{M} + \tilde{M}^T(I_N \otimes F_2) - (I_N \otimes F_2^T)\tilde{Q}_2(I_N \otimes F_2) > 0, \tag{3.50}$$

then the network (3.42) *is output strictly passive under the adaptive law* (3.41).

3.5.2 Synchronization Criteria

Theorem 3.23 *The CDNMWs* (3.1) *under the edges-based pinning adaptive control strategy* (3.41) *is synchronized if the network* (3.42) *is output strictly passive with respect to storage function* $\bar{S}(t) = \frac{V_3(t)}{2}$ *and matrix* $F_1 \in \mathbb{R}^{n\times n}$ *is nonsingular.*

Proof If the network (3.42) is output strictly passive with respect to storage function $\bar{S}(t)$, we can easily obtain

$$\dot{\bar{S}}(t) \leqslant -\lambda_m\Big((I_N \otimes F_1^T)\tilde{Q}_2(I_N \otimes F_1)\Big)\|\tilde{z}(t)\|^2,$$

where $0 < \tilde{Q}_2 \in \mathbb{R}^{nN\times nN}$.

Obviously, $\bar{S}(t)$ is non-increasing and bounded. Consequently, both the error vector $\tilde{z}(t)$ and the coupling weights $A^s_{ij}(t)$, $(i, j) \in \hat{\mathcal{B}}$, are bounded. Since $A^s_{ij}(t)((i, j) \in \hat{\mathcal{B}})$ is monotonically increasing (see (3.41)), we have $\lim_{t\to+\infty} A^s_{ij}(t) = \breve{A}^s_{ij} > 0$. On the other hand, according to the definition of $\bar{S}(t)$, we can conclude that $\lim_{t\to+\infty} \sum_{i=1}^N \tilde{z}_i^T(t) P\tilde{z}_i(t)$ exists and is a non-negative real number. Then, similar to the proof of Theorem 3.17, we can easily get that $\lim_{t\to+\infty} \|\tilde{z}(t)\| = 0$. Therefore, the network (3.1) under edges-based pinning adaptive control strategy (3.41) is synchronized.

From the proof of Theorems 3.11 and 3.23, we can get the following conclusion.

Corollary 3.24 *If there exist matrices* $\tilde{M} \in \mathbb{R}^{nN\times mN}$, $0 < \tilde{Q}_2 \in \mathbb{R}^{nN\times nN}$ *and* $F_2 \in \mathbb{R}^{n\times m}$ *such that*

$$(I_N \otimes F_2^T)\tilde{M} + \tilde{M}^T(I_N \otimes F_2) - (I_N \otimes F_2^T)\tilde{Q}_2(I_N \otimes F_2) > 0, \tag{3.51}$$

then the network (3.1) *under edges-based pinning adaptive control strategy* (3.41) *is synchronized.*

Remark 3.25 In this section, by choosing suitable adaptive strategy to adjust a fraction of coupling weights, some sufficient conditions for ensuring passivity, input strict passivity and output strict passivity of the CDNMWs (3.1) are presented (see Theorems 3.20, 3.21, and 3.22). In addition, a synchronization criterion is also developed for CDNMWs (3.1) by employing output strict passivity and edges-based pinning adaptive control strategy (3.41).

Remark 3.26 More recently, very few researchers have considered the synchronization for CDNMWs [29–31]. In [29, 30], the authors studied the synchronization problem for a kind of CDNMWs by employing the designed state feedback controllers and Lyapunov functional method. Wang and Wu [31] considered a CRDNNs with state coupling and spatial diffusion coupling. In this paper, several synchronization criteria are presented for CDNMWs (3.1) by utilizing the output strict passivity, and two nodes-based and edges-based pinning adaptive control strategies are also developed. Obviously, these existing results [29–31] are totally different from our work.

3.6 Numerical Examples

Example 3.27 Consider the following CDNMWs:

$$
\begin{aligned}
\dot{w}_i(t) &= g(w_i(t)) + a_1 \sum_{j=1}^{6} A_{ij}^1 H_1 w_j(t) + a_2 \sum_{j=1}^{6} A_{ij}^2 H_2 w_j(t) \\
&\quad + a_3 \sum_{j=1}^{6} A_{ij}^3 H_3 w_j(t) + B u_i(t),
\end{aligned}
$$

where $i = 1, 2, \ldots, 6$, $g_l(\xi) = \frac{|\xi+1|-|\xi-1|}{4}$, $l = 1, 2, 3$, $H_1 = \mathrm{diag}(0.6, 0.5, 0.6)$, $H_2 = \mathrm{diag}(0.8, 0.4, 0.7)$, $H_3 = \mathrm{diag}(0.8, 0.7, 0.5)$, $a_1 = 3$, $a_2 = 4$, $a_3 = 5$, $\Delta = I_3$, $P = I_3$, $\beta = 0.5$, the matrices B, A^1, A^2 and A^3 are chosen as, respectively

$$
B = \begin{pmatrix} 0.1 & 0.2 \\ 0.1 & 0.3 \\ 0.3 & 0.1 \end{pmatrix},
$$

$$
A^1 = \begin{pmatrix}
-0.5 & 0.1 & 0 & 0.2 & 0.1 & 0.1 \\
0.1 & -0.4 & 0.1 & 0.1 & 0.1 & 0 \\
0 & 0.1 & -0.8 & 0 & 0.5 & 0.2 \\
0.2 & 0.1 & 0 & -0.6 & 0.3 & 0 \\
0.1 & 0.1 & 0.5 & 0.3 & -1.1 & 0.1 \\
0.1 & 0 & 0.2 & 0 & 0.1 & -0.4
\end{pmatrix},
$$

$$A^2 = \begin{pmatrix} -0.6 & 0.2 & 0 & 0.2 & 0.1 & 0.1 \\ 0.2 & -0.7 & 0.3 & 0.1 & 0.1 & 0 \\ 0 & 0.3 & -1 & 0 & 0.5 & 0.2 \\ 0.2 & 0.1 & 0 & -0.6 & 0.3 & 0 \\ 0.1 & 0.1 & 0.5 & 0.3 & -1.1 & 0.1 \\ 0.1 & 0 & 0.2 & 0 & 0.1 & -0.4 \end{pmatrix},$$

$$A^3 = \begin{pmatrix} -0.8 & 0.3 & 0 & 0.3 & 0.1 & 0.1 \\ 0.3 & -0.9 & 0.3 & 0.2 & 0.1 & 0 \\ 0 & 0.3 & -0.8 & 0 & 0.3 & 0.2 \\ 0.3 & 0.2 & 0 & -0.8 & 0.3 & 0 \\ 0.1 & 0.1 & 0.3 & 0.3 & -1 & 0.2 \\ 0.1 & 0 & 0.2 & 0 & 0.2 & -0.5 \end{pmatrix}.$$

We select the nodes 1, 2, 3, 4 and 5 as pinned nodes.
Case 1: Choose

$$F_1 = \begin{pmatrix} 0.8 & 0 & 0 \\ 0 & 0.6 & 0 \\ 0 & 0 & 0.8 \end{pmatrix}, \quad F_2 = \begin{pmatrix} 0.1 & 0.3 \\ 0.3 & 0.2 \\ 0.1 & 0.4 \end{pmatrix},$$

$\hat{k}_1^1 = 0.2$, $\hat{k}_2^1 = 0.3$, $\hat{k}_3^1 = 0.4$, $\hat{k}_4^1 = 0.3$, $\hat{k}_5^1 = 0.2$, $\hat{k}_1^2 = 0.4$, $\hat{k}_2^2 = 0.3$, $\hat{k}_3^2 = 0.2$, $\hat{k}_4^2 = 0.5$, $\hat{k}_5^2 = 0.4$, $\hat{k}_1^3 = 0.2$, $\hat{k}_2^3 = 0.4$, $\hat{k}_3^3 = 0.2$, $\hat{k}_4^3 = 0.3$, $\hat{k}_5^3 = 0.2$. Utilizing the MATLAB YALMIP Toolbox, we can get the following matrices $\hat{M}$ and $\hat{Q}_2$ satisfying (3.32):

$$\hat{M} = I_6 \otimes \begin{pmatrix} 0.9129 & 2.3412 \\ 2.4663 & 2.2370 \\ 0.9673 & 2.1245 \end{pmatrix},$$

$$\hat{Q}_2 = I_6 \otimes \begin{pmatrix} 1.3972 & 0.0125 & -0.3016 \\ 0.0125 & 1.0689 & 0.0873 \\ -0.3016 & 0.0873 & 0.9016 \end{pmatrix}.$$

According to Theorem 3.16, the network (3.21) is output strictly passive. The simulation results are displayed in Figs. 3.1 and 3.2.
Case 2: Take the same matrices F_1, F_2, $\hat{M}$, $\hat{Q}_2$, $\hat{k}^s (s = 1, 2, 3)$ as in Case 1. Obviously, F_1 is a nonsingular matrix, and the condition of Corollary 3.18 can also be satisfied. Therefore, the network (3.1) under nodes-based pinning adaptive controller (3.20) is synchronized. The simulation results are displayed in Figs. 3.3 and 3.4.

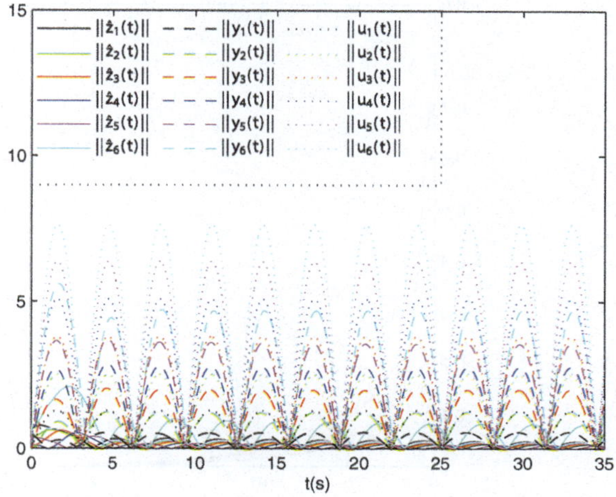

Fig. 3.1 $\|\hat{z}_i(t)\|$, $\|y_i(t)\|$, $\|u_i(t)\|$, $i = 1, 2, \ldots, 6$

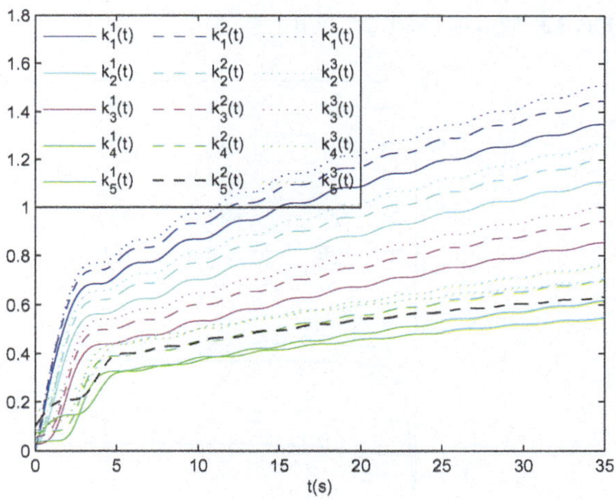

Fig. 3.2 $k_i^s(t)$, $s = 1, 2, 3$, $i = 1, 2, \ldots, 5$

Example 3.28 Consider the following CDNMWs:

$$
\begin{aligned}
\dot{w}_i(t) = {} & g(w_i(t)) + a_1 \sum_{j=1}^{6} A_{ij}^1(t) H_1 w_j(t) + a_2 \sum_{j=1}^{6} A_{ij}^2(t) H_2 w_j(t) \\
& + a_3 \sum_{j=1}^{6} A_{ij}^3(t) H_3 w_j(t) + B u_i(t),
\end{aligned}
$$

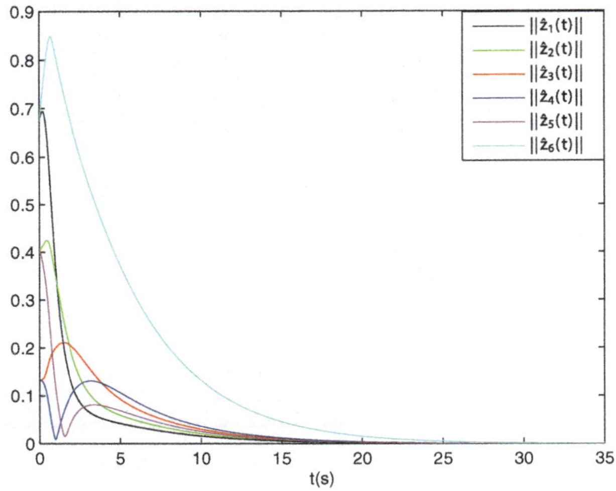

Fig. 3.3 $\|\hat{z}_i(t)\|, i = 1, 2, \ldots, 6$

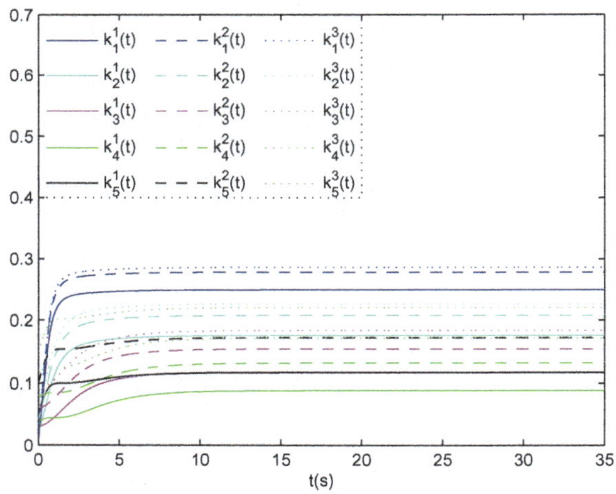

Fig. 3.4 $k_i^s(t), s = 1, 2, 3, i = 1, 2, \ldots, 5$

where $i = 1, 2, \ldots, 6$, $g_1(\xi) = -0.3\tanh(\xi)$, $g_2(\xi) = -0.4\tanh(\xi)$, $g_3(\xi) = -0.5\tanh(\xi)$, $H_1 = \text{diag}(0.6, 0.5, 0.6)$, $H_2 = \text{diag}(0.8, 0.4, 0.7)$, $H_3 = \text{diag}(0.8, 0.7, 0.5)$, $a_1 = 3$, $a_2 = 4$, $a_3 = 5$, $\Delta = 1.5I_3$, $P = I_3$, $\beta = 0.5$, the matrices B, $A^1(0)$, $A^2(0)$ and $A^3(0)$ are chosen as, respectively

$$B = \begin{pmatrix} 0.1 & 0.2 \\ 0.1 & 0.3 \\ 0.3 & 0.1 \end{pmatrix},$$

$$A^1(0) = \begin{pmatrix} -0.08 & 0.02 & 0.03 & 0.02 & 0.01 & 0 \\ 0.02 & -0.07 & 0.02 & 0 & 0.01 & 0.02 \\ 0.03 & 0.02 & -0.1 & 0.02 & 0.03 & 0 \\ 0.02 & 0 & 0.02 & -0.05 & 0.01 & 0 \\ 0.01 & 0.01 & 0.03 & 0.01 & -0.07 & 0.01 \\ 0 & 0.02 & 0 & 0 & 0.01 & -0.03 \end{pmatrix},$$

$$A^2(0) = \begin{pmatrix} -0.08 & 0.01 & 0.02 & 0.03 & 0.02 & 0 \\ 0.01 & -0.05 & 0.01 & 0 & 0.01 & 0.02 \\ 0.02 & 0.01 & -0.09 & 0.03 & 0.03 & 0 \\ 0.03 & 0 & 0.03 & -0.07 & 0.01 & 0 \\ 0.02 & 0.01 & 0.03 & 0.01 & -0.1 & 0.03 \\ 0 & 0.02 & 0 & 0 & 0.03 & -0.05 \end{pmatrix},$$

$$A^3(0) = \begin{pmatrix} -0.06 & 0.02 & 0.01 & 0.02 & 0.01 & 0 \\ 0.02 & -0.06 & 0.01 & 0 & 0.01 & 0.02 \\ 0.01 & 0.01 & -0.06 & 0.02 & 0.02 & 0 \\ 0.02 & 0 & 0.02 & -0.05 & 0.01 & 0 \\ 0.01 & 0.01 & 0.02 & 0.01 & -0.07 & 0.02 \\ 0 & 0.02 & 0 & 0 & 0.02 & -0.04 \end{pmatrix}.$$

Select undirected edges $(1, 2)$, $(2, 3)$, $(3, 4)$, $(4, 5)$, $(5, 6)$ as pinned edges.
Case 1: Take

$$F_1 = \begin{pmatrix} 0.2 & 0 & 0 \\ 0 & 0.1 & 0 \\ 0 & 0 & 0.3 \end{pmatrix}, \quad F_2 = \begin{pmatrix} 0.1 & 0.3 \\ 0.3 & 0.2 \\ 0.1 & 0.4 \end{pmatrix}.$$

We can easily find the following matrices $\tilde{M}$ and $\tilde{Q}_2$ satisfying (3.50):

$$\tilde{M} = I_6 \otimes \begin{pmatrix} -0.3463 & 0.9983 \\ 2.6689 & -0.5704 \\ -0.8760 & 1.5076 \end{pmatrix},$$

$$\tilde{Q}_2 = I_6 \otimes \begin{pmatrix} 1.2224 & 0 & 0 \\ 0 & 1.2224 & 0 \\ 0 & 0 & 1.2224 \end{pmatrix}.$$

According to Theorem 3.22, the network (3.42) under adaptive law (3.41) is output strictly passive. The simulation results are displayed in Figs. 3.5 and 3.6.

Case 2: Take the same matrices F_1, F_2, $\tilde{M}$, $\tilde{Q}_2$ as in Case 1. Obviously, the condition of Corollary 3.24 can also be satisfied. Therefore, the network (3.1) under edges-based pinning adaptive control strategy (3.41) is synchronized. The simulation results are displayed in Figs. 3.7 and 3.8.

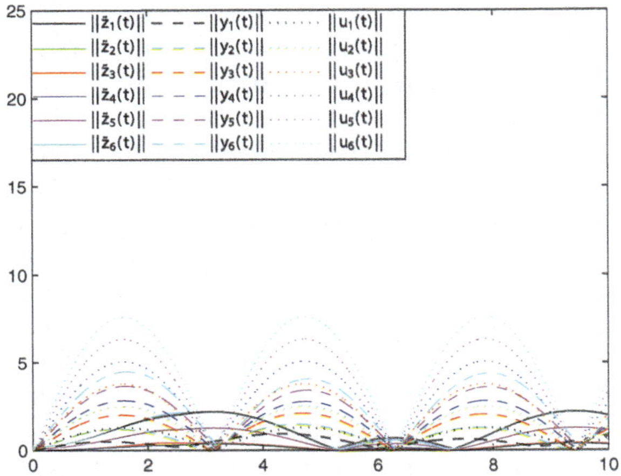

Fig. 3.5 $\|\tilde{z}_i(t)\|$, $\|y_i(t)\|$, $\|u_i(t)\|$, $i = 1, 2, \ldots, 6$

Fig. 3.6 $A^s_{ij}(t)$, $s = 1, 2, 3$

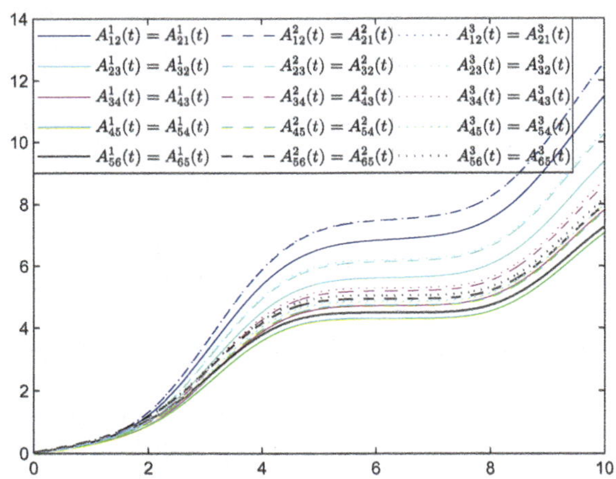

3.7 Conclusion

In this chapter, the passivity problem for a CDNMWs with different dimensions of input and output vectors has been discussed. By employing Lyapunov functional method and some inequality techniques, several passivity criteria have been presented for such network model. Furthermore, we also have studied the passivity of CDNMWs by pinning a fraction of nodes or edges with adaptive strategies. In addition, some synchronization criteria have also been established by exploiting the output strict passivity. Finally, two illustrative examples have been provided to demonstrate the correctness of these obtained results.

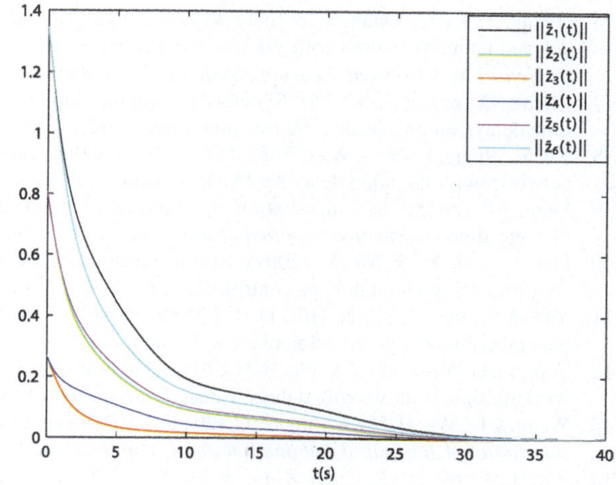

Fig. 3.7 $\|\tilde{z}_i(t)\|$, $i = 1, 2, \ldots, 6$

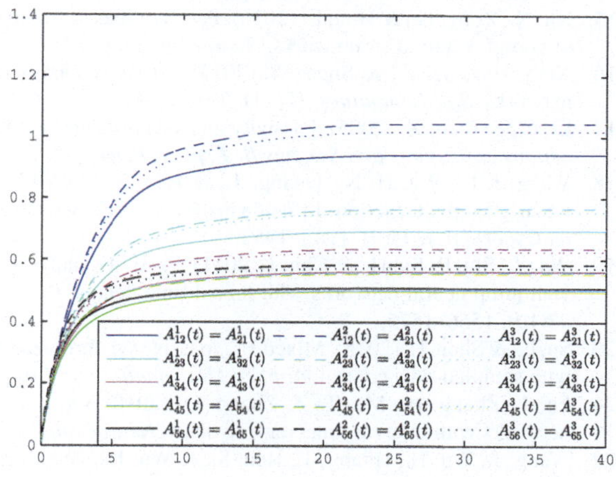

Fig. 3.8 $A^s_{ij}(t)$, $s = 1, 2, 3$

References

1. Chen, W. H., Jiang, Z., Lu, X., & Luo, S. (2015). $\mathcal{H}_\infty$ synchronization for complex dynamical networks with coupling delays using distributed impulsive control. *Nonlinear Analysis: Hybrid Systems*, *17*, 111–127.
2. Wang, X., She, K., Zhong, S., & Yang, H. (2016). New result on synchronization of complex dynamical networks with time-varying coupling delay and sampled-data control. *Neurocomputing*, *214*(19), 508–515.
3. Li, Z. X., Park, J. H., Wu, Z. G. (2013). Synchronization of complex networks with nonhomogeneous Markov jump topology. *Nonlinear Dynamics 74*, 65–75.
4. Wang, Y., & Li, T. (2015). Synchronization of fractional order complex dynamical networks. *Physica A*, *428*, 1–12.
5. Yang, L., Jiang, J., & Liu, X. (2016). Synchronization of fractional-order colored dynamical networks via open-plus-closed-loop control. *Physica A*, *443*, 200–211.

6. Huang, T., Li, C., Duan, S., & Starzyk, J. A. (2012). Robust exponential stability of uncertain delayed neural networks with stochastic perturbation and impulse effects. *IEEE Transactions on Neural Networks and Learning Systems*, *23*(6), 866–875.
7. Cheng, Q., & Cao, J. (2015). Synchronization of complex dynamical networks with discrete time delays on time scales. *Neurocomputing*, *151*(2), 729–736.
8. Xu, Y., Zhang, J., Zhou, W., & Tong, D. (2017). Adaptive synchronization of complex dynamical networks with bounded delay feedback controller. *Optik*, *131*(1), 467–474.
9. Fang, M. (2015). Synchronization for complex dynamical networks with time delay and discrete-time information. *Applied Mathematics and Computation*, *258*, 1–11.
10. Dai, Y., Cai, Y., & Xu, X. (2009). Synchronization of complex networks with time-varying coupling delay via impulsive control. *Complex Sciences*, *4*, 912–923.
11. Yao, J., Guan, Z. H., & Hill, D. J. (2009). Passivity-based control and synchronization of general complex dynamical networks. *Automatica*, *45*(9), 2107–2113.
12. Wang, J. L., Yang, Z. C., & Wu, H. N. (2012). Passivity analysis of complex dynamical networks with multiple time-varying delays. *Journal of Engineering Mathematics*, *74*(1), 175–188.
13. Wang, J. L., Wu, H. N., & Yang, Z. C. (2011). Passivity analysis of impulsive complex networks. *International Journal of Automation and Computing*, *8*(4), 484–489.
14. Yao, J., Wang, H. O., Guan, Z. H., & Xu, W. (2009). Passive stability and synchronization of complex spatio-temporal switching networks with time delays. *Automatica*, *45*(7), 1721–1728.
15. Xu, X., Zong, G., & Hou, L. (2016). Passivity-based stabilization and passive synchronization of complex nonlinear networks. *Neurocomputing*, *175*, 101–109.
16. Xiao, J., Zeng, Z., & Shen, W. (2013). Passivity analysis for delayed discontinuous neural networks. *Soft Computing*, *17*(11), 2033–2041.
17. Li, C., & Liao, X. (2005). Passivity analysis of neural networks with time delay. *IEEE Transactions on Circuits and Systems II: Express Briefs*, *52*(8), 471–475.
18. Wang, J. L., Wu, H. N., Huang, T., & Ren, S. Y. (2015). Passivity and synchronization of linearly coupled reaction-diffusion neural networks with adaptive coupling. *IEEE Transactions on Cybernetics*, *45*(9), 1942–1952.
19. Wu, Z., Shi, P., Su, H., & Chu, J. (2011). Passivity analysis for discrete-time stochastic Markovian jump neural networks with mixed time delays. *IEEE Transactions on Neural Networks*, *22*(10), 1566–1575.
20. Su, L., & Shen, H. (2015). Mixed $\mathcal{H}_\infty$/passive synchronization for complex dynamical networks with sampled-data control. *Applied Mathematics and Computation*, *259*, 931–942.
21. Fu, J., Zhang, H., Ma, T., & Zhang, Q. (2010). On passivity analysis for stochastic neural networks with interval time-varying delay. *Neurocomputing*, *73*(4–6), 795–801.
22. Wang, J., Wu, H., Huang, T., Ren, S., & Wu, J. (2018). Passivity and output synchronization of complex dynamical networks with fixed and adaptive coupling strength. *IEEE Transactions on Neural Networks and Learning Systems*, *29*(2), 364–376.
23. Lou, X., & Cui, B. (2007). Passivity analysis of integro-differential neural networks with time-varying delays. *Neurocomputing*, *70*(4), 1071–1078.
24. Song, Q., Liang, J., & Wang, Z. (2009). Passivity analysis of discrete-time stochastic neural networks with time-varying delays. *Neurocomputing*, *72*(7–9), 1782–1788.
25. Zeng, H. B., Park, J. H., Shen, H. (2015). Robust passivity analysis of neural networks with discrete and distributed delays. *Neurocomputing* *149*, 1092–1097.
26. Zhang, J., Ma, L., & Liu, Y. (2016). Passivity analysis for discrete-time neural networks with mixed time-delays and randomly occurring quantization effects. *Neurocomputing*, *216*, 657–665.
27. Wang, J., Wu, H., Huang, T., Ren, S., & Wu, J. (2017). Passivity analysis of coupled reaction-diffusion neural networks with Dirichlet boundary conditions. *IEEE Transactions on Systems, Man, and Cybernetics: Systems*, *47*(8), 2148–2159.
28. Wang, J. L., Wu, H. N., Huang, T., Ren, S. Y., & Wu, J. (2017). Passivity of directed and undirected complex dynamical networks with adaptive coupling weights. *IEEE Transactions on Neural Networks and Learning Systems*, *28*(8), 1827–1839.

29. An, X., Zhang, L., & Zhang, J. (2015). Research on urban public traffic network with multi-weights based on single bus transfer junction. *Physica A*, *436*, 748–755.
30. An, X., Zhang, L., Li, Y., & Zhang, J. (2014). Synchronization analysis of complex networks with multi-weights and its application in public traffic network. *Physica A*, *412*, 149–156.
31. Wang, J., & Wu, H. (2014). Synchronization and adaptive control of an array of linearly coupled reaction-diffusion neural networks with hybrid coupling. *IEEE Transactions on Cybernetics*, *44*(8), 1350–1361.
32. Yang, X., Cao, J., & Yang, Z. (2013). Synchronization of coupled reaction-diffusion neural networks with time-varying delays via pinning-impulsive controller. *SIAM Journal on Control and Optimization*, *51*(5), 3486–3510.
33. Tang, Y., Gao, H., Lu, J., & Kurths, J. (2014). Pinning distributed synchronization of stochastic dynamical networks: A mixed optimization approach. *IEEE Transactions on Neural Networks and Learning Systems*, *25*(10), 1804–1815.
34. Tang, Y., Gao, H., & Kurths, J. (2014). Distributed robust synchronization of dynamical networks with stochastic coupling. *IEEE Transactions on Circuits and Systems I: Regular Papers*, *61*(5), 1508–1519.
35. Tang, Y., Qian, F., Gao, H., & Kurths, J. (2014). Synchronization in complex networks and its application—A survey of recent advances and challenges. *Annual Reviews in Control*, *38*(2), 184–198.
36. Wang, J. L., Wu, H. N., & Guo, L. (2014). Novel adaptive strategies for synchronization of linearly coupled neural networks with reaction-diffusion terms. *IEEE Transactions on Neural Networks and Learning Systems*, *25*(2), 429–440.
37. Yu, W., DeLellis, P., Chen, G., di Bernardo, M., & Kurths, J. (2012). Distributed adaptive control of synchronization in complex networks. *IEEE Transactions on Automatic Control*, *57*(8), 2153–2158.
38. Willems, J. C. (1972). Dissipative dynamical systems part I: General theory. *Archive for Rational Mechanics and Analysis*, *45*(5), 321–351.
39. Ren, S. Y., Wu, J., & Xu, B. B. (2017). Passivity and pinning passivity of complex dynamical networks with spatial diffusion coupling. *Neurocomputing*, *227*, 3–9.
40. Ren, S. Y., Wu, J., & Wei, P. C. (2017). Passivity and pinning passivity of coupled delayed reaction–diffusion neural networks with Dirichlet boundary conditions. *Neural Processing Letters*, *45*(3), 869–885.

Chapter 4
FTP and FTS of CDNs with State and Derivative Coupling

4.1 Introduction

In recent years, numerous researchers have turned their attention to dynamical behaviors of CDNs, particularly the synchronization and passivity of CDNs [1–17]. In [4], Xu et al. gave several synchronization criteria for CDNs by taking advantage of the designed pinning control schemes and adaptive strategies. Wang et al. [6] researched the global synchronization of CDNs under digital communication with limited bandwidth. In [8], the authors studied the robust H_∞ synchronization for Markov jump stochastic neural networks with decentralized event triggered scheme and mixed time varying delays. In [15], Wang et al. took into account the output and input strict passivity for a generalized complex network with delayed coupling, and several sufficient conditions for assuring the output and input strict passivity were obtained by choosing some suitable Lyapunov functionals. In [17], the authors respectively analyzed the passivity for multi-weighted CDNs with switching and fixed topologies by employing Lyapunov functional approach, and a pinning adaptive state feedback controller was presented to guarantee the passivity of the multi-weighted CDNs.

However, in these existing works [1–17], the network models with non-derivative coupling were discussed. As a matter of fact, in many real networks, different time derivatives of node state are likely to result in different changes of other nodes [18]. Therefore, some researchers have taken the synchronization and passivity into consideration for CDNs with derivative coupling [18–22]. Xu et al. [18] investigated the adaptive synchronization of CDNs with derivative and non-derivative coupling, and several criteria for synchronization were obtained by taking advantage of the LaSalle's invariance principle. In [20], the authors studied the pinning synchronization of a complex network with derivative and non-derivative coupling, and several synchronization criteria were obtained by exploiting the devised state feedback controllers and the adaptive laws. In [23], Wang et al. analyzed the passivity for two types

The original version of this book was revised: The belated corrections have been incorporated. The corrections to this book are available at https://doi.org/10.1007/978-981-33-4287-3_10

J.-L. Wang et al., *Passivity of Complex Dynamical Networks*,
https://doi.org/10.1007/978-981-33-4287-3_4

of CDNs with multiple derivative couplings, and an adaptive state feedback control strategy was developed in order to ensure the passivity of these network models.

Obviously, it may be better if the CDNs can achieve synchronization over the finite time interval in a number of actual applications. But, in these existing works [1–12], the authors always require the synchronization to be achieved in CDNs only while time approaches infinity. More recently, some investigators have considered the FTS for CDNs [24–31]. Qiu et al. [24] respectively discussed the FTS of multi-weighted CDNs without and with coupling delay, and several FTS criteria were presented for these network models with switching and fixed topologies by means of state feedback controllers and Lyapunov functionals. In [25], the authors considered a nonlinear CDN, and a FTS criterion was given by exploiting the mathematical induction approach and pinning impulsive control strategy. But, very few authors have considered the FTS of CDNs with derivative coupling [31]. Xu et al. [31] studied the FTS for a CDN with derivative and non-derivative coupling by utilizing a new FTS theory, and some sufficient conditions were obtained to ensure the FTS. Therefore, it is very significative to further study the FTS of CDNs with derivative coupling.

To the best of our knowledge, the passivity is an advantageous tool to investigate the synchronization of CDNs. Therefore, in order to research the FTS of CDNs better, it's essential to further consider the FTP for CDNs. Regretfully, very few results about FTP of CDNs have been reported [32–35]. In [32], Wang et al. presented some new concepts about the FTP on the basis of the existing passivity definitions, and gave several FTP criteria for multi-weighted coupled neural networks without and with coupling delays by selecting suitable controllers. Huang et al. [33] discussed the FTP for multi-weighted CDNs with different dimensional nodes, obtained some sufficient conditions to ensure the FTP of the network models with switching and fixed topologies by choosing suitable controllers, and analyzed the finite-time stability of the network models when they are finite-time passive. To our knowledge, the FTP for CDNs with derivative coupling has not yet been researched. Therefore, it's very interesting and important to further investigate the FTP of CDNs with derivative coupling.

In the present chapter , two classes of CDNs with derivative coupling and state coupling are put forward. By exploiting Lyapunov functional method and devising suitable controllers, we derive some sufficient conditions that can make sure the FTP of these network models. Moreover, we also research the FTS of these network models when they are finite-time passive.

4.2 Definitions

Definition 4.1 A system with input $w(t) \in \mathbb{R}^n$ and output $y(t) \in \mathbb{R}^n$ achieves FTP if

$$y^T(t)w(t) \geqslant \dot{V}(t) + \upsilon V^{\phi}(t)$$

for any $t \in [t^*, +\infty)$, where $\mathbb{R} \ni t^* \geqslant 0, \mathbb{R} \ni \upsilon > 0, 0 < \phi < 1$, V is a nonnegative function.

Definition 4.2 A system with input $w(t) \in \mathbb{R}^n$ and output $y(t) \in \mathbb{R}^n$ achieves finite-time input strict passivity (FTISP) if

$$y^T(t)w(t) - r_1 w^T(t)w(t) \geqslant \dot{V}(t) + \upsilon V^{\phi}(t)$$

for any $t \in [t^*, +\infty)$, where $\mathbb{R} \ni t^* \geqslant 0$, $\mathbb{R} \ni r_1 > 0$, $\mathbb{R} \ni \upsilon > 0$, $0 < \phi < 1$, V is a nonnegative function.

Definition 4.3 A system with input $w(t) \in \mathbb{R}^n$ and output $y(t) \in \mathbb{R}^n$ achieves finite-time output strict passivity (FTOSP) if

$$y^T(t)w(t) - r_2 y^T(t)y(t) \geqslant \dot{V}(t) + \upsilon V^{\phi}(t)$$

for any $t \in [t^*, +\infty)$ where $\mathbb{R} \ni t^* \geqslant 0, \mathbb{R} \ni r_2 > 0, \mathbb{R} \ni \upsilon > 0, 0 < \phi < 1$, V is a nonnegative function.

4.3 FTP and FTS of CDNs with State Coupling

4.3.1 Network Model

The network model is given by:

$$\dot{z}_i(t) = p(z_i(t)) + a_1 \sum_{j=1}^{N} A^1_{ij} \Gamma^1 z_j(t) + w_i(t) + v_i(t), \tag{4.1}$$

where $i = 1, 2, \ldots, N$; $w_i(t) \in \mathbb{R}^n$ is the external input; $v_i(t) \in \mathbb{R}^n$ is the control input; $z_i(t) = (z_{i1}(t), z_{i2}(t), \ldots, z_{in}(t))^T \in \mathbb{R}^n$ is the state vector of the ith node; $p(z_i(t)) = (p_1(z_i(t)), p_2(z_i(t)), \ldots, p_n(z_i(t)))^T \in \mathbb{R}^n$ is a continuously differentiable vector function; $0 < a_1 \in \mathbb{R}$ denotes the coupling strength; $\Gamma^1 \in \mathbb{R}^{n\times n} > 0$ is the inner coupling matrix; $A^1 = (A^1_{ij})_{N\times N} \in \mathbb{R}^{N\times N}$ is the outer coupling matrix, where $\mathbb{R} \ni A^1_{ij}$ is defined as follows: if node i and node j are connected, then $\mathbb{R} \ni A^1_{ij} = A^1_{ji} > 0$ $(i \neq j)$; otherwise, $\mathbb{R} \ni A^1_{ij} = A^1_{ji} = 0$ $(i \neq j)$; and $A^1_{ii} = -\sum_{\substack{j=1\\ j\neq i}}^{N} A^1_{ij}, i = 1, 2, \ldots, N$.

In this chapter, the network (4.1) is connected and $p(\cdot)$ meets the Lipschitz condition, i.e., there exists $\mathbb{R} \ni \delta > 0$ which satisfies

$$\|p(\psi_1) - p(\psi_2)\| \leqslant \delta \|\psi_1 - \psi_2\| \tag{A1}$$

for any $\psi_1, \psi_2 \in \mathbb{R}^n$.

4.3.2 FTP of CDNs

Define $\mathbb{R}^n \ni \bar{z}(t) = (\bar{z}_1(t), \bar{z}_2(t), \ldots, \bar{z}_n(t))^T = \frac{1}{N}\sum_{i=1}^{N} z_i(t)$. Then, we have

$$\begin{aligned}\dot{\bar{z}}(t) &= \frac{1}{N}\sum_{i=1}^{N} p(z_i(t)) + \frac{a_1}{N}\sum_{j=1}^{N}\left(\sum_{i=1}^{N} A^1_{ij}\right)\Gamma^1 z_j(t) + \frac{1}{N}\sum_{i=1}^{N} w_i(t) + \frac{1}{N}\sum_{i=1}^{N} v_i(t)\\ &= \frac{1}{N}\sum_{i=1}^{N} p(z_i(t)) + \frac{1}{N}\sum_{i=1}^{N} w_i(t) + \frac{1}{N}\sum_{i=1}^{N} v_i(t).\end{aligned}$$

Letting $c_i(t) = z_i(t) - \bar{z}(t)$, we get

$$\begin{aligned}\dot{c}_i(t) = {} & p(z_i(t)) + a_1\sum_{j=1}^{N} A^1_{ij}\Gamma^1 c_j(t) + w_i(t) + v_i(t) - \frac{1}{N}\sum_{h=1}^{N} p(z_h(t))\\ & -\frac{1}{N}\sum_{h=1}^{N} w_h(t) - \frac{1}{N}\sum_{h=1}^{N} v_h(t), i = 1, 2, \ldots, N.\end{aligned} \tag{4.2}$$

Choose the output vector $y_i(t) \in \mathbb{R}^n$ for the system (4.2) as follows:

$$y_i(t) = D_1 c_i(t) + D_2 w_i(t), i = 1, 2, \ldots, N,$$

where $D_1 \in \mathbb{R}^{n\times n}$, $D_2 \in \mathbb{R}^{n\times n}$.

A. State feedback controller

Construct the following controller for the system (4.2):

$$v_i(t) = -K_i(z_i(t) - \bar{z}(t)) - \varphi \mathrm{sign}(z_i(t) - \bar{z}(t))|z_i(t) - \bar{z}(t)|^{\alpha}, \tag{4.3}$$

where $K_i \in \mathbb{R}^{n\times n}$, $0 < \alpha < 1$, $\mathrm{sign}(z_i(t) - \bar{z}(t)) = \mathrm{diag}(\mathrm{sign}(z_{i1}(t) - \bar{z}_1(t)),$ $\mathrm{sign}(z_{i2}(t) - \bar{z}_2(t)), \ldots, \mathrm{sign}(z_{in}(t) - \bar{z}_n(t)))$, $|z_i(t) - \bar{z}(t)|^{\alpha} = (|z_{i1}(t) - \bar{z}_1(t)|^{\alpha},$ $|z_{i2}(t) - \bar{z}_2(t)|^{\alpha}, \ldots, |z_{in}(t) - \bar{z}_n(t)|^{\alpha})^T$, $\mathbb{R} \ni \varphi > 0$.

Theorem 4.4 *If there is a matrix* $K = \mathrm{diag}(K_1, K_2, \ldots, K_N) \in \mathbb{R}^{nN\times nN}$ *such that*

$$\begin{pmatrix} Q_1 & \Theta_1 \\ \Theta_1^T & -I_N \otimes \frac{D_2+D_2^T}{2}\end{pmatrix} \leqslant 0, \tag{4.4}$$

where $Q_1 = (1+\delta^2)I_{Nn} - K - K^T + 2a_1 A^1 \otimes \Gamma^1$, $\Theta_1 = I_{Nn} - \frac{I_N \otimes D_1^T}{2}$, *then the model* (4.2) *under the controller* (4.3) *achieves FTP.*

Proof The following Lyapunov functional can be chosen for the model (4.2):

$$V_1(t) = \sum_{i=1}^{N} c_i^T(t)c_i(t). \tag{4.5}$$

Then, we have

$$\begin{aligned}\dot{V}_1(t) = 2\sum_{i=1}^{N} c_i^T(t)\Big(&p(z_i(t)) - p(\bar{z}(t)) + p(\bar{z}(t)) - \frac{1}{N}\sum_{h=1}^{N} p(z_h(t)) \\ &+a_1\sum_{j=1}^{N} A_{ij}^1 \Gamma^1 c_j(t) + w_i(t) - \frac{1}{N}\sum_{h=1}^{N} w_h(t) - \frac{1}{N}\sum_{h=1}^{N} v_h(t) \\ &-K_i c_i(t) - \varphi \text{sign}(c_i(t))|c_i(t)|^{\alpha}\Big).\end{aligned} \tag{4.6}$$

Obviously,

$$\begin{aligned}&2c_i^T(t)(p(z_i(t)) - p(\bar{z}(t))) \\ &\leqslant c_i^T(t)c_i(t) + (p(z_i(t)) - p(\bar{z}(t)))^T(p(z_i(t)) - p(\bar{z}(t))) \\ &\leqslant (1+\delta^2)c_i^T(t)c_i(t).\end{aligned} \tag{4.7}$$

Moreover,

$$\begin{aligned}\sum_{i=1}^{N} c_i(t) &= \sum_{i=1}^{N}(z_i(t) - \bar{z}(t)) \\ &= \sum_{i=1}^{N}\left(z_i(t) - \frac{1}{N}\sum_{h=1}^{N} z_h(t)\right) \\ &= \sum_{i=1}^{N} z_i(t) - \sum_{h=1}^{N} z_h(t) \\ &= 0.\end{aligned} \tag{4.8}$$

From (4.8), we have

$$\sum_{i=1}^{N} c_i^T(t)\left(p(\bar{z}(t)) - \frac{1}{N}\sum_{h=1}^{N} p(z_h(t))\right) = 0, \tag{4.9}$$

$$\sum_{i=1}^{N} c_i^T(t)\left(\frac{1}{N}\sum_{h=1}^{N} w_h(t) + \frac{1}{N}\sum_{h=1}^{N} v_h(t)\right) = 0. \tag{4.10}$$

From (4.6), (4.7), (4.9), (4.10), one derives

$$
\begin{aligned}
\dot{V}_1(t) \leqslant\ & (1+\delta^2)\sum_{i=1}^{N} c_i^T(t)c_i(t) + 2\sum_{i=1}^{N}\sum_{j=1}^{N} a_1 A_{ij}^1 c_i^T(t)\Gamma^1 c_j(t) + 2\sum_{i=1}^{N} c_i^T(t)w_i(t) \\
& -2\sum_{i=1}^{N} c_i^T(t)K_i c_i(t) - 2\varphi\sum_{i=1}^{N} c_i^T(t)\mathrm{sign}(c_i(t))|c_i(t)|^{\alpha} \\
=\ & c^T(t)\Big[(1+\delta^2)I_{Nn} - K - K^T + 2a_1 A^1\otimes\Gamma^1\Big]c(t) + 2c^T(t)w(t) \\
& -2\varphi\sum_{i=1}^{N} c_i^T(t)\mathrm{sign}(c_i(t))|c_i(t)|^{\alpha},
\end{aligned} \tag{4.11}
$$

in which $c(t) = (c_1^T(t), c_2^T(t), \ldots, c_N^T(t))^T$, $w(t) = (w_1^T(t), w_2^T(t), \ldots, w_N^T(t))^T$.

In addition,

$$
\begin{aligned}
\sum_{i=1}^{N} c_i^T(t)\mathrm{sign}(c_i(t))|c_i(t)|^{\alpha} &= \sum_{i=1}^{N}\sum_{j=1}^{n} |c_{ij}(t)|^{\alpha+1} \\
&\geqslant \sum_{i=1}^{N}\left(\sum_{j=1}^{n} c_{ij}^2(t)\right)^{\frac{\alpha+1}{2}} \\
&= \sum_{i=1}^{N}(c_i^T(t)c_i(t))^{\frac{\alpha+1}{2}}.
\end{aligned} \tag{4.12}
$$

By (4.11) and (4.12), one obtains

$$
\begin{aligned}
& \dot{V}_1(t) - y^T(t)w(t) \\
=\ & \dot{V}_1(t) - c^T(t)(I_N\otimes D_1^T)w(t) - w^T(t)(I_N\otimes D_2^T)w(t) \\
\leqslant\ & -2\varphi\sum_{i=1}^{N}(c_i^T(t)c_i(t))^{\frac{\alpha+1}{2}} + \xi^T(t)\begin{pmatrix} Q_1 & \Theta_1 \\ \Theta_1^T & -I_N\otimes\frac{D_2+D_2^T}{2}\end{pmatrix}\xi(t) \\
\leqslant\ & -2\varphi\sum_{i=1}^{N}(c_i^T(t)c_i(t))^{\frac{\alpha+1}{2}} \\
\leqslant\ & -2\varphi\left(\sum_{i=1}^{N} c_i^T(t)c_i(t)\right)^{\frac{\alpha+1}{2}} \\
=\ & -2\varphi V_1^{\frac{\alpha+1}{2}}(t),
\end{aligned}
$$

where $\xi(t) = (c^T(t), w^T(t))^T$ and $y(t) = (y_1^T(t), y_2^T(t), \ldots, y_N^T(t))^T$.

Then, we can derive

$$y^T(t)w(t) \geqslant \dot{V}_1(t) + 2\varphi V_1^{\frac{\alpha+1}{2}}(t)$$

for any $t \in [0, +\infty)$.

Therefore, the network (4.2) under the controller (4.3) achieves FTP.

Similar to the proof of the Theorem 4.4, one could derive the following conclusions.

Theorem 4.5 *If there are a matrix* $K = \mathrm{diag}(K_1, K_2, \ldots, K_N) \in \mathbb{R}^{nN \times nN}$ *and a positive real number* r_1 *which satisfy*

$$\begin{pmatrix} Q_1 & \Theta_1 \\ \Theta_1^T & r_1 I_{Nn} - I_N \otimes \frac{D_2 + D_2^T}{2} \end{pmatrix} \leqslant 0, \tag{4.13}$$

where Q_1 *and* Θ_1 *have the same meanings as these in Theorem 4.4 , then the model* (4.2) *under the controller* (4.3) *achieves FTISP.*

Theorem 4.6 *If there are a matrix* $K = \mathrm{diag}(K_1, K_2, \ldots, K_N) \in \mathbb{R}^{nN \times nN}$ *and a positive real number* r_2 *which satisfy*

$$\begin{pmatrix} Q_2 & \Theta_2 \\ \Theta_2^T & Q_3 \end{pmatrix} \leqslant 0, \tag{4.14}$$

where $Q_2 = Q_1 + r_2(I_N \otimes D_1^T)(I_N \otimes D_1)$, $\Theta_2 = \Theta_1 + r_2(I_N \otimes D_1^T)(I_N \otimes D_2)$, $Q_3 = r_2(I_N \otimes D_2^T)(I_N \otimes D_2) - I_N \otimes \frac{D_2 + D_2^T}{2}$, Q_1 *and* Θ_1 *have the same meanings as these in Theorem 4.4, then the model* (4.2) *under the controller* (4.3) *achieves FTOSP.*

B. Adaptive state feedback controller:

Construct the following controller for the system (4.2):

$$v_i(t) = a_1 \sum_{j=1}^{N} \bar{A}_{ij}^1(t)\Gamma^1 z_j(t) - \varphi \mathrm{sign}(z_i(t) - \bar{z}(t))|z_i(t) - \bar{z}(t)|^{\alpha},$$

$$\dot{\bar{A}}_{ij}^1(t) = \omega_{ij}^1 (z_i(t) - z_j(t))^T \Gamma^1 (z_i(t) - z_j(t)) + \omega_{ij}^1, \qquad j \in \mathcal{N}_i, \tag{4.15}$$

where $\omega_{ij}^1 = \omega_{ji}^1$ are positive constants, $0 < \alpha < 1$, $\mathbb{R} \ni \varphi > 0$, $\mathbb{R} \ni \bar{A}_{ij}^1(t)$ is defined as follows: if there exists a connection between node i and node j $(i \neq j)$, then $\bar{A}_{ij}^1(t) = \bar{A}_{ji}^1(t) > 0$; otherwise, $\bar{A}_{ij}^1(t) = \bar{A}_{ji}^1(t) = 0$, and

$$\bar{A}^1_{ii}(t) = -\sum_{\substack{j=1\\ j\neq i}}^{N} \bar{A}^1_{ij}(t),\ i = 1, 2, \ldots, N.$$

Theorem 4.7 *If the following condition holds:*

$$D_2 + D_2^T > 0, \tag{4.16}$$

then the model (4.2) *realizes FTP under the adaptive controller* (4.15).

Proof Devise the following Lyapunov functional for the model (4.2):

$$\begin{aligned} V(t) &= V_1(t) + V_2(t), \\ V_1(t) &= \sum_{i=1}^{N} c_i^T(t)c_i(t), \\ V_2(t) &= \sum_{i=1}^{N}\sum_{j\in\mathcal{N}_i} \frac{a_1}{2\omega^1_{ij}}\left(\bar{A}^1_{ij}(t) - \tilde{A}^1_{ij}\right)^2, \end{aligned}$$

where $\tilde{A}^1_{ij} = \tilde{A}^1_{ji}(i \neq j)$ are nonnegative constants, and $\tilde{A}^1_{ij} = 0(i \neq j)$ if and only if $\bar{A}^1_{ij}(t) = 0$.

Therefore,

$$\begin{aligned} \dot{V}(t) &= 2\sum_{i=1}^{N} c_i^T(t)\dot{c}_i(t) + \sum_{i=1}^{N}\sum_{j\in\mathcal{N}_i} \frac{a_1}{\omega^1_{ij}}\left(\bar{A}^1_{ij}(t) - \tilde{A}^1_{ij}\right)\dot{\bar{A}}^1_{ij}(t) \\ &= 2\sum_{i=1}^{N} c_i^T(t)\Big(p(z_i(t)) - p(\bar{z}(t)) + p(\bar{z}(t)) - \frac{1}{N}\sum_{h=1}^{N} p(z_h(t)) \\ &\quad +a_1\sum_{j=1}^{N} A^1_{ij}\Gamma^1 c_j(t) + w_i(t) - \frac{1}{N}\sum_{h=1}^{N} w_h(t) + a_1\sum_{j=1}^{N}\bar{A}^1_{ij}(t)\Gamma^1 c_j(t) \\ &\quad -\frac{1}{N}\sum_{h=1}^{N} v_h(t) - \varphi \mathrm{sign}(c_i(t))|c_i(t)|^{\alpha}\Big) + a_1\sum_{i=1}^{N}\sum_{j\in\mathcal{N}_i}\left(\bar{A}^1_{ij}(t) - \tilde{A}^1_{ij}\right) \\ &\quad +\sum_{i=1}^{N}\sum_{j\in\mathcal{N}_i} a_1\left(\bar{A}^1_{ij}(t) - \tilde{A}^1_{ij}\right)(c_i(t) - c_j(t))^T\Gamma^1(c_i(t) - c_j(t)) \\ &\leqslant (1+\delta^2)\sum_{i=1}^{N} c_i^T(t)c_i(t) + 2\sum_{i=1}^{N} c_i^T(t)w_i(t) + 2a_1\sum_{i=1}^{N}\sum_{j=1}^{N} A^1_{ij}c_i^T(t)\Gamma^1 c_j(t) \\ &\quad +2a_1\sum_{i=1}^{N}\sum_{j=1}^{N}\bar{A}^1_{ij}(t)c_i^T(t)\Gamma^1 c_j(t) - 2\varphi\sum_{i=1}^{N} c_i^T(t)\mathrm{sign}(c_i(t))|c_i(t)|^{\alpha} \end{aligned}$$

$$+\sum_{i=1}^{N}\sum_{j\in\mathcal{N}_i} a_1\big(\bar{A}_{ij}^1(t)-\tilde{A}_{ij}^1\big)(c_i(t)-c_j(t))^T\Gamma^1(c_i(t)-c_j(t))$$
$$+a_1\sum_{i=1}^{N}\sum_{j\in\mathcal{N}_i}\big(\bar{A}_{ij}^1(t)-\tilde{A}_{ij}^1\big).$$

Define $\tilde{A}^1=(\tilde{A}_{ij}^1)_{N\times N}\in\mathbb{R}^{N\times N}$, where $\tilde{A}_{ii}^1=-\sum_{\substack{j=1\\ j\neq i}}^{N}\tilde{A}_{ij}^1$, $i=1,2,\ldots,N$. Then, one gets

$$\sum_{i=1}^{N}\sum_{j\in\mathcal{N}_i}\big(\bar{A}_{ij}^1(t)-\tilde{A}_{ij}^1\big)(c_i(t)-c_j(t))^T\Gamma^1(c_i(t)-c_j(t))$$
$$=-2\sum_{i=1}^{N}\sum_{j=1}^{N}(\bar{A}_{ij}^1(t)-\tilde{A}_{ij}^1)c_i^T(t)\Gamma^1c_j(t). \tag{4.17}$$

From (4.12), (4.17) and Lemma 1.7, one obtains

$$\dot{V}(t)\leqslant(1+\delta^2)\sum_{i=1}^{N}c_i^T(t)c_i(t)-2\varphi\sum_{i=1}^{N}(c_i^T(t)c_i(t))^{\frac{\alpha+1}{2}}+2\sum_{i=1}^{N}c_i^T(t)w_i(t)$$
$$+2a_1\sum_{i=1}^{N}\sum_{j=1}^{N}A_{ij}^1c_i^T(t)\Gamma^1c_j(t)+2a_1\sum_{i=1}^{N}\sum_{j=1}^{N}\tilde{A}_{ij}^1c_i^T(t)\Gamma^1c_j(t)$$
$$+a_1\sum_{i=1}^{N}\sum_{j\in\mathcal{N}_i}\big(\bar{A}_{ij}^1(t)-\tilde{A}_{ij}^1\big)$$
$$\leqslant c^T(t)\Big[(1+\delta^2)I_{Nn}+2a_1A^1\otimes\Gamma^1+2a_1\tilde{A}^1\otimes\Gamma^1\Big]c(t)+2c^T(t)w(t)$$
$$+a_1\sum_{i=1}^{N}\sum_{j\in\mathcal{N}_i}\big(\bar{A}_{ij}^1(t)-\tilde{A}_{ij}^1\big)-2\varphi V_1^{\frac{\alpha+1}{2}}(t). \tag{4.18}$$

By (4.18), we can derive

$$\dot{V}(t)-y^T(t)w(t)$$
$$=\dot{V}(t)-c^T(t)(I_N\otimes D_1^T)w(t)-w^T(t)(I_N\otimes D_2^T)w(t)$$
$$\leqslant c^T(t)\Big[(1+\delta^2)I_{Nn}+2a_1A^1\otimes\Gamma^1+2a_1\tilde{A}^1\otimes\Gamma^1\Big]c(t)+2c^T(t)w(t)$$
$$-2\varphi V_1^{\frac{\alpha+1}{2}}(t)-c^T(t)(I_N\otimes D_1^T)w(t)-w^T(t)\Psi w(t)$$
$$+a_1\sum_{i=1}^{N}\sum_{j\in\mathcal{N}_i}\big(\bar{A}_{ij}^1(t)-\tilde{A}_{ij}^1\big)$$

$$
\begin{aligned}
&\leqslant c^T(t)\Big[(1+\delta^2)I_{Nn} + 2a_1 A^1 \otimes \Gamma^1 + 2a_1\tilde{A}^1 \otimes \Gamma^1 + \Xi\Big]c(t) - 2\varphi V_1^{\frac{\alpha+1}{2}}(t)\\
&\quad + a_1 \sum_{i=1}^{N}\sum_{j\in\mathcal{N}_i}\big(\bar{A}^1_{ij}(t) - \tilde{A}^1_{ij}\big),
\end{aligned}
$$

where $\Psi = I_N \otimes \frac{D_2+D_2^T}{2}$, $\Xi = \Big(I_{Nn} - I_N \otimes \frac{D_1^T}{2}\Big)\Psi^{-1}\big(I_{Nn} - I_N \otimes \frac{D_1}{2}\big)$.

By the definition of $\tilde{A}^1$, there exists a unitary matrix $\kappa = (\kappa_1, \kappa_2, \cdots, \kappa_N) \in \mathbb{R}^{N\times N}$ such that $\kappa^T\tilde{A}^1\kappa = \Lambda = \mathrm{diag}(\lambda_1, \lambda_2, \cdots, \lambda_N)$, where λ_i are the eigenvalues of $\tilde{A}^1$ and $0 = \lambda_1 > \lambda_2 \geqslant \lambda_3 \geqslant \cdots \geqslant \lambda_N$. Let $\mathbb{R}^{Nn} \ni m(t) = (m_1^T(t), m_2^T(t), \cdots, m_N^T(t))^T = (\kappa^T \otimes I_n)c(t)$. Since $\kappa_1 = \frac{1}{\sqrt{N}}(1, 1, \cdots, 1)^T$, we have $m_1(t) = (\kappa_1^T \otimes I_n)c(t) = 0$. Then, we can derive

$$
\begin{aligned}
&c^T(t)\Big[(1+\delta^2)I_{Nn} + 2a_1 A^1 \otimes \Gamma^1 + \Xi\Big]c(t)\\
&+2a_1 c^T(t)(\kappa\otimes I_n)(\Lambda\otimes\Gamma^1)(\kappa^T\otimes I_n)c(t)\\
\leqslant\ & c^T(t)\Big[(1+\delta^2)I_{Nn} + 2a_1 A^1 \otimes \Gamma^1 + \Xi\Big]c(t) + 2a_1\lambda_2 m^T(t)(I_N\otimes\Gamma^1)m(t)\\
=\ & c^T(t)\Big[(1+\delta^2)I_{Nn} + 2a_1 A^1 \otimes \Gamma^1 + \Xi + 2a_1\lambda_2(I_N\otimes\Gamma^1)\Big]c(t).
\end{aligned}
$$

Selecting $\tilde{A}^1_{ij}$ large enough such that

$$
1 + \delta^2 + \lambda_M(\Xi) + 2a_1\lambda_2\lambda_m(\Gamma^1) \leqslant 0.
$$

From (4.15), a positive real number t^* could be found, which satisfies

$$
\bar{A}^1_{ij}(t) \geqslant \tilde{A}^1_{ij}
$$

for any $t \geqslant t^*$ and $j \in \mathcal{N}_i$.

Thus, when $t \geqslant t^*$, we have

$$
\begin{aligned}
&\dot{V}_1(t) - y^T(t)w(t)\\
=\ & 2\sum_{i=1}^{N} c_i^T(t)\Big(p(z_i(t)) - p(\bar{z}(t)) + p(\bar{z}(t)) - \frac{1}{N}\sum_{h=1}^{N} p(z_h(t))\\
&+a_1\sum_{j=1}^{N} A^1_{ij}\Gamma^1 c_j(t) + w_i(t) - \frac{1}{N}\sum_{h=1}^{N} w_h(t) + a_1\sum_{j=1}^{N}\bar{A}^1_{ij}(t)\Gamma^1 c_j(t)\\
&-\frac{1}{N}\sum_{h=1}^{N} v_h(t) - \varphi\,\mathrm{sign}(c_i(t))|c_i(t)|^{\alpha}\Big) - c^T(t)(I_N\otimes D_1^T)w(t) - w^T(t)\Psi w(t)\\
\leqslant\ & c^T(t)\Big[(1+\delta^2)I_{Nn} + 2a_1 A^1\otimes\Gamma^1 + 2a_1(\bar{A}^1(t) - \tilde{A}^1)\otimes\Gamma^1
\end{aligned}
$$

$$
\begin{aligned}
&+2a_1\tilde{A}^1\otimes\Gamma^1+\Xi\Big]c(t)-2\varphi V_1^{\frac{\alpha+1}{2}}(t)\\
&\leqslant c^T(t)\Big[(1+\delta^2)I_{Nn}+2a_1A^1\otimes\Gamma^1+2a_1\tilde{A}^1\otimes\Gamma^1+\Xi\Big]c(t)-2\varphi V_1^{\frac{\alpha+1}{2}}(t)\\
&\leqslant -2\varphi V_1^{\frac{\alpha+1}{2}}(t),
\end{aligned}
\tag{4.19}
$$

where $\bar{A}^1(t)=(\bar{A}^1_{ij}(t))_{N\times N}\in\mathbb{R}^{N\times N}$.

By (4.19), we have

$$
y^T(t)w(t)\geqslant \dot{V}_1(t)+2\varphi V_1^{\frac{\alpha+1}{2}}(t),\quad t\geqslant t^*.
$$

Therefore, the model (4.2) realizes FTP under the adaptive controller (4.15).

Similar to the proof of the Theorem 4.7, one could derive the following conclusions.

Theorem 4.8 *If there is an* $\mathbb{R}\ni r_1>0$ *which satisfies*

$$
D_2+D_2^T-r_1I_n>0,
\tag{4.20}
$$

then the model (4.2) *under the adaptive controller* (4.15) *achieves FTISP.*

Theorem 4.9 *If there is an* $\mathbb{R}\ni r_2>0$ *which satisfies*

$$
D_2+D_2^T-r_2D_2^TD_2>0,
\tag{4.21}
$$

then the model (4.2) *realizes FTOSP under the adaptive controller* (4.15).

4.3.3 FTS of CDNs

Definition 4.10 The model (4.1) achieves FTS if there is $0<\rho\in\mathbb{R}$ which meets

$$
\lim_{t\to\rho^-}\left\|z_i(t)-\frac{1}{N}\sum_{l=1}^{N}z_l(t)\right\|=0,\ i=1,2,\ldots,N,
$$

$$
\sum_{i=1}^{N}\left\|z_i(t)-\frac{1}{N}\sum_{l=1}^{N}z_l(t)\right\|\neq 0,\ 0\leqslant t<\rho
$$

under the condition that $w_i(t)=0, i=1,2,\ldots,N$.

Theorem 4.11 *Assume that* $V(t):[0,+\infty)\to[0,+\infty)$ *is continuously differentiable and meets the following condition:*

$$
\kappa_1(\|c(t)\|)\leqslant V(t),
$$

where $\kappa_1 : [0, +\infty) \to [0, +\infty)$ is strictly monotonically increasing and continuous function, $\kappa_1(\varsigma)$ is positive for $\varsigma > 0$ with $\kappa_1(0) = 0$. If the network (4.2) *under the controller* (4.3) *or* (4.15) *achieves FTP in regard to $V(t)$, then the model* (4.1) *under the controller* (4.3) *or* (4.15) *realizes FTS.*

Proof If the model (4.2) under the controller (4.3) or (4.15) achieves FTP in regard to $V(t)$, there obviously exist $0 < \iota < 1$, $\mathbb{R} \ni t^* \geqslant 0$ and $\mathbb{R} \ni \zeta > 0$ which satisfy

$$y^T(t)w(t) \geqslant \dot{V}(t) + \zeta V^{\iota}(t)$$

for any $t \geqslant t^*$.

Letting $w(t) = 0$, one gets

$$\dot{V}(t) \leqslant -\zeta V^{\iota}(t)$$

for any $t \geqslant t^*$.

From Lemma 1.6, we have

$$V(t) = 0 \ \text{ for } \ t \geqslant t^* + \mu,$$

where $\mu = \frac{V^{1-\iota}(t^*)}{\zeta(1-\iota)}$.

Then, we gain

$$\|c(t)\| = 0 \ \text{ for } \ t \geqslant t^* + \mu.$$

Evidently, a ρ which meets $0 < \rho \leqslant t^* + \mu$ can be found such that

$$\lim_{t \to \rho^-} \|c_i(t)\| = 0, \ i = 1, 2, \ldots, N,$$

$$\sum_{i=1}^{N} \|c_i(t)\| \neq 0, \ 0 \leqslant t < \rho.$$

Thus, the model (4.1) under the controller (4.3) or (4.15) achieves FTS.

On the basis of Theorems 4.4, 4.7 and 4.11, one could obtain the following corollaries.

Corollary 4.12 *If there are matrices $K = \text{diag}(K_1, K_2, \ldots, K_N) \in \mathbb{R}^{nN \times nN}$, $D_1 \in \mathbb{R}^{n \times n}$ and $D_2 \in \mathbb{R}^{n \times n}$ such that*

$$\begin{pmatrix} Q_1 & \Theta_1 \\ \Theta_1^T & -I_N \otimes \frac{D_2 + D_2^T}{2} \end{pmatrix} \leqslant 0,$$

where $Q_1 = (1 + \delta^2)I_{Nn} - K - K^T + 2a_1 A^1 \otimes \Gamma^1$, $\Theta_1 = I_{Nn} - \frac{I_N \otimes D_1^T}{2}$, then the network (4.1) *under the controller* (4.3) *achieves FTS.*

Corollary 4.13 *If there is a matrix* $D_2 \in \mathbb{R}^{n\times n}$ *such that*

$$D_2 + D_2^T > 0, \tag{4.22}$$

then the model (4.1) *realizes FTS under the adaptive state feedback controller* (4.15).

4.4 FTP and FTS of CDNs with Derivative Coupling

4.4.1 Network Model

The network model is given by:

$$\dot{z}_i(t) = p(z_i(t)) + a_2 \sum_{j=1}^{N} A_{ij}^2 \Gamma^2 \dot{z}_j(t) + w_i(t) + v_i(t), \tag{4.23}$$

where $i = 1, 2, \ldots, N$, $w_i(t), v_i(t), z_i(t), p(\cdot)$ have the same meanings as these in the model (4.1), a_2, Γ^2, $A^2 = (A_{ij}^2)_{N\times N} \in \mathbb{R}^{N\times N}$ have the similar definitions as $a_1, \Gamma^1, A^1 = (A_{ij}^1)_{N\times N} \in \mathbb{R}^{N\times N}$ in the model (4.1). In this chapter, the network (4.23) is also connected.

4.4.2 FTP of CDNs

Define $\mathbb{R}^n \ni \hat{z}(t) = (\hat{z}_1(t), \hat{z}_2(t), \ldots, \hat{z}_n(t))^T = \frac{1}{N}\sum_{i=1}^{N} z_i(t)$. Then, we have

$$\begin{aligned}
\dot{\hat{z}}(t) &= \frac{1}{N}\sum_{i=1}^{N} p(z_i(t)) + \frac{a_2}{N}\sum_{j=1}^{N}\left(\sum_{i=1}^{N} A_{ij}^2\right)\Gamma^2 \dot{z}_j(t) + \frac{1}{N}\sum_{i=1}^{N} w_i(t) + \frac{1}{N}\sum_{i=1}^{N} v_i(t) \\
&= \frac{1}{N}\sum_{i=1}^{N} p(z_i(t)) + \frac{1}{N}\sum_{i=1}^{N} w_i(t) + \frac{1}{N}\sum_{i=1}^{N} v_i(t).
\end{aligned}$$

Letting $\hat{c}_i(t) = z_i(t) - \hat{z}(t)$, we have

$$\begin{aligned}
\dot{\hat{c}}_i(t) &= p(z_i(t)) + a_2 \sum_{j=1}^{N} A_{ij}^2 \Gamma^2 \dot{\hat{c}}_j(t) + w_i(t) + v_i(t) - \frac{1}{N}\sum_{h=1}^{N} p(z_h(t)) \\
&\quad - \frac{1}{N}\sum_{h=1}^{N} w_h(t) - \frac{1}{N}\sum_{h=1}^{N} v_h(t), \quad i = 1, 2, \ldots, N.
\end{aligned} \tag{4.24}$$

Select the following output vector $\hat{y}_i(t) \in \mathbb{R}^n$ for the model (4.24)

$$\hat{y}_i(t) = \hat{D}_1\hat{c}_i(t) + \hat{D}_2 w_i(t), \tag{4.25}$$

where $\hat{D}_1 \in \mathbb{R}^{n\times n}$, $\hat{D}_2 \in \mathbb{R}^{n\times n}$.

Denote

$$\begin{aligned}
\hat{c}(t) &= (\hat{c}_1^T(t), \hat{c}_2^T(t), \ldots, \hat{c}_N^T(t))^T,\\
w(t) &= (w_1^T(t), w_2^T(t), \ldots, w_N^T(t))^T,\\
\hat{y}(t) &= (\hat{y}_1^T(t), \hat{y}_2^T(t), \ldots, \hat{y}_N^T(t))^T.
\end{aligned}$$

A. State feedback controller

Construct the following controller for the model (4.24):

$$v_i(t) = -K_i(z_i(t) - \hat{z}(t)) - \varphi \text{sign}(z_i(t) - \hat{z}(t))|z_i(t) - \hat{z}(t)|^{\alpha}, \tag{4.26}$$

where $K_i \in \mathbb{R}^{n\times n}, 0 < \alpha < 1, \mathbb{R} \ni \varphi > 0$.

Theorem 4.14 *If there is a matrix* $K = \text{diag}(K_1, K_2, \ldots, K_N) \in \mathbb{R}^{nN\times nN}$ *such that*

$$\begin{pmatrix} \hat{Q}_1 & \hat{\Theta}_1 \\ \hat{\Theta}_1^T & -I_N \otimes \frac{\hat{D}_2+\hat{D}_2^T}{2} \end{pmatrix} \leqslant 0, \tag{4.27}$$

where $\hat{Q}_1 = (1+\delta^2)I_{Nn} - K - K^T$, $\hat{\Theta}_1 = I_{Nn} - \frac{I_N\otimes\hat{D}_1^T}{2}$, *then the network* (4.24) *under the controller* (4.26) *achieves FTP.*

Proof The following Lyapunov functional can be chosen for the model (4.24):

$$\hat{V}_1(t) = \sum_{i=1}^{N} \hat{c}_i^T(t)\hat{c}_i(t) - a_2\hat{c}^T(t)(A^2 \otimes \Gamma^2)\hat{c}(t). \tag{4.28}$$

Then, we have

$$\begin{aligned}
\dot{\hat{V}}_1(t) &= 2\sum_{i=1}^{N} \hat{c}_i^T(t)\dot{\hat{c}}_i(t) - 2a_2\hat{c}^T(t)(A^2 \otimes \Gamma^2)\dot{\hat{c}}(t)\\
&= 2\sum_{i=1}^{N} \hat{c}_i^T(t)\Big(p(z_i(t)) - p(\hat{z}(t)) + p(\hat{z}(t)) + a_2\sum_{j=1}^{N} A_{ij}^2\Gamma^2\dot{\hat{c}}_j(t) + w_i(t)\\
&\quad -K_i\hat{c}_i(t) - \frac{1}{N}\sum_{h=1}^{N} p(z_h(t)) - \frac{1}{N}\sum_{h=1}^{N} w_h(t) - \frac{1}{N}\sum_{h=1}^{N} v_h(t)
\end{aligned}$$

$$-\varphi \mathrm{sign}(\hat{c}_i(t))|\hat{c}_i(t)|^{\alpha}\Big) - 2a_2\hat{c}^T(t)(A^2 \otimes \Gamma^2)\dot{\hat{c}}(t)$$
$$\leqslant (1+\delta^2)\hat{c}^T(t)\hat{c}(t) + 2\hat{c}^T(t)w(t) - 2\varphi\sum_{i=1}^{N}(\hat{c}_i^T(t)\hat{c}_i(t))^{\frac{\alpha+1}{2}}$$
$$-\hat{c}^T(t)(K+K^T)\hat{c}(t). \tag{4.29}$$

From (4.28), we can get

$$\hat{V}_1(t) \leqslant \lambda_M(P)\hat{c}^T(t)\hat{c}(t), \tag{4.30}$$

where $P = I_{Nn} - a_2A^2 \otimes \Gamma^2$.

By (4.29) and (4.30), one derives

$$\begin{aligned}
&\dot{\hat{V}}_1(t) - \hat{y}^T(t)w(t)\\
&= \dot{\hat{V}}_1(t) - \hat{c}^T(t)(I_N \otimes \hat{D}_1^T)w(t) - w^T(t)(I_N \otimes \hat{D}_2^T)w(t)\\
&\leqslant -2\varphi\sum_{i=1}^{N}(\hat{c}_i^T(t)\hat{c}_i(t))^{\frac{\alpha+1}{2}} + \hat{\xi}^T(t)\begin{pmatrix} \hat{Q}_1 & \hat{\Theta}_1 \\ \hat{\Theta}_1^T & -I_N \otimes \frac{\hat{D}_2+\hat{D}_2^T}{2} \end{pmatrix}\hat{\xi}(t)\\
&\leqslant -2\varphi(\hat{c}^T(t)\hat{c}(t))^{\frac{\alpha+1}{2}}\\
&\leqslant -2\varphi\left(\frac{\hat{V}_1(t)}{\lambda_M(P)}\right)^{\frac{\alpha+1}{2}}\\
&= \frac{-2\varphi}{\lambda_M^{\frac{\alpha+1}{2}}(P)}\hat{V}_1^{\frac{\alpha+1}{2}}(t),
\end{aligned}$$

where $\hat{\xi}(t) = (\hat{c}^T(t), w^T(t))^T$.

Then, one has

$$\hat{y}^T(t)w(t) \geqslant \dot{\hat{V}}_1(t) + \frac{2\varphi}{\lambda_M^{\frac{\alpha+1}{2}}(P)}\hat{V}_1^{\frac{\alpha+1}{2}}(t).$$

Therefore, the network (4.24) under the controller (4.26) reaches FTP.

Theorem 4.15 *If there are a matrix* $K = \mathrm{diag}(K_1, K_2, \ldots, K_N) \in \mathbb{R}^{nN \times nN}$ *and a positive real number* $\hat{r}_1$ *such that*

$$\begin{pmatrix} \hat{Q}_1 & \hat{\Theta}_1 \\ \hat{\Theta}_1^T & \hat{r}_1 I_{Nn} - I_N \otimes \frac{\hat{D}_2+\hat{D}_2^T}{2} \end{pmatrix} \leqslant 0, \tag{4.31}$$

where $\hat{Q}_1$, $\hat{\Theta}_1$ *have the same meanings as these in Theorem 4.14, then the model* (4.24) *under the controller* (4.26) *achieves FTISP.*

Theorem 4.16 *If there are a matrix* $K = \mathrm{diag}(K_1, K_2, \ldots, K_N) \in \mathbb{R}^{nN\times nN}$ *and a positive real number* $\hat{r}_2$ *such that*

$$\begin{pmatrix} \hat{Q}_2 & \hat{\Theta}_2 \\ \hat{\Theta}_2^T & \hat{Q}_3 \end{pmatrix} \leqslant 0, \tag{4.32}$$

where $\hat{Q}_2 = \hat{Q}_1 + \hat{r}_2(I_N \otimes \hat{D}_1^T)(I_N \otimes \hat{D}_1)$, $\hat{\Theta}_2 = \hat{\Theta}_1 + \hat{r}_2(I_N \otimes \hat{D}_1^T)(I_N \otimes \hat{D}_2)$, $\hat{Q}_3 = \hat{r}_2(I_N \otimes \hat{D}_2^T)(I_N \otimes \hat{D}_2) - I_N \otimes \frac{\hat{D}_2+\hat{D}_2^T}{2}$, $\hat{Q}_1$ *and* $\hat{\Theta}_1$ *have the same meanings as these in Theorem 4.14, then the model* (4.24) *under the controller* (4.26) *achieves FTOSP.*

B. Adaptive state feedback controller

Construct the following controller for the model (4.24):

$$\begin{aligned} v_i(t) &= a_2 \sum_{j=1}^{N} \bar{A}_{ij}^2(t)\Gamma^2 z_j(t) - \varphi \mathrm{sign}(z_i(t) - \hat{z}(t))|z_i(t) - \hat{z}(t)|^{\alpha}, \\ \dot{\bar{A}}_{ij}^2(t) &= \omega_{ij}^2 (z_i(t) - z_j(t))^T \Gamma^2 (z_i(t) - z_j(t)) + \omega_{ij}^2, \qquad j \in \mathcal{N}_i, \end{aligned} \tag{4.33}$$

where $\mathbb{R} \ni \omega_{ij}^2 = \omega_{ji}^2 > 0$, $\bar{A}^2(t) = (\bar{A}_{ij}^2(t))_{N\times N} \in \mathbb{R}^{N\times N}$ has the similar meaning as $\bar{A}^1(t) = (\bar{A}_{ij}^1(t))_{N\times N} \in \mathbb{R}^{N\times N}$ in the model (4.15).

Theorem 4.17 *If the following condition holds:*

$$\hat{D}_2 + \hat{D}_2^T > 0, \tag{4.34}$$

then the model (4.24) *realizes FTP under the adaptive controller* (4.33).

Proof Devise the following Lyapunov functional for the model (4.24):

$$\begin{aligned} \hat{V}(t) &= \hat{V}_1(t) + \hat{V}_2(t), \\ \hat{V}_1(t) &= \sum_{i=1}^{N} \hat{c}_i^T(t)\hat{c}_i(t) - a_2 \hat{c}^T(t)(A^2 \otimes \Gamma^2)\hat{c}(t), \\ \hat{V}_2(t) &= \sum_{i=1}^{N} \sum_{j\in\mathcal{N}_i} \frac{a_2}{2\omega_{ij}^2} (\bar{A}_{ij}^2(t) - \tilde{A}_{ij}^2)^2, \end{aligned}$$

where $\tilde{A}_{ij}^2 = \tilde{A}_{ji}^2 (i \neq j)$ are nonnegative constants, and $\tilde{A}_{ij}^2 = 0 (i \neq j)$ if and only if $\bar{A}_{ij}^2(t) = 0$.

Therefore,

$$
\begin{aligned}
\dot{V}(t) &= 2\sum_{i=1}^{N}\hat{c}_i^T(t)\dot{\hat{c}}_i(t) - 2a_2\hat{c}^T(t)(A^2\otimes\Gamma^2)\dot{\hat{c}}(t)\\
&\quad+\sum_{i=1}^{N}\sum_{j\in\mathcal{N}_i}\frac{a_2}{\omega_{ij}^2}\big(\bar{A}_{ij}^2(t)-\tilde{A}_{ij}^2\big)\dot{\bar{A}}_{ij}^2(t)\\
&= 2\sum_{i=1}^{N}\hat{c}_i^T(t)\bigg(p(z_i(t)) - p(\hat{z}(t)) + p(\hat{z}(t)) + a_2\sum_{j=1}^{N}A_{ij}^2\Gamma^2\dot{\hat{c}}_j(t) + w_i(t)\\
&\quad+a_2\sum_{j=1}^{N}\bar{A}_{ij}^2(t)\Gamma^2\hat{c}_j(t) - \frac{1}{N}\sum_{h=1}^{N}p(z_h(t)) - \frac{1}{N}\sum_{h=1}^{N}w_h(t) - \frac{1}{N}\sum_{h=1}^{N}v_h(t)\\
&\quad-\varphi\,\mathrm{sign}(\hat{c}_i(t))|\hat{c}_i(t)|^{\alpha}\bigg) - 2a_2\hat{c}^T(t)(A^2\otimes\Gamma^2)\dot{\hat{c}}(t)\\
&\quad+\sum_{i=1}^{N}\sum_{j\in\mathcal{N}_i}a_2\big(\bar{A}_{ij}^2(t)-\tilde{A}_{ij}^2\big)(\hat{c}_i(t)-\hat{c}_j(t))^T\Gamma^2(\hat{c}_i(t)-\hat{c}_j(t))\\
&\quad+a_2\sum_{i=1}^{N}\sum_{j\in\mathcal{N}_i}\big(\bar{A}_{ij}^2(t)-\tilde{A}_{ij}^2\big)\\
&\leqslant (1+\delta^2)\hat{c}^T(t)\hat{c}(t) + 2\hat{c}^T(t)w(t) + 2a_2\sum_{i=1}^{N}\sum_{j=1}^{N}\bar{A}_{ij}^2(t)\hat{c}_i^T(t)\Gamma^2\hat{c}_j(t)\\
&\quad-2\varphi\sum_{i=1}^{N}(\hat{c}_i^T(t)\hat{c}_i(t))^{\frac{\alpha+1}{2}} - 2a_2\sum_{i=1}^{N}\sum_{j=1}^{N}\big(\bar{A}_{ij}^2(t)-\tilde{A}_{ij}^2\big)\hat{c}_i^T(t)\Gamma^2\hat{c}_j(t)\\
&\quad+a_2\sum_{i=1}^{N}\sum_{j\in\mathcal{N}_i}\big(\bar{A}_{ij}^2(t)-\tilde{A}_{ij}^2\big)\\
&= (1+\delta^2)\hat{c}^T(t)\hat{c}(t) + 2\hat{c}^T(t)w(t) + 2a_2\hat{c}^T(t)(\tilde{A}^2\otimes\Gamma^2)\hat{c}(t)\\
&\quad-2\varphi\sum_{i=1}^{N}(\hat{c}_i^T(t)\hat{c}_i(t))^{\frac{\alpha+1}{2}} + a_2\sum_{i=1}^{N}\sum_{j\in\mathcal{N}_i}\big(\bar{A}_{ij}^2(t)-\tilde{A}_{ij}^2\big)\\
&= \hat{c}^T(t)\Big[(1+\delta^2)I_{Nn} + 2a_2\tilde{A}^2\otimes\Gamma^2\Big]\hat{c}(t) + 2\hat{c}^T(t)w(t)\\
&\quad-2\varphi\sum_{i=1}^{N}(\hat{c}_i^T(t)\hat{c}_i(t))^{\frac{\alpha+1}{2}} + a_2\sum_{i=1}^{N}\sum_{j\in\mathcal{N}_i}\big(\bar{A}_{ij}^2(t)-\tilde{A}_{ij}^2\big),
\end{aligned}
\tag{4.35}
$$

where $\tilde{A}_{ii}^2 = -\sum_{\substack{j=1\\ j\neq i}}^{N}\tilde{A}_{ij}^2$, $i = 1, 2, \ldots, N$ and $\tilde{A}^2 = (\tilde{A}_{ij}^2)_{N\times N} \in \mathbb{R}^{N\times N}$.

By (4.35), one obtains

$$\begin{aligned}
&\dot{\hat{V}}(t) - \hat{y}^T(t)w(t) \\
&= \dot{\hat{V}}(t) - \hat{c}^T(t)(I_N \otimes \hat{D}_1^T)w(t) - w^T(t)(I_N \otimes \hat{D}_2^T)w(t) \\
&\leqslant \hat{c}^T(t)\Big[(1+\delta^2)I_{Nn} + 2a_2\tilde{A}^2 \otimes \Gamma^2\Big]\hat{c}(t) + 2\hat{c}^T(t)w(t) - \hat{c}^T(t)(I_N \otimes \hat{D}_1^T)w(t) \\
&\quad - w^T(t)\hat{\Psi}w(t) - 2\varphi(\hat{c}^T(t)\hat{c}(t))^{\frac{\alpha+1}{2}} + a_2\sum_{i=1}^{N}\sum_{j\in\mathcal{N}_i}\left(\bar{A}_{ij}^2(t) - \tilde{A}_{ij}^2\right) \\
&\leqslant \hat{c}^T(t)\Big[(1+\delta^2)I_{Nn} + 2a_2\tilde{A}^2 \otimes \Gamma^2 + \hat{\Xi}\Big]\hat{c}(t) - 2\varphi(\hat{c}^T(t)\hat{c}(t))^{\frac{\alpha+1}{2}} \\
&\quad + a_2\sum_{i=1}^{N}\sum_{j\in\mathcal{N}_i}\left(\bar{A}_{ij}^2(t) - \tilde{A}_{ij}^2\right),
\end{aligned}$$

where $\hat{\Psi} = I_N \otimes \frac{\hat{D}_2 + \hat{D}_2^T}{2}$, $\hat{\Xi} = \left(I_{Nn} - I_N \otimes \frac{\hat{D}_1^T}{2}\right)\hat{\Psi}^{-1}\left(I_{Nn} - I_N \otimes \frac{\hat{D}_1}{2}\right)$.

By the definition of $\tilde{A}^2$, there exists a unitary matrix $\hat{\kappa} = (\hat{\kappa}_1, \hat{\kappa}_2, \cdots, \hat{\kappa}_N) \in \mathbb{R}^{N\times N}$ such that $\hat{\kappa}^T\tilde{A}^2\hat{\kappa} = \hat{\Lambda} = \text{diag}(\hat{\lambda}_1, \hat{\lambda}_2, \cdots, \hat{\lambda}_N)$, where $\hat{\lambda}_i$ are the eigenvalues of $\tilde{A}^2$ and $0 = \hat{\lambda}_1 > \hat{\lambda}_2 \geqslant \hat{\lambda}_3 \geqslant \cdots \geqslant \hat{\lambda}_N$. Let $\mathbb{R}^{Nn} \ni \hat{m}(t) = (\hat{m}_1^T(t), \hat{m}_2^T(t), \cdots, \hat{m}_N^T(t))^T = (\hat{\kappa}^T \otimes I_n)\hat{c}(t)$. Since $\hat{\kappa}_1 = \frac{1}{\sqrt{N}}(1, 1, \cdots, 1)^T$, we have $\hat{m}_1(t) = (\hat{\kappa}_1^T \otimes I_n)\hat{c}(t) = 0$. Then, we can derive

$$\begin{aligned}
&\hat{c}^T(t)\Big[(1+\delta^2)I_{Nn} + \hat{\Xi}\Big]\hat{c}(t) + 2a_2\hat{c}^T(t)(\hat{\kappa} \otimes I_n)(\hat{\Lambda} \otimes \Gamma^2)(\hat{\kappa}^T \otimes I_n)\hat{c}(t) \\
&\leqslant \hat{c}^T(t)\Big[(1+\delta^2)I_{Nn} + \hat{\Xi}\Big]\hat{c}(t) + 2a_2\hat{\lambda}_2\hat{m}^T(t)(I_N \otimes \Gamma^2)\hat{m}(t) \\
&= \hat{c}^T(t)\Big[(1+\delta^2)I_{Nn} + \hat{\Xi} + 2a_2\hat{\lambda}_2(I_N \otimes \Gamma^2)\Big]\hat{c}(t).
\end{aligned}$$

Selecting $\tilde{A}_{ij}^2$ large enough such that

$$1 + \delta^2 + \lambda_M(\hat{\Xi}) + 2a_2\hat{\lambda}_2\lambda_m(\Gamma^2) \leqslant 0.$$

From (4.33), a positive real number t^* could be found, which satisfies

$$\bar{A}_{ij}^2(t) \geqslant \tilde{A}_{ij}^2$$

for any $t \geqslant t^*$ and $j \in \mathcal{N}_i$.

Thus, when $t \geqslant t^*$, we have

$$\dot{\hat{V}}_1(t) - \hat{y}^T(t)w(t)$$

$$
\begin{aligned}
&= 2\sum_{i=1}^{N}\hat{c}_i^T(t)\Big(p(z_i(t)) - p(\hat{z}(t)) + p(\hat{z}(t)) + a_2\sum_{j=1}^{N}A_{ij}^2\Gamma^2\dot{\hat{c}}_j(t) + w_i(t) \\
&\quad + a_2\sum_{j=1}^{N}\tilde{A}_{ij}^2(t)\Gamma^2\hat{c}_j(t) - \frac{1}{N}\sum_{h=1}^{N}p(z_h(t)) - \frac{1}{N}\sum_{h=1}^{N}w_h(t) - \frac{1}{N}\sum_{h=1}^{N}v_h(t) \\
&\quad - \varphi\,\mathrm{sign}(\hat{c}_i(t))|\hat{c}_i(t)|^{\alpha}\Big) - 2a_2\hat{c}^T(t)(A^2\otimes\Gamma^2)\dot{\hat{c}}(t) - \hat{c}^T(t)(I_N\otimes\hat{D}_1^T)w(t) \\
&\quad - w^T(t)(I_N\otimes\hat{D}_2^T)w(t) \\
&\leqslant \hat{c}^T(t)\Big[(1+\delta^2)I_{Nn} + 2a_2(\bar{A}^2(t) - \tilde{A}^2)\otimes\Gamma^2 + 2a_2\tilde{A}^2\otimes\Gamma^2 + \hat{\Xi}\Big]\hat{c}(t) \\
&\quad - 2\varphi\sum_{i=1}^{N}(\hat{c}_i^T(t)\hat{c}_i(t))^{\frac{\alpha+1}{2}} \\
&\leqslant \hat{c}^T(t)\Big[(1+\delta^2)I_{Nn} + 2a_2\tilde{A}^2\otimes\Gamma^2 + \hat{\Xi}\Big]\hat{c}(t) - 2\varphi(\hat{c}^T(t)\hat{c}(t))^{\frac{\alpha+1}{2}} \\
&\leqslant -2\varphi(\hat{c}^T(t)\hat{c}(t))^{\frac{\alpha+1}{2}} \\
&\leqslant \frac{-2\varphi}{\lambda_M^{\frac{\alpha+1}{2}}(P)}\hat{V}_1^{\frac{\alpha+1}{2}}(t).
\end{aligned} \tag{4.36}
$$

By (4.36), one has

$$
\hat{y}^T(t)w(t) \geqslant \dot{\hat{V}}_1(t) + \frac{2\varphi}{\lambda_M^{\frac{\alpha+1}{2}}(P)}\hat{V}_1^{\frac{\alpha+1}{2}}(t),\ t \geqslant t^*.
$$

Therefore, the model (4.24) under the adaptive controller (4.33) realizes FTP.

Similar to the proof of the Theorem 4.17, one could derive the following conclusions.

Theorem 4.18 *If there is an* $\mathbb{R} \ni \hat{r}_1 > 0$ *which satisfies*

$$
\hat{D}_2 + \hat{D}_2^T - \hat{r}_1 I_n > 0, \tag{4.37}
$$

then the network (4.24) *under the adaptive controller* (4.33) *reaches FTISP.*

Theorem 4.19 *If there is an* $\mathbb{R} \ni \hat{r}_2 > 0$ *which satisfies*

$$
\hat{D}_2 + \hat{D}_2^T - \hat{r}_2\hat{D}_2^T\hat{D}_2 > 0, \tag{4.38}
$$

then the model (4.24) *reaches FTOSP under the adaptive controller* (4.33).

4.4.3 FTS of CDNs

Definition 4.20 The model (4.23) achieves FTS if there is $0 < \rho \in \mathbb{R}$ such that

$$\lim_{t\to\rho^-}\left\| z_i(t) - \frac{1}{N}\sum_{l=1}^{N} z_l(t)\right\| = 0,\ i = 1, 2, \ldots, N,$$

$$\sum_{i=1}^{N}\left\| z_i(t) - \frac{1}{N}\sum_{l=1}^{N} z_l(t)\right\| \neq 0,\ 0 \leqslant t < \rho$$

under the condition that $w_i(t) = 0, i = 1, 2, \ldots, N$.

Theorem 4.21 *Assume that $\hat{V}(t) : [0, +\infty) \to [0, +\infty)$ is continuously differentiable and meets the following condition:*

$$\kappa_2(\|\hat{c}(t)\|) \leqslant \hat{V}(t),$$

where $\kappa_2 : [0, +\infty) \to [0, +\infty)$ is strictly monotonically increasing and continuous function, $\kappa_2(\varsigma)$ is positive for $\varsigma > 0$ with $\kappa_2(0) = 0$. If the network (4.24) *under the controller* (4.26) *or* (4.33) *achieves FTP in regard to $\hat{V}(t)$, then the model* (4.23) *achieves FTS under controller* (4.26) *or* (4.33).

Proof If the network (4.24) under the controller (4.26) or (4.33) achieves FTP in regard to $\hat{V}(t)$, there obviously exist $0 < \iota < 1$, $\mathbb{R} \ni t^* \geqslant 0$ and $\mathbb{R} \ni \zeta > 0$ which satisfy

$$\hat{y}^T(t)w(t) \geqslant \dot{\hat{V}}(t) + \zeta\hat{V}^{\iota}(t)$$

for any $t \geqslant t^*$.

Letting $w(t) = 0$, we can derive

$$\dot{\hat{V}}(t) \leqslant -\zeta\hat{V}^{\iota}(t)$$

for any $t \geqslant t^*$.

From Lemma 1.6, we have

$$\hat{V}(t) = 0 \ \text{ for } \ t \geqslant t^* + \mu,$$

in which $\mu = \frac{\hat{V}^{1-\iota}(t^*)}{\zeta(1-\iota)}$.

Then, we can conclude that

$$\|\hat{c}(t)\| = 0 \ \text{ for } \ t \geqslant t^* + \mu.$$

Evidently, a ρ which meets $0 < \rho \leqslant t^* + \mu$ could be found such that

$$\lim_{t\to\rho^-} \|\hat{c}_i(t)\| = 0, i = 1, 2, \ldots, N,$$

$$\sum_{i=1}^{N} \|\hat{c}_i(t)\| \neq 0,\ 0 \leqslant t < \rho.$$

Therefore, the model (4.23) reaches FTS under the controller (4.26) or (4.33).

On the basis of Theorem 4.14, 4.17 and 4.21, we can obtain the following corollaries.

Corollary 4.22 *If there are matrices* $K = \text{diag}(K_1, K_2, \ldots, K_N) \in \mathbb{R}^{nN\times nN}$, $\hat{D}_1 \in \mathbb{R}^{n\times n}$ *and* $\hat{D}_2 \in \mathbb{R}^{n\times n}$ *such that*

$$\begin{pmatrix} \hat{Q}_1 & \hat{\Theta}_1 \\ \hat{\Theta}_1^T & -I_N \otimes \frac{\hat{D}_2+\hat{D}_2^T}{2} \end{pmatrix} \leqslant 0, \tag{4.39}$$

where $\hat{Q}_1 = (1+\delta^2)I_{Nn} - K - K^T$, $\hat{\Theta}_1 = I_{Nn} - \frac{I_N\otimes\hat{D}_1^T}{2}$, *then the network* (4.23) *under the controller* (4.26) *achieves FTS.*

Corollary 4.23 *If there is a matrix* $\hat{D}_2 \in \mathbb{R}^{n\times n}$ *such that*

$$\hat{D}_2 + \hat{D}_2^T > 0, \tag{4.40}$$

then the network (4.23) *reaches FTS under the adaptive controller* (4.33).

Remark 4.24 In a number of actual applications, the CDNs are required to achieve synchronization over the finite time interval. Moreover, in order to better study the FTS problem, it's also essential to consider the FTP for CDNs. Therefore, we respectively investigate the FTP and FTS problems for CDNs with state and derivative coupling, and several FTP and FTS criteria are derived by exploiting the Lyapunov functional method, and devised state feedback controllers and adaptive state feedback controllers.

Remark 4.25 In this chapter, two adaptive state feedback controllers are devised to ensure the FTP and FTS of the CDNs with derivative and state coupling (see (4.15) and (4.33)). From the proofs of the Theorems 4.7 and 4.17 , when $t \geqslant t^*$, we can stop adjusting the feedback gains $\bar{A}_{ij}^1(t)$ and $\bar{A}_{ij}^2(t) (j \in \mathcal{N}_i)$.

4.5 Numerical Examples

Example 4.26 Discuss the following network model:

$$\dot{z}_i(t) = p(z_i(t)) + 0.1 \sum_{j=1}^{6} A_{ij}^1 \Gamma^1 z_j(t) + w_i(t) + v_i(t),$$

where $p_s(z_i(t)) = \frac{1}{8}\left(|z_{is}(t)+1| - |z_{is}(t)-1|\right)$, $s = 1, 2, 3$, $\Gamma^1 = \text{diag}(0.7, 0.5, 0.3)$,

$$A^1 = \begin{pmatrix} -0.7 & 0.2 & 0 & 0.3 & 0 & 0.2 \\ 0.2 & -0.6 & 0.2 & 0 & 0.2 & 0 \\ 0 & 0.2 & -0.7 & 0.5 & 0 & 0 \\ 0.3 & 0 & 0.5 & -1.1 & 0.3 & 0 \\ 0 & 0.2 & 0 & 0.3 & -1 & 0.5 \\ 0.2 & 0 & 0 & 0 & 0.5 & -0.7 \end{pmatrix}.$$

$$\bar{A}^1(0) = \begin{pmatrix} -0.8 & 0.2 & 0 & 0.2 & 0 & 0.4 \\ 0.2 & -0.5 & 0.1 & 0 & 0.2 & 0 \\ 0 & 0.1 & -0.4 & 0.3 & 0 & 0 \\ 0.2 & 0 & 0.3 & -0.9 & 0.4 & 0 \\ 0 & 0.2 & 0 & 0.4 & -0.9 & 0.3 \\ 0.4 & 0 & 0 & 0 & 0.3 & -0.7 \end{pmatrix}.$$

Evidently, $p(\cdot)$ meets the Lipschitz condition with $\delta = 0.25$.

Case 1: Take

$$D_1 = \begin{pmatrix} 0.3 & 0.5 & 0.7 \\ 0.4 & 0.6 & 0.5 \\ 0.3 & 0.7 & 0.4 \end{pmatrix},$$

$$D_2 = \begin{pmatrix} 0.4 & 0 & 0 \\ 0 & 0.2 & 0 \\ 0 & 0 & 0.6 \end{pmatrix}.$$

From the Theorem 4.7 , the model (4.2) under the adaptive state feedback controller (4.15) achieves FTP. Figs. 4.1 and 4.2 display the simulation results.

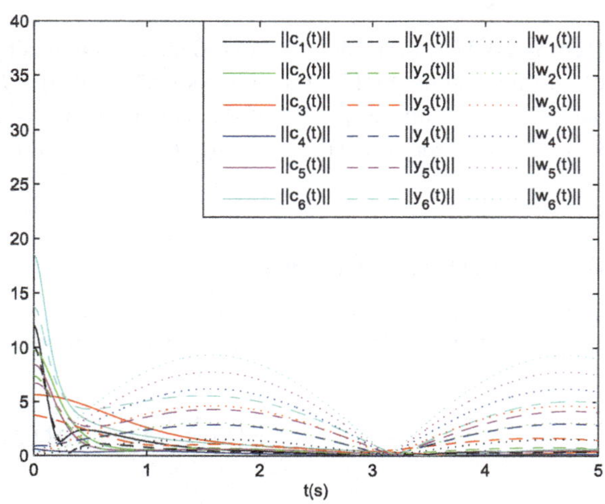

Fig. 4.1 $\|c_i(t)\|$, $\|y_i(t)\|$, $\|w_i(t)\|$, $i = 1, 2, \ldots, 6$

Fig. 4.2 $\bar{A}^1_{ij}(t),\ j \in \mathcal{N}_i$

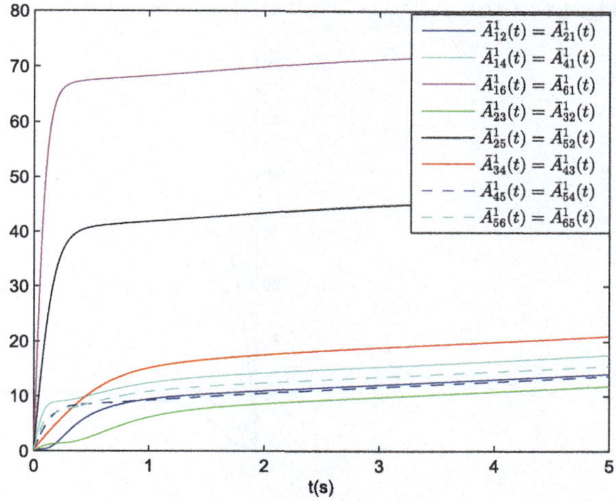

Fig. 4.3 $\|c_i(t)\|,\ i = 1, 2, \ldots, 6$

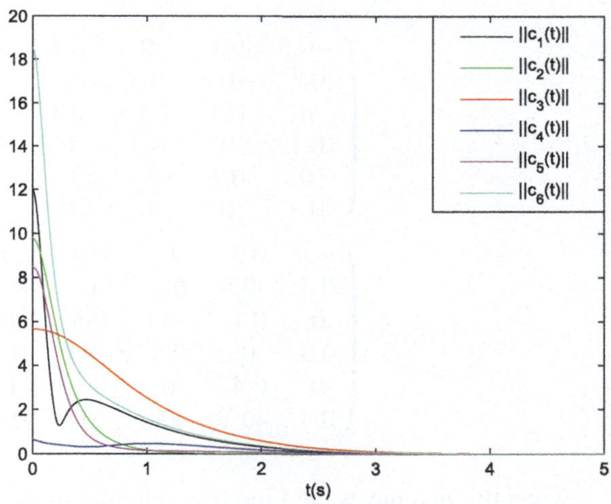

Case 2: By Corollary 4.12, the model (4.1) reaches FTS under the adaptive state feedback controller (4.15). Figures 4.3 and 4.4 show the simulation results.

Example 4.27 Discuss the following network model:

$$\dot{z}_i(t) = p(z_i(t)) + 0.2\sum_{j=1}^{6} A^2_{ij}\Gamma^2\dot{z}_j(t) + w_i(t) + v_i(t),$$

where $p_s(z_i(t)) = \frac{1}{8}\left(|z_{is}(t) + 1| - |z_{is}(t) - 1|\right),\ s = 1, 2, 3,\ \Gamma^2 = \text{diag}(0.4, 0.6, 0.9)$,

Fig. 4.4 $\bar{A}^1_{ij}(t),\ j \in \mathcal{N}_i$

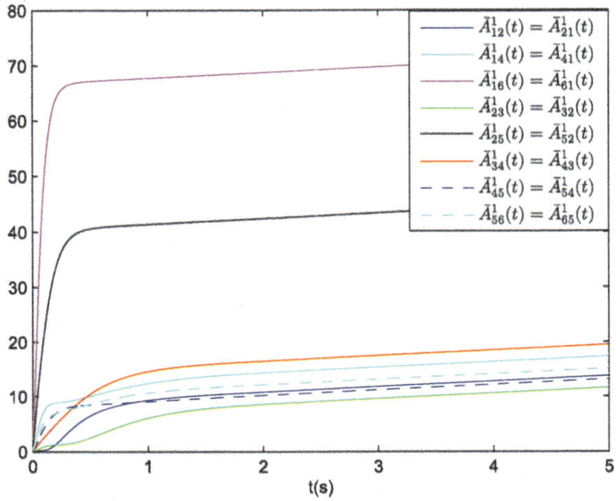

$$A^2 = \begin{pmatrix} -0.5 & 0.1 & 0 & 0.1 & 0 & 0.3 \\ 0.1 & -0.6 & 0.4 & 0 & 0.1 & 0 \\ 0 & 0.4 & -0.5 & 0.1 & 0 & 0 \\ 0.1 & 0 & 0.1 & -0.5 & 0.3 & 0 \\ 0 & 0.1 & 0 & 0.3 & -0.8 & 0.4 \\ 0.3 & 0 & 0 & 0 & 0.4 & -0.7 \end{pmatrix}.$$

$$\bar{A}^2(0) = \begin{pmatrix} -1 & 0.3 & 0 & 0.3 & 0 & 0.4 \\ 0.3 & -0.8 & 0.1 & 0 & 0.4 & 0 \\ 0 & 0.1 & -0.6 & 0.5 & 0 & 0 \\ 0.3 & 0 & 0.5 & -0.9 & 0.1 & 0 \\ 0 & 0.4 & 0 & 0.1 & -0.9 & 0.4 \\ 0.4 & 0 & 0 & 0 & 0.4 & -0.8 \end{pmatrix}.$$

Evidently, $p(\cdot)$ meets the Lipschitz condition with $\delta = 0.25$.

Case 1: Take

$$\hat{D}_1 = \begin{pmatrix} 0.2 & 0.3 & 0.6 \\ 0.3 & 0.5 & 0.7 \\ 0.8 & 0.6 & 0.2 \end{pmatrix},$$

$$\hat{D}_2 = \begin{pmatrix} 0.1 & 0 & 0 \\ 0 & 0.7 & 0 \\ 0 & 0 & 0.3 \end{pmatrix}.$$

Fig. 4.5 $\|\hat{c}_i(t)\|$, $\|\hat{y}_i(t)\|$, $\|w_i(t)\|$, $i = 1, 2, \ldots, 6$

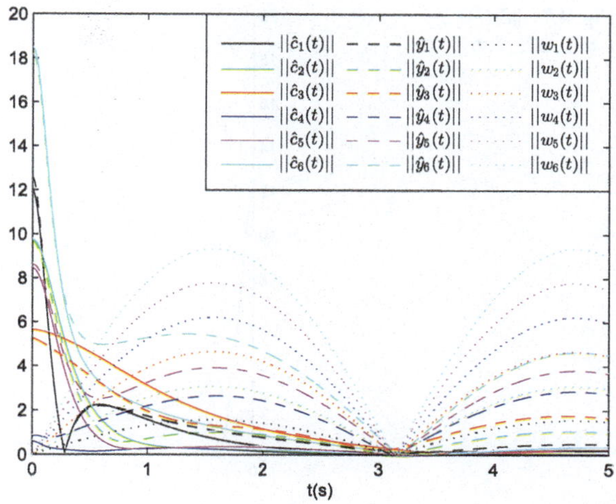

Fig. 4.6 $\bar{A}^2_{ij}(t)$, $j \in \mathcal{N}_i$

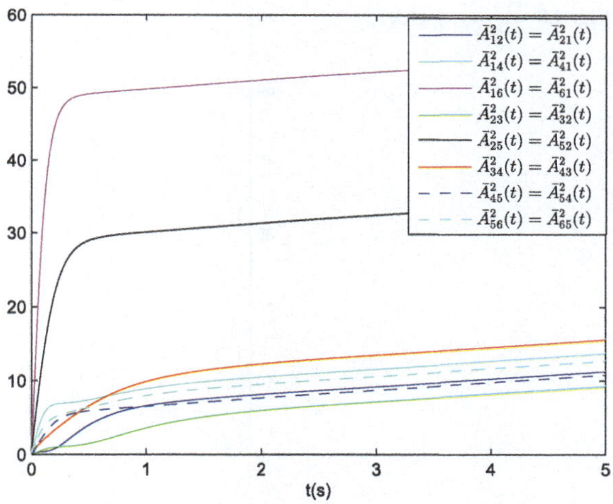

From the Theorem 4.17 , the model (4.24) under the adaptive state feedback controller (4.33) achieves FTP. Figs. 4.5 and 4.6 display the simulation results.

Case 2: By Corollary 4.22, the network (4.23) under the adaptive state feedback controller (4.33) achieves FTS. Figs. 4.7 and 4.8 display the simulation results.

Fig. 4.7 $\|\hat{c}_i(t)\|$, $i = 1, 2, \ldots, 6$

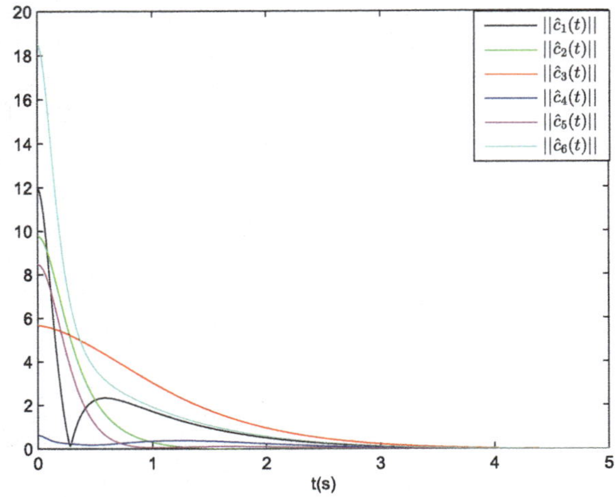

Fig. 4.8 $\bar{A}^2_{ij}(t)$, $j \in \mathcal{N}_i$

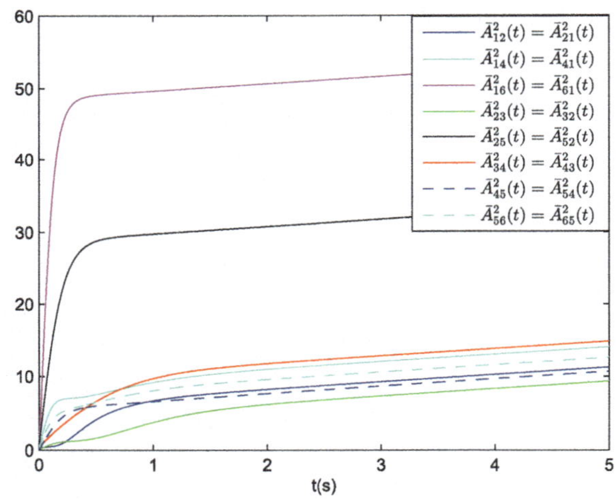

4.6 Conclusion

In this chapter, two complex network models have been presented, in which nodes are coupled by the states or derivatives of the states. On one hand, we have discussed the FTP, FTISP and FTOSP of the presented network models by making use of the adaptive state feedback controllers and state feedback controllers. Moreover, some FTS criteria for these network models have been derived on the basis of the obtained FTP results. Finally, we have provided two numeral examples to verify the correctness of the derived FTP and FTS criteria.

References

1. Wang, A., Dong, T., & Liao, X. (2016). Event-triggered synchronization strategy for complex dynamical networks with the Markovian switching topologies. *Neural Networks, 74*, 52–57.
2. Liu, T., Zhao, J., & Hill, D. J. (2010). Exponential synchronization of complex delayed dynamical networks with switching topology. *IEEE Transactions on Circuits and Systems I: Regular Papers, 57*(11), 2967–2980.
3. Xu, R., Kao, Y., & Gao, C. (2015). Exponential synchronization of delayed Markovian jump complex networks with generally uncertain transition rates. *Applied Mathematics and Computation, 271*, 682–693.
4. Xu, M., Wang, J. L., Huang, Y. L., Wei, P. C., & Wang, S. X. (2017). Pinning synchronization of complex dynamical networks with and without time-varying delay. *Neurocomputing, 266*, 263–273.
5. Li, X. J., & Yang, G. H. (2016). FLS-based adaptive synchronization control of complex dynamical networks with nonlinear couplings and state-dependent uncertainties. *IEEE Transactions on Cybernetics, 46*(1), 171–180.
6. Wang, Y. W., Bian, T., Xiao, J. W., & Wen, C. (2015). Global synchronization of complex dynamical networks through digital communication with limited data rate. *IEEE Transactions on Neural Networks and Learning Systems, 26*(10), 2487–2499.
7. Ali, M. S., Usha, M., Cao, J., & Lu, G. (2019). Synchronisation analysis for stochastic T-S fuzzy complex networks with coupling delay. *International Journal of Systems Science, 50*, 585–598.
8. Vadivel, R., Ali, M. S., & Alzahrani, F. (2019). Robust $\mathcal{H}_\infty$ synchronization of Markov jump stochastic uncertain neural networks with decentralized event-triggered mechanism. *Chinese Journal of Physics, 60*, 68–87.
9. Li, C., Yu, W., & Huang, T. (2014). Impulsive synchronization schemes of stochastic complex networks with switching topology: Average time approach. *Neural Networks, 54*, 85–94.
10. Liu, H., Cao, M., Wu, C. W., Lu, J. A., & Tse, C. K. (2015). Synchronization in directed complex networks using graph comparison tools. *IEEE Transactions on Circuits and Systems I: Regular Papers, 62*(4), 1185–1194.
11. Liu, X., & Chen, T. (2015). Synchronization of complex networks via aperiodically intermittent pinning control. *IEEE Transactions on Automatic Control 60*(12), 3316–3321.
12. Lu, R., Yu, W., Lü, J., & Xue, A. (2014). Synchronization on complex networks of networks. *IEEE Transactions on Neural Networks and Learning Systems, 25*(11), 2110–2118.
13. Yao, J., Wang, H. O., Guan, Z. H., & Xu, W. (2009). Passive stability and synchronization of complex spatio-temporal switching networks with time delays. *Automatica, 45*(7), 1721–1728.
14. Wang, J., Wu, H., Huang, T., Ren, S., & Wu, J. (2018). Passivity and output synchronization of complex dynamical networks with fixed and adaptive coupling strength. *IEEE Transactions on Neural Networks and Learning Systems, 29*(2), 364–376.
15. Wang, J. L., Yang, Z. C., & Wu, H. N. (2012). Passivity analysis of complex dynamical networks with multiple time-varying delays. *Journal of Engineering Mathematics, 74*(1), 175–188.
16. Yao, J., Guan, Z. H., & Hill, D. J. (2009). Passivity-based control and synchronization of general complex dynamical networks. *Automatica, 45*(9), 2107–2113.
17. Zhang, X. X., Wang, J. L., Huang, Y. L., & Ren, S. Y. (2018). Analysis and pinning control for passivity of multi-weighted complex dynamical networks with fixed and switching topologies. *Neurocomputing, 275*, 958–968.
18. Xu, Y. H., Zhou, W. N., Fang, J. A., & Sun, W. (2010). Adaptive synchronization of the complex dynamical network with non-derivative and derivative coupling. *Physics Letters A, 374*(15–16), 1673–1677.
19. Zhu, Y., Luo, X., Li, S., & Guan, X. (2012). Exponential synchronization of complex dynamical systems with delay and derivative coupling via impulse. In *Proceedings of the 10th World Congress on Intelligent Control and Automation* (pp. 875–880).
20. Deng, L. P., Wu, Z. Y., & Wu, Q. C. (2013). Pinning synchronization of complex network with non-derivative and derivative coupling. *Nonlinear Dynamics, 73*(1–2), 775–782.

21. Xu, Y. H., Zhou, W. N., Fang, J. A., Sun, W., & Pan, L. (2010). Topology identification and adaptive synchronization of uncertain complex networks with non-derivative and derivative coupling. *Journal of the Franklin Institute*, *347*(8), 1566–1576.
22. Wang, D. Y., Wang, J. L., Ren, S. Y., & Huang, Y. L. (2019). Output synchronization and H_∞ output synchronization of complex dynamical networks with multiple derivative couplings. *Journal of the Franklin Institute*, *356*(1), 407–440.
23. Wang, D. Y., Wang, J. L., Ren, S. Y., & Huang, Y. L. (2020). Passivity and synchronisation of complex dynamical networks with multiple derivative couplings. *International Journal of Control*, *93*(7), 1710–1724.
24. Qiu, S. H., Huang, Y. L., & Ren, S. Y. (2018). Finite-time synchronization of multi-weighted complex dynamical networks with and without coupling delay. *Neurocomputing*, *275*, 1250–1260.
25. Lu, X. D., Zhang, X. F., & Liu, Q. R. (2018). Finite-time synchronization of nonlinear complex dynamical networks on time scales via pinning impulsive control. *Neurocomputing*, *275*, 2104–2110.
26. Li, J. R., Jiang, H. J., Hu, C., & Yu, J. (2018). Analysis and discontinuous control for finite-time synchronization of delayed complex dynamical networks. *Chaos, Solitons and Fractals*, *114*, 291–305.
27. Li, H. L., Cao, J. D., Jiang, H. J., & Alsaedi, A. (2018). Graph theory-based finite-time synchronization of fractional-order complex dynamical networks. *Journal of the Franklin Institute*, *355*(13), 5711–5789.
28. He, G., Fang, J. A., & Li, Z. (2017). Finite-time synchronization of cyclic switched complex networks under feedback control. *Journal of the Franklin Institute*, *354*(9), 3780–3796.
29. Ren, H. W., Deng, F. Q., & Peng, Y. J. (2018). Finite time synchronization of Markovian jumping stochastic complex dynamical systems with mix delays via hybrid control strategy. *Neurocomputing*, *272*, 683–693.
30. Mei, J., Jiang, M. H., Xu, W. M., & Wang, B. (2013). Finite-time synchronization control of complex dynamical networks with time delay. *Communications in Nonlinear Science and Numerical Simulation*, *18*(9), 2462–2478.
31. Xu, Y. H., Zhou, W. N., Fang, J. A., Xie, C. R., & Tong, D. B. (2016). Finite-time synchronization of the complex dynamical network with non-derivative and derivative coupling. *Neurocomputing*, *173*, 1356–1361.
32. Wang, J. L., Xu, M., Wu, H. N., & Huang, T. (2018). Finite-time passivity of coupled neural networks with multiple weights. *IEEE Transactions on Network Science and Engineering*, *5*(3), 184–197.
33. Huang, Y. L., Chen, W. Z., & Wang, J. M. (2018). Finite-time passivity of delayed multi-weighted complex dynamical networks with different dimensional nodes. *Neurocomputing*, *312*, 74–89.
34. Wang, J., Zhang, X., Wu, H., Huang, T., & Wang, Q. (2019). Finite-time passivity and synchronization of coupled reaction-diffusion neural networks with multiple weights. *IEEE Transactions on Cybernetics*, *49*(9), 3385–3397.
35. Rajavel, S., Samidurai, R., Cao, J. D., Alsaedi, A., & Ahmad, B. (2017). Finite-time non-fragile passivity control for neural networks with time-varying delay. *Applied Mathematics and Computation*, *297*, 145–158.

Chapter 5
FTP of Adaptive Coupled Neural Networks with Undirected and Directed Topologies

5.1 Introduction

During the last few decades, CDNs have roused widespread attention of researchers. As one kind of CDNs, CNNs have been comprehensively studied on account of their extensive applications in secure communication, brain science, etc. To our knowledge, such applications rely on the dynamical behaviors, especially the synchronization, of CNNs. Therefore, the synchronization problem of CNNs has become a hot issue in recent years [1–8]. In [1], the synchronization problem for CNNs with random coupling strength and Markovian jumping was studied, and several delay-dependent synchronization criteria were established. Wang et al. [2] analyzed the output synchronization of directed and undirected CNNs, and two adaptive control schemes for ensuring the output synchronization were developed. In [3], the authors put forward a fractional-order CNNs model, and explored the synchronization and robust synchronization for the fractional-order CNNs model. Furthermore, as a valid approach to study synchronization, passivity of CNNs also has been investigated by some authors [9, 10]. In [9], the authors presented a switched CNNs with uncertain parameters and the same dimension of output and input, and respectively considered the passivity and the relatedness of synchronization and passivity of the presented switched CNNs. Ren et al. [10] concerned the passivity and pinning passivity for two kinds of CNNs.

However, in these existing results about synchronization of CNNs [1–8], the authors always require that the networks can achieve synchronization over the infinite time interval, that is, CNNs are synchronized only when time approaches infinity. But, in many nature phenomena and practical applications, CDNs need to realize synchronization in a finite time. Therefore, few authors have dealt with the FTS for CNNs [11–20]. In [11], Wu et al. discussed the FTS of an uncertain switched CNNs with the help of the average dwell time technique. In [12], several delay-dependent

The original version of this book was revised: The belated corrections have been incorporated. The corrections to this book are available at https://doi.org/10.1007/978-981-33-4287-3_10

J.-L. Wang et al., *Passivity of Complex Dynamical Networks*,
https://doi.org/10.1007/978-981-33-4287-3_5

FTS criteria were put forward for a coupled hierarchical hybrid delayed neural networks (DNNs) by exploiting weighted integral inequality and selecting appropriate Lyapunov functionals. In [13], a CNNs with mixed delays, Markovian jumping parameters and discontinuous activation functions was introduced, and several FTS conditions for such network model were derived by utilizing M-matrix method. For all we know, only a small group of authors have studied the FTP problem of CNNs [21]. Wang et al. [21] presented two multi-weighted CNNs without and with coupling delays, not only obtained some FTP criteria with the help of the designed controllers but also illustrated the relatedness of FTP and FTS for these network models. On the other hand, in most of the existing results about passivity and FTP of CDNs, the dimensions of output and input are the same [9, 21]. In fact, the dimension of output differs from the dimension of input in various cases. Regrettably, very few authors have considered the passivity of CDNs with different dimensions of output and input [10, 22, 23]. Especially, the FTP of CNNs with different dimensions of output and input has not yet been considered.

As is well known, on various occasions, CDNs with given coupling weights are not passive and synchronized. Even though the networks are passive and synchronized, the given coupling weights are generally larger than the needed values. Moreover, in many real networks, for instance, wireless sensor networks, NNs, CNNs, etc., the coupling weights between nodes are adaptively tuned. Therefore, it is necessary to discuss the adaptive coupling weights when we investigate the passivity and synchronization for CDNs [24–27]. In [24], Xu et al. discussed the synchronization for a fractional-order CDNs with adaptive coupling weights, and two synchronization criteria were obtained by employing Laplace transform. Moreover, several criteria were given to ensure the passivity and synchronization of the directed and undirected CDNs with adaptive coupling weights in [25]. Obviously, it is also meaningful and challenging to explore the FTP and FTS of CNNs with adaptive coupling weights.

In the present chapter, two classes of adaptive coupled neural networks with undirected and directed topologies are proposed. Taking advantage of Lyapunov functional method, designing appropriate controllers and adaptive laws, we obtain some sufficient conditions to make sure the FTP of these CNNs. In addition, we also study the FTS of these network models on the basis of the FTP.

5.2 Definitions

Definition 5.1 A system with output $y(t) \in \mathbb{R}^p$ and input $u(t) \in \mathbb{R}^n$ realizes FTP if

$$y^T(t)Wu(t) \geqslant \dot{V}(t) + \varphi V^{\theta}(t)$$

for $W \in \mathbb{R}^{p\times n}$, $t \in [t^*, +\infty)$, $\mathbb{R} \ni \varphi > 0$ and $0 < \theta < 1$, where V is a nonnegative function, $\mathbb{R} \ni t^* \geqslant 0$.

Definition 5.2 A system with output $y(t) \in \mathbb{R}^p$ and input $u(t) \in \mathbb{R}^n$ realizes FTISP if

$$y^T(t)Wu(t) - u^T(t)W_1u(t) \geqslant \dot{V}(t) + \varphi V^\theta(t)$$

for $W \in \mathbb{R}^{p\times n}, 0 < W_1 \in \mathbb{R}^{n\times n}, t \in [t^*, +\infty), \mathbb{R} \ni \varphi > 0$ and $0 < \theta < 1$, where V is a nonnegative function, $\mathbb{R} \ni t^* \geqslant 0$.

Definition 5.3 A system with output $y(t) \in \mathbb{R}^p$ and input $u(t) \in \mathbb{R}^n$ realizes FTOSP if

$$y^T(t)Wu(t) - y^T(t)W_2y(t) \geqslant \dot{V}(t) + \varphi V^\theta(t)$$

for $W \in \mathbb{R}^{p\times n}, 0 < W_2 \in \mathbb{R}^{p\times p}, t \in [t^*, +\infty), \mathbb{R} \ni \varphi > 0$ and $0 < \theta < 1$, where V is a nonnegative function, $\mathbb{R} \ni t^* \geqslant 0$.

Remark 5.4 In recent years, the passivity has been widely studied for various complex network models, and many significant results have been reported. Especially, some authors have investigated the synchronization of CDNs by employing the passivity property. In order to better study the FTS problem of CDNs, three FTP concepts are presented in this chapter.

5.3 FTP of Adaptive Coupled Neural Networks with Undirected Topology

5.3.1 Network Model

In what follows, the undirected network model is discussed:

$$\begin{aligned}\dot{x}_i(t) = &-Bx_i(t) + Cg(x_i(t)) + J + d\sum_{j=1}^{N} A_{ij}(t)Hx_j(t)\\ &+Eu_i(t) + v_i(t),\ i = 1, 2, \ldots, N,\end{aligned} \tag{5.1}$$

where $x_i(t) = (x_{i1}(t), x_{i2}(t), \ldots, x_{in}(t))^T \in \mathbb{R}^n$ is the state vector of the ith node; $u_i(t) \in \mathbb{R}^n$ represents the external input; $B = \text{diag}(b_1, b_2, \ldots, b_n) \in \mathbb{R}^{n\times n} > 0, C = (c_{lp})_{n\times n} \in \mathbb{R}^{n\times n}$ and $E \in \mathbb{R}^{n\times n}$ are constant matrices; $g(x_i(t)) = (g_1(x_{i1}(t)), g_2(x_{i2}(t)), \ldots, g_n(x_{in}(t)))^T \in \mathbb{R}^n, J = (J_1, J_2, \ldots, J_n)^T \in \mathbb{R}^n; \mathbb{R} \ni d > 0$ denotes coupling strength; $H \in \mathbb{R}^{n\times n} > 0$ is the inner coupling matrix; $v_i(t) \in \mathbb{R}^n$ is the control input; $\mathbb{R}^{N\times N} \ni A(t) = (A_{ij}(t))_{N\times N}$ represents coupling weights, where $\mathbb{R} \ni A_{ij}(t)$ is defined as follows: if there exists a connection between node i and node j $(i \neq j)$ at time t, then $\mathbb{R} \ni A_{ij}(t) = A_{ji}(t) > 0$; if not, $\mathbb{R} \ni A_{ij}(t) = A_{ji}(t) = 0$, and

$$A_{ii}(t) = -\sum_{\substack{j=1\\ j\neq i}}^{N} A_{ij}(t), i = 1, 2, \ldots, N.$$

Remark 5.5 In recent years, the dynamical behaviors of CNNs have been comprehensively studied on account of their extensive applications in secure communication, brain science, etc. Therefore, a CNNs model (see (5.1)) is presented and studied in this section, in which each node is an n-dimensional NN. The single NN is described by:

$$\dot{z}_l(t) = -b_l z_l(t) + J_l + \sum_{p=1}^{n} c_{lp} g_p(z_p(t)),$$

where $l = 1, 2, \ldots, n$, $z_l(t) \in \mathbb{R}$ represents the state of the lth neuron, b_l is the rate with which the lth neuron will reset its potential to the resting state when disconnected from the network and external input, J_l is a constant external input, c_{lp} represents the strength of the pth neuron on the lth neuron, $g_p(\cdot)$ is the activation function of the pth neuron. Then, one has

$$\dot{z}(t) = -Bz(t) + J + Cg(z(t)),$$

where $z(t) = (z_1(t), z_2(t), \ldots, z_n(t))^T \in \mathbb{R}^n$.

In this section, the topology of network (5.1) is assumed to be fixed and connected, and $g_s(\cdot)(s = 1, 2, \ldots, n)$ fulfills the Lipschitz condition, that is, there exists $\mathbb{R} \ni \vartheta_s > 0$ satisfying

$$|g_s(\beta_1) - g_s(\beta_2)| \leqslant \vartheta_s |\beta_1 - \beta_2| \qquad (A1)$$

for any $\beta_1, \beta_2 \in \mathbb{R}$. Let $\vartheta = \mathrm{diag}(\vartheta_1^2, \vartheta_2^2, \ldots, \vartheta_n^2) \in \mathbb{R}^{n\times n}$.

Design the following adaptive law:

$$\dot{A}_{ij}(t) = \begin{cases} \xi_{ij}(x_i(t) - x_j(t))^T H(x_i(t) - x_j(t)) + \xi_{ij}, & (i, j) \in \mathcal{B}, \\ -\sum_{\substack{\varsigma=1 \\ \varsigma\neq i}}^{N} \dot{A}_{i\varsigma}(t), & i = j, \\ 0, & \text{otherwise}, \end{cases} \tag{5.2}$$

where $\xi_{ij} = \xi_{ji}$ are positive constants.

5.3.2 FTP Criteria

Letting $\bar{x}(t) = \frac{1}{N}\sum_{r=1}^{N} x_r(t)$, one has

$$\dot{\bar{x}}(t) = \frac{1}{N}\sum_{r=1}^{N} \dot{x}_r(t)$$

$$
\begin{aligned}
&= -B\bar{x}(t) + \frac{1}{N}\sum_{r=1}^{N} Cg(x_r(t)) + J \\
&\quad + \frac{d}{N}\sum_{j=1}^{N}\left(\sum_{r=1}^{N} A_{rj}(t)\right) Hx_j(t) + \frac{1}{N}\sum_{r=1}^{N} Eu_r(t) + \frac{1}{N}\sum_{r=1}^{N} v_r(t) \\
&= -B\bar{x}(t) + \frac{1}{N}\sum_{r=1}^{N} Cg(x_r(t)) + J + \frac{1}{N}\sum_{r=1}^{N} Eu_r(t) + \frac{1}{N}\sum_{r=1}^{N} v_r(t).
\end{aligned}
$$

Taking $w_i(t) = (w_{i1}(t), w_{i2}(t), \ldots, w_{in}(t))^T = x_i(t) - \bar{x}(t), i = 1, 2, \ldots, N$, we have

$$
\begin{aligned}
\dot{w}_i(t) &= -B(x_i(t) - \bar{x}(t)) + Cg(x_i(t)) - \frac{1}{N}\sum_{r=1}^{N} Cg(x_r(t)) \\
&\quad + d\sum_{j=1}^{N} A_{ij}(t)Hw_j(t) + d\left(\sum_{j=1}^{N} A_{ij}(t)\right) H\bar{x}(t) + Eu_i(t) \\
&\quad - \frac{1}{N}\sum_{r=1}^{N} Eu_r(t) + v_i(t) - \frac{1}{N}\sum_{r=1}^{N} v_r(t) \\
&= -Bw_i(t) + Cg(x_i(t)) + d\sum_{j=1}^{N} A_{ij}(t)Hw_j(t) - \frac{1}{N}\sum_{r=1}^{N} Cg(x_r(t)) \\
&\quad + Eu_i(t) - \frac{1}{N}\sum_{r=1}^{N} Eu_r(t) + v_i(t) - \frac{1}{N}\sum_{r=1}^{N} v_r(t),
\end{aligned}
\tag{5.3}
$$

where $i = 1, 2, \ldots, N$.

Define the output vector $y_i(t) \in \mathbb{R}^p$ of network (5.3) as follows:

$$
y_i(t) = M_1 w_i(t) + M_2 u_i(t),
$$

where $M_1 \in \mathbb{R}^{p\times n}$ and $M_2 \in \mathbb{R}^{p\times n}$.

The controller for network (5.1) is selected as

$$
v_i(t) = -\alpha \text{sign}(w_i(t))|w_i(t)|^{\delta}, \tag{5.4}
$$

where $\text{sign}(w_i(t)) = \text{diag}(\text{sign}(w_{i1}(t)), \text{sign}(w_{i2}(t)), \ldots, \text{sign}(w_{in}(t))), 0 < \alpha \in \mathbb{R}$, $|w_i(t)|^{\delta} = (|w_{i1}(t)|^{\delta}, |w_{i2}(t)|^{\delta}, \ldots, |w_{in}(t)|^{\delta})^T$, $\delta \in (0, 1)$.

Theorem 5.6 *If there exits a* $W \in \mathbb{R}^{pN\times nN}$ *satisfying*

$$
W^T(I_N \otimes M_2) + (I_N \otimes M_2^T)W > 0, \tag{5.5}
$$

the network (5.3) *under the adaptive law* (5.2) *and controller* (5.4) *realizes FTP.*

Proof The Lyapunov functional for network (5.3) is constructed as

$$
\begin{aligned}
V(t) &= V_1(t) + V_2(t),\\
V_1(t) &= \sum_{i=1}^{N} w_i^T(t) w_i(t),\\
V_2(t) &= \sum_{i=1}^{N} \sum_{(i,j)\in\mathcal{B}} \frac{d}{2\xi_{ij}} (A_{ij}(t) - \gamma_{ij})^2,
\end{aligned}
$$

where $\gamma_{ij} = \gamma_{ji} (i \neq j)$ are nonnegative constants, and $\gamma_{ij} = 0 (i \neq j)$ if and only if $A_{ij}(t) = 0$.

Then, one gets

$$
\begin{aligned}
\dot{V}(t) &= 2\sum_{i=1}^{N} w_i^T(t)\dot{w}_i(t) + \sum_{i=1}^{N}\sum_{(i,j)\in\mathcal{B}} \frac{d}{\xi_{ij}}(A_{ij}(t) - \gamma_{ij})\dot{A}_{ij}(t)\\
&= 2\sum_{i=1}^{N} w_i^T(t)\Big[- Bw_i(t) + Cg(x_i(t)) - Cg(\bar{x}(t)) + Cg(\bar{x}(t))\\
&\quad - \frac{1}{N}\sum_{r=1}^{N} Cg(x_r(t)) + d\sum_{j=1}^{N} A_{ij}(t)Hw_j(t) + Eu_i(t)\\
&\quad - \frac{1}{N}\sum_{r=1}^{N} Eu_r(t) - \alpha \mathrm{sign}(w_i(t))|w_i(t)|^{\delta} - \frac{1}{N}\sum_{r=1}^{N} v_r(t)\Big]\\
&\quad + \sum_{i=1}^{N}\sum_{(i,j)\in\mathcal{B}} d(A_{ij}(t) - \gamma_{ij})(w_i(t) - w_j(t))^T H(w_i(t) - w_j(t))\\
&\quad + \sum_{i=1}^{N}\sum_{(i,j)\in\mathcal{B}} d(A_{ij}(t) - \gamma_{ij}).
\end{aligned}
$$

Since

$$
\begin{aligned}
\sum_{i=1}^{N} w_i(t) &= \sum_{i=1}^{N}\Big(x_i(t) - \frac{1}{N}\sum_{r=1}^{N} x_r(t)\Big)\\
&= \sum_{i=1}^{N} x_i(t) - \sum_{r=1}^{N} x_r(t)\\
&= 0,
\end{aligned}
$$

one obtains

$$\sum_{i=1}^{N} w_i^T(t)\left(Cg(\bar{x}(t)) - \frac{1}{N}\sum_{r=1}^{N} Cg(x_r(t))\right) = 0, \tag{5.6}$$

$$\sum_{i=1}^{N} w_i^T(t)\left(\frac{1}{N}\sum_{r=1}^{N} Eu_r(t) + \frac{1}{N}\sum_{r=1}^{N} v_r(t)\right) = 0. \tag{5.7}$$

Furthermore,

$$2w_i^T(t)C\left(g(x_i(t)) - g(\bar{x}(t))\right) \leqslant w_i^T(t)(CC^T + \vartheta)w_i(t). \tag{5.8}$$

Defining $\Gamma = (\gamma_{ij})_{N\times N} \in \mathbb{R}^{N\times N}$ and $\gamma_{ii} = -\sum_{\substack{j=1\\ j\neq i}}^{N} \gamma_{ij}, i = 1, 2, \ldots, N$, one has

$$\begin{aligned}
&\sum_{i=1}^{N}\sum_{(i,j)\in\mathcal{B}} d(A_{ij}(t) - \gamma_{ij})(w_i(t) - w_j(t))^T H(w_i(t) - w_j(t)) \\
&= -2\sum_{i=1}^{N}\sum_{j=1}^{N} d(A_{ij}(t) - \gamma_{ij}) w_i^T(t) H w_j(t).
\end{aligned} \tag{5.9}$$

On the basis of (5.6)–(5.9) and Lemma 1.7 we have

$$\begin{aligned}
\dot{V}(t) \leqslant &\sum_{i=1}^{N} w_i^T(t)\left(-2B + CC^T + \vartheta\right) w_i(t) + 2\sum_{i=1}^{N} w_i^T(t)Eu_i(t) \\
&-2\alpha\sum_{i=1}^{N}\sum_{l=1}^{n} |w_{il}(t)|^{\delta+1} + 2d\sum_{i=1}^{N}\sum_{j=1}^{N}\gamma_{ij} w_i^T(t)Hw_j(t) \\
&+\sum_{i=1}^{N}\sum_{(i,j)\in\mathcal{B}} d(A_{ij}(t) - \gamma_{ij}) \\
= &\sum_{i=1}^{N} w_i^T(t)\left(-2B + CC^T + \vartheta\right) w_i(t) + 2\sum_{i=1}^{N} w_i^T(t)Eu_i(t) \\
&-2\alpha\sum_{i=1}^{N}\sum_{l=1}^{n} \left(w_{il}^2(t)\right)^{\frac{\delta+1}{2}} + 2d\sum_{i=1}^{N}\sum_{j=1}^{N}\gamma_{ij} w_i^T(t)Hw_j(t) \\
&+\sum_{i=1}^{N}\sum_{(i,j)\in\mathcal{B}} d(A_{ij}(t) - \gamma_{ij}) \\
\leqslant &\, w^T(t)\Big[I_N \otimes (-2B + CC^T + \vartheta) + 2d\Gamma \otimes H\Big]w(t) + 2w^T(t)(I_N \otimes E)u(t) \\
&+\sum_{i=1}^{N}\sum_{(i,j)\in\mathcal{B}} d(A_{ij}(t) - \gamma_{ij}) - 2\alpha\left(\sum_{i=1}^{N}\sum_{l=1}^{n} w_{il}^2(t)\right)^{\frac{\delta+1}{2}}
\end{aligned}$$

$$= w^T(t)\Big[I_N \otimes (-2B+CC^T+\vartheta)+2d\Gamma \otimes H\Big]w(t)+2w^T(t)(I_N \otimes E)u(t)$$
$$+\sum_{i=1}^{N}\sum_{(i,j)\in\mathcal{B}} d(A_{ij}(t)-\gamma_{ij})-2\alpha V_1^{\frac{\delta+1}{2}}(t), \tag{5.10}$$

where $w(t)=(w_1^T(t), w_2^T(t), \ldots, w_N^T(t))^T$, $u(t)=(u_1^T(t), u_2^T(t), \ldots, u_N^T(t))^T$.

According to (5.5) and (5.10), one gets

$$\begin{aligned}
&\dot{V}(t)-2y^T(t)Wu(t)\\
&=\dot{V}(t)-2w^T(t)(I_N \otimes M_1^T)Wu(t)-2u^T(t)(I_N \otimes M_2^T)Wu(t)\\
&\leqslant w^T(t)\Big[I_N \otimes (-2B+CC^T+\vartheta)+2d\Gamma \otimes H\Big]w(t)\\
&\quad +2w^T(t)(I_N \otimes E)u(t)+\sum_{i=1}^{N}\sum_{(i,j)\in\mathcal{B}} d(A_{ij}(t)-\gamma_{ij})\\
&\quad -2\alpha V_1^{\frac{\delta+1}{2}}(t)-2w^T(t)(I_N \otimes M_1^T)Wu(t)-u^T(t)\Upsilon u(t)\\
&\leqslant w^T(t)\Big[I_N \otimes (-2B+CC^T+\vartheta)+2d\Gamma \otimes H+\Xi\Big]w(t)\\
&\quad -2\alpha V_1^{\frac{\delta+1}{2}}(t)+\sum_{i=1}^{N}\sum_{(i,j)\in\mathcal{B}} d(A_{ij}(t)-\gamma_{ij}),
\end{aligned}$$

where $y(t)=(y_1^T(t), y_2^T(t), \ldots, y_N^T(t))^T$, $\Upsilon = W^T(I_N \otimes M_2)+(I_N \otimes M_2^T)W$, $\Xi=[I_N \otimes E-(I_N \otimes M_1^T)W]\Upsilon^{-1}[I_N \otimes E^T - W^T(I_N \otimes M_1)]$.

By the definition of Γ, there exists a unitary matrix $\kappa=(\kappa_1, \kappa_2, \ldots, \kappa_N) \in \mathbb{R}^{N\times N}$ such that $\kappa^T\Gamma\kappa=\Lambda=\mathrm{diag}(\lambda_1, \lambda_2, \ldots, \lambda_N)$, where $\lambda_i, i=1,2,\ldots,N$, are the eigenvalues of Γ and $0=\lambda_1>\lambda_2\geqslant\lambda_3\geqslant\cdots\geqslant\lambda_N$. Let $m(t)=(m_1^T(t), m_2^T(t), \ldots, m_N^T(t))^T=(\kappa^T \otimes I_n)w(t)$. Since $\kappa_1=\frac{1}{\sqrt{N}}(1,1,\ldots,1)^T$, one gets $m_1(t)=(\kappa_1^T \otimes I_n)w(t)=0$. Then, we can derive

$$\begin{aligned}
&w^T(t)\Big[I_N \otimes (-2B+CC^T+\vartheta)+\Xi\Big]w(t)\\
&+2dw^T(t)(\kappa \otimes I_n)(\Lambda \otimes H)(\kappa^T \otimes I_n)w(t)\\
&\leqslant w^T(t)\Big[I_N \otimes (-2B+CC^T+\vartheta)+\Xi\Big]w(t)+2d\lambda_2 m^T(t)(I_N \otimes H)m(t)\\
&= w^T(t)\Big[I_N \otimes (-2B+CC^T+\vartheta)+\Xi+2d\lambda_2(I_N \otimes H)\Big]w(t).
\end{aligned}$$

Selecting γ_{ij} large enough such that

$$\lambda_M(-2B+CC^T+\vartheta)+\lambda_M(\Xi)+2d\lambda_2\lambda_m(H)\leqslant 0.$$

According to (5.2) , we can find a t^* ($0<t^*\in\mathbb{R}$) such that $A_{ij}(t)\geqslant\gamma_{ij}$ for any $t\geqslant t^*$ and $(i,j)\in\mathcal{B}$.

Therefore, when $t \geqslant t^*$, one gets

$$\begin{aligned}
&\dot{V}_1(t) - 2y^T(t)Wu(t) \\
&= 2\sum_{i=1}^{N} w_i^T(t)\Big[- Bw_i(t) + Cg(x_i(t)) - Cg(\bar{x}(t)) + d\sum_{j=1}^{N} A_{ij}(t)Hw_j(t) \\
&+Eu_i(t) - \alpha\text{sign}(w_i(t))|w_i(t)|^{\delta}\Big] - 2w^T(t)(I_N \otimes M_1^T)Wu(t) \\
&-2u^T(t)(I_N \otimes M_2^T)Wu(t) \\
&\leqslant w^T(t)\Big[I_N \otimes (-2B + CC^T + \vartheta) + 2d(A(t) - \Gamma) \otimes H \\
&+2d\Gamma \otimes H + \Xi\Big]w(t) - 2\alpha V_1^{\frac{\delta+1}{2}}(t) \\
&\leqslant w^T(t)\Big[I_N \otimes (-2B + CC^T + \vartheta) + 2d\Gamma \otimes H + \Xi\Big]w(t) - 2\alpha V_1^{\frac{\delta+1}{2}}(t) \\
&\leqslant -2\alpha V_1^{\frac{\delta+1}{2}}(t).
\end{aligned} \tag{5.11}$$

From (5.11) , we can derive

$$y^T(t)Wu(t) \geqslant \dot{\hat{V}}_1(t) + \hat{\alpha}\hat{V}_1^{\frac{\delta+1}{2}}(t), \quad t \geqslant t^*,$$

where $\hat{V}_1(t) = \frac{V_1(t)}{2}$, $\hat{\alpha} = 2^{\frac{\delta+1}{2}}\alpha$.

Similarly, according to the proof method in Theorem 5.6 , the following conclusions can be easily derived.

Theorem 5.7 *If there exist* $W \in \mathbb{R}^{pN \times nN}$ *and* $0 < W_1 \in \mathbb{R}^{nN \times nN}$ *satisfying*

$$W^T(I_N \otimes M_2) + (I_N \otimes M_2^T)W - W_1 > 0, \tag{5.12}$$

the network (5.3) *under the adaptive law* (5.2) *and controller* (5.4) *realizes FTISP.*

Theorem 5.8 *If there exist* $W \in \mathbb{R}^{pN \times nN}$ *and* $0 < W_2 \in \mathbb{R}^{pN \times pN}$ *satisfying*

$$W^T(I_N \otimes M_2) + (I_N \otimes M_2^T)W - (I_N \otimes M_2^T)W_2(I_N \otimes M_2) > 0, \tag{5.13}$$

the network (5.3) *under the adaptive law* (5.2) *and controller* (5.4) *realizes FTOSP.*

5.3.3 FTS Criteria

Definition 5.9 The network (5.1) realizes FTS under the condition that $u_i(t) = 0, i = 1, 2, \ldots, N$, if there exists $0 < \rho \in \mathbb{R}$ satisfying

$$\lim_{t\to\rho^-}\left\|x_i(t)-\frac{1}{N}\sum_{r=1}^{N}x_r(t)\right\|=0,\ i=1,2,\ldots,N,$$

$$\sum_{i=1}^{N}\left\|x_i(t)-\frac{1}{N}\sum_{r=1}^{N}x_r(t)\right\|\neq 0,\ 0\leqslant t<\rho.$$

Theorem 5.10 *Suppose that* $V(t):[0,+\infty)\to[0,+\infty)$ *is continuously differentiable and satisfies*

$$\upsilon_1(\|w(t)\|)\leqslant V(t),$$

where $\upsilon_1:[0,+\infty)\to[0,+\infty)$ *is continuous and strictly monotonically increasing function with* $\upsilon_1(0)=0$. *If network* (5.3) *under the adaptive law* (5.2) *and controller* (5.4) *realizes FTP (FTISP, FTOSP) with respect to* $V(t)$, *the network* (5.1) *under the adaptive law* (5.2) *and controller* (5.4) *achieves FTS.*

Proof If network (5.3) under the adaptive law (5.2) and controller (5.4) realizes FTP, there apparently exist $W\in\mathbb{R}^{p\times n}$, $0<\theta<1$, $\mathbb{R}\ni t^*\geqslant 0$ and $\mathbb{R}\ni\varphi>0$ such that

$$y^T(t)Wu(t)\geqslant\dot V(t)+\varphi V^\theta(t)$$

for any $t\geqslant t^*$.

Taking $u(t)=0$, one has

$$\dot V(t)\leqslant-\varphi V^\theta(t).$$

for any $t\geqslant t^*$.

Based on Lemma 1.6 , one obtains

$$V(t)=0\ \text{ for }\ t\geqslant t^*+\mu,$$

where $\mu=\frac{V^{1-\theta}(t^*)}{\varphi(1-\theta)}$.

Then, one has

$$\|w(t)\|=0\ \text{ for }\ t\geqslant t^*+\mu.$$

Obviously, we can find a positive constant $0<\rho\leqslant t^*+\mu$ such that

$$\lim_{t\to\rho^-}\|w(t)\|=0,$$

$$\|w(t)\|>0,\ 0\leqslant t<\rho.$$

Therefore, network (5.1) achieves FTS under the adaptive law (5.2) and controller (5.4).

Similarly, when network (5.3) achieves FTISP or FTOSP, it can be proved that the network (5.1) achieves FTS under the adaptive law (5.2) and controller (5.4).

On the basis of Theorems 5.6 and 5.10, the following conclusion can be obtained.

Corollary 5.11 *The network* (5.1) *under the adaptive law* (5.2) *and controller* (5.4) *realizes FTS if there exists a* $W \in \mathbb{R}^{pN\times nN}$ *satisfying*

$$W^T(I_N \otimes M_2) + (I_N \otimes M_2^T)W > 0. \tag{5.14}$$

5.4 FTP of Adaptive Coupled Neural Networks with Directed Topology

5.4.1 Network Model

In what follows, we consider the case that the topology structure of network (5.1) is directed:

$$\begin{aligned}\dot{x}_i(t) = &-Bx_i(t) + Cg(x_i(t)) + J + d\sum_{j=1}^{N} A_{ij}(t)Hx_j(t)\\ &+Eu_i(t) + v_i(t),\ i = 1, 2, \ldots, N,\end{aligned} \tag{5.15}$$

in which the definitions of B, C, J, d, H, E, $x_i(t)$, $g(\cdot)$, $u_i(t)$, $v_i(t)$ are the same as these parameters in network (5.1), $\mathbb{R}^{N\times N} \ni A(t) = (A_{ij}(t))_{N\times N}$ denotes coupling weights, where $A_{ij}(t) \in \mathbb{R}$ is defined as follows: if there exists a connection from node i to node j $(i \neq j)$, then $\mathbb{R} \ni A_{ij}(t) > 0$; if not, $\mathbb{R} \ni A_{ij}(t) = 0$, and

$$A_{ii}(t) = -\sum_{\substack{j=1\\ j\neq i}}^{N} A_{ij}(t), i = 1, 2, \ldots, N.$$

According to Lemma 1.3, there exists a positive vector $\phi = (\phi_1, \phi_2, \ldots, \phi_N)^T \in \mathbb{R}^N$ such that $A^T(0)\phi = 0$ and

$$\sum_{j=1}^{N} \hat{A}_{ij} = \sum_{j=1}^{N} \hat{A}_{ji} = 0,$$

where $\hat{A} = (\hat{A}_{ij})_{N\times N} = \Phi A(0) + A^T(0)\Phi$, $\Phi = \text{diag}(\phi_1, \phi_2, \ldots, \phi_N)$.

Letting $\eta_i = \phi_i/(\phi_1 + \phi_2 + \cdots + \phi_N), i = 1, 2, \ldots, N$, one has $(\eta_1, \eta_2, \ldots, \eta_N)$ $A(0) = 0$ and

$$A_{ii}(0) = \frac{-\sum_{\substack{j=1\\ j\neq i}}^{N}(\eta_i A_{ij}(0) + \eta_j A_{ji}(0))}{2\eta_i}. \tag{5.16}$$

Design the following adaptive law:

$$\dot{A}_{ij}(t) = \begin{cases} \xi_{ij}\eta_j(x_i(t)-x_j(t))^T H(x_i(t)-x_j(t)) & \\ +\xi_{ij}\eta_j, & (i,j)\in\mathcal{B} \text{ and } (j,i)\in\mathcal{B}, \\ -\sum_{\substack{\zeta=1\\ \zeta\neq i}}^{N}\dot{A}_{i\zeta}(t), & i=j, \\ 0, & \text{otherwise,} \end{cases} \tag{5.17}$$

where $\xi_{ij} = \xi_{ji}$ are positive constants.

Then,

$$\sum_{j=1}^{N}\eta_i\dot{A}_{ij}(t) = \sum_{j=1}^{N}\eta_j\dot{A}_{ji}(t) = 0. \tag{5.18}$$

By (5.18), one gets

$$\dot{A}_{ii}(t) = \frac{-\sum_{\substack{j=1\\ j\neq i}}^{N}\left(\eta_i\dot{A}_{ij}(t) + \eta_j\dot{A}_{ji}(t)\right)}{2\eta_i}, \tag{5.19}$$

$$\sum_{j=1}^{N}\eta_j A_{ji}(t) = 0 \tag{5.20}$$

for all t.

Furthermore, based on (5.16) and (5.19) , we can derive

$$A_{ii}(t) = \frac{-\sum_{\substack{j=1\\ j\neq i}}^{N}(\eta_i A_{ij}(t) + \eta_j A_{ji}(t))}{2\eta_i}$$

for all t.

5.4.2 FTP Criteria

Letting $\bar{x}(t) = \sum_{r=1}^{N}\eta_r x_r(t)$, one obtains

$$\dot{\bar{x}}(t) = \sum_{r=1}^{N}\eta_r\dot{x}_r(t)$$

$$
\begin{aligned}
&= -B\bar{x}(t) + \sum_{r=1}^{N} \eta_r C g(x_r(t)) + J + d \sum_{j=1}^{N} \left(\sum_{r=1}^{N} \eta_r A_{rj}(t) \right) H x_j(t) \\
&\quad + \sum_{r=1}^{N} \eta_r E u_r(t) + \sum_{r=1}^{N} \eta_r v_r(t) \\
&= -B\bar{x}(t) + \sum_{r=1}^{N} \eta_r C g(x_r(t)) + J + \sum_{r=1}^{N} \eta_r E u_r(t) + \sum_{r=1}^{N} \eta_r v_r(t).
\end{aligned}
$$

Taking $w_i(t) = (w_{i1}(t), w_{i2}(t), \ldots, w_{in}(t))^T = x_i(t) - \bar{x}(t)$, $i = 1, 2, \ldots, N$, we have

$$
\begin{aligned}
\dot{w}_i(t) = &-Bw_i(t) + Cg(x_i(t)) + d\sum_{j=1}^{N} A_{ij}(t) H w_j(t) - \sum_{r=1}^{N} \eta_r C g(x_r(t)) + E u_i(t) \\
&- \sum_{r=1}^{N} \eta_r E u_r(t) + v_i(t) - \sum_{r=1}^{N} \eta_r v_r(t),
\end{aligned} \tag{5.21}
$$

where $i = 1, 2, \ldots, N$.

Define the output vector $y_i(t) \in \mathbb{R}^p$ of network (5.21) as follows:

$$
y_i(t) = M_1 w_i(t) + M_2 u_i(t),
$$

where $M_1 \in \mathbb{R}^{p\times n}$ and $M_2 \in \mathbb{R}^{p\times n}$.

The controller for network (5.21) is selected as:

$$
v_i(t) = -\alpha \eta_i^{\frac{\delta-1}{2}} \operatorname{sign}(w_i(t)) |w_i(t)|^{\delta}, \tag{5.22}
$$

where $0 < \alpha \in \mathbb{R}$, $\delta \in (0, 1)$.

As a matter of convenience, we define

$$
K_{ij}(t) = \begin{cases}
\eta_i A_{ij}(t) + \eta_j A_{ji}(t), & (i, j) \in \mathcal{B} \text{ and } (j, i) \in \mathcal{B}, \\
-\sum_{\substack{\varsigma=1 \\ \varsigma\neq i}}^{N} K_{i\varsigma}(t), & i = j, \\
0, & \text{otherwise},
\end{cases}
$$

$$
F_{ij} = \begin{cases}
\eta_i A_{ij}(0), & (i, j) \in \mathcal{B} \text{ and } (j, i) \notin \mathcal{B}, \\
\eta_j A_{ji}(0), & (j, i) \in \mathcal{B} \text{ and } (i, j) \notin \mathcal{B}, \\
-\sum_{\substack{\varsigma=1 \\ \varsigma\neq i}}^{N} F_{i\varsigma}, & i = j, \\
0, & \text{otherwise},
\end{cases}
$$

$$
\eta = \operatorname{diag}(\eta_1, \eta_2, \ldots, \eta_N).
$$

Theorem 5.12 *The network* (5.21) *under the adaptive law* (5.17) *and controller* (5.22) *realizes FTP if there exist* $W \in \mathbb{R}^{pN\times nN}$ *and* $\mathbb{R}^{N\times N} \ni \Gamma = (\gamma_{ij})_{N\times N}$ *satisfying*

$$\begin{pmatrix} Q_1 & Z_1 \\ Z_1^T & -W^T(I_N \otimes M_2) - (I_N \otimes M_2^T)W \end{pmatrix} \leqslant 0, \tag{5.23}$$

where $Q_1 = \eta \otimes (-2B + CC^T + \vartheta) + d(F + \Gamma) \otimes H$, $Z_1 = \eta \otimes E - (I_N \otimes M_1^T)W$, $\mathbb{R}^{N\times N} \ni F = (F_{ij})_{N\times N}$, $\gamma_{ii} = -\sum_{\substack{j=1\\ j\neq i}}^{N} \gamma_{ij}$, $\gamma_{ij} = \gamma_{ji}(i \neq j)$ *are nonnegative constants, and* $\gamma_{ij} = 0(i \neq j)$ *if and only if* $K_{ij}(t) = 0$.

Proof The Lyapunov functional is constructed as:

$$\tilde{V}(t) = \sum_{i=1}^{N} \eta_i w_i^T(t) w_i(t).$$

Then,

$$\begin{aligned}
\dot{\tilde{V}}(t) &= 2\sum_{i=1}^{N} \eta_i w_i^T(t)\dot{w}_i(t) \\
&= 2\sum_{i=1}^{N} \eta_i w_i^T(t)\Big[- Bw_i(t) + Cg(x_i(t)) - Cg(\bar{x}(t)) + Cg(\bar{x}(t)) \\
&\quad - \sum_{r=1}^{N} \eta_r Cg(x_r(t)) + d\sum_{j=1}^{N} A_{ij}(t)Hw_j(t) + Eu_i(t) - \sum_{r=1}^{N} \eta_r Eu_r(t) \\
&\quad - \alpha\eta_i^{\frac{\delta-1}{2}} \operatorname{sign}(w_i(t))|w_i(t)|^{\delta} - \sum_{r=1}^{N} \eta_r v_r(t)\Big].
\end{aligned}$$

Since

$$\begin{aligned}
\sum_{i=1}^{N} \eta_i w_i(t) &= \sum_{i=1}^{N} \eta_i \Big(x_i(t) - \sum_{r=1}^{N} \eta_r x_r(t)\Big) \\
&= \sum_{i=1}^{N} \eta_i x_i(t) - \sum_{r=1}^{N} \left(\sum_{i=1}^{N} \eta_i\right) \eta_r x_r(t) \\
&= \sum_{i=1}^{N} \eta_i x_i(t) - \sum_{r=1}^{N} \eta_r x_r(t) \\
&= 0,
\end{aligned}$$

one obtains

$$\sum_{i=1}^{N}\eta_i w_i^T(t)\left(Cg(\bar{x}(t))-\sum_{r=1}^{N}\eta_r Cg(x_r(t))\right)=0, \tag{5.24}$$

$$\sum_{i=1}^{N}\eta_i w_i^T(t)\left(\sum_{r=1}^{N}\eta_r Eu_r(t)+\sum_{r=1}^{N}\eta_r v_r(t)\right)=0. \tag{5.25}$$

On the basis of (5.8), (5.24), (5.25) and Lemma 1.7, we can derive

$$\begin{aligned}
\dot{\tilde{V}}(t) \leqslant{}& \sum_{i=1}^{N}\eta_i w_i^T(t)\left(-2B+CC^T+\vartheta\right)w_i(t)+2\sum_{i=1}^{N}\eta_i w_i^T(t)Eu_i(t)\\
&-2\alpha\sum_{i=1}^{N}\sum_{l=1}^{n}\eta_i^{\frac{\delta+1}{2}}|w_{il}(t)|^{\delta+1}+\sum_{i=1}^{N}\sum_{j=1}^{N}d(\eta_i A_{ij}(t)+\eta_j A_{ji}(t))w_i^T(t)Hw_j(t)\\
={}& \sum_{i=1}^{N}\eta_i w_i^T(t)\left(-2B+CC^T+\vartheta\right)w_i(t)+2\sum_{i=1}^{N}\eta_i w_i^T(t)Eu_i(t)\\
&-2\alpha\sum_{i=1}^{N}\sum_{l=1}^{n}\eta_i^{\frac{\delta+1}{2}}|w_{il}(t)|^{\delta+1}+\sum_{i=1}^{N}\sum_{j=1}^{N}d\Big[(F_{ij}+\gamma_{ij})\\
&+(K_{ij}(t)-\gamma_{ij})\Big]w_i^T(t)Hw_j(t)\\
={}& w^T(t)[\eta\otimes(-2B+CC^T+\vartheta)+d(F+\Gamma)\otimes H]w(t)+2w^T(t)(\eta\otimes E)u(t)\\
&-2\alpha\sum_{i=1}^{N}\sum_{l=1}^{n}\left(\eta_i w_{il}^2(t)\right)^{\frac{\delta+1}{2}}+dw^T(t)[(K(t)-\Gamma)\otimes H]w(t)\\
\leqslant{}& w^T(t)[\eta\otimes(-2B+CC^T+\vartheta)+d(F+\Gamma)\otimes H]w(t)+2w^T(t)(\eta\otimes E)u(t)\\
&+dw^T(t)[(K(t)-\Gamma)\otimes H]w(t)-2\alpha\tilde{V}^{\frac{\delta+1}{2}}(t),
\end{aligned} \tag{5.26}$$

where $w(t)=(w_1^T(t),w_2^T(t),\ldots,w_N^T(t))^T$, $u(t)=(u_1^T(t),u_2^T(t),\ldots,u_N^T(t))^T$, $\mathbb{R}^{N\times N}\ni K(t)=(K_{ij}(t))_{N\times N}$.

According to (5.23) and (5.26), one gets

$$\begin{aligned}
&\dot{\tilde{V}}(t)-2y^T(t)Wu(t)\\
={}& \dot{\tilde{V}}(t)-2w^T(t)(I_N\otimes M_1^T)Wu(t)-2u^T(t)(I_N\otimes M_2^T)Wu(t)\\
\leqslant{}& w^T(t)[\eta\otimes(-2B+CC^T+\vartheta)+d(F+\Gamma)\otimes H]w(t)+2w^T(t)(\eta\otimes E)u(t)\\
&-2\alpha\tilde{V}^{\frac{\delta+1}{2}}(t)-2w^T(t)(I_N\otimes M_1^T)Wu(t)-u^T(t)[W^T(I_N\otimes M_2)\\
&+(I_N\otimes M_2^T)W]u(t)+dw^T(t)[(K(t)-\Gamma)\otimes H]w(t)\\
={}& \chi^T(t)\begin{pmatrix} Q_1 & Z_1\\ Z_1^T & Q_2\end{pmatrix}\chi(t)-2\alpha\tilde{V}^{\frac{\delta+1}{2}}(t)+dw^T(t)[(K(t)-\Gamma)\otimes H]w(t)
\end{aligned}$$

$$\leqslant -2\alpha \tilde{V}^{\frac{\delta+1}{2}}(t) + dw^T(t)[(K(t) - \Gamma) \otimes H]w(t),$$

where $y(t) = (y_1^T(t), y_2^T(t), \ldots, y_N^T(t))^T$, $Q_2 = -W^T(I_N \otimes M_2) - (I_N \otimes M_2^T)W$, $\chi(t) = (w^T(t), u^T(t))^T$.

According to (5.17) , we can find a positive real number t^* such that $K_{ij}(t) \geqslant \gamma_{ij}$ for any $t \geqslant t^*$, (i, j) and $(j, i) \in \mathcal{B}$.

Therefore, when $t \geqslant t^*$, one gets

$$\dot{\tilde{V}}(t) \leqslant 2y^T(t)Wu(t) - 2\alpha \tilde{V}^{\frac{\delta+1}{2}}(t).$$

Then, we can derive

$$y^T(t)Wu(t) \geqslant \dot{\hat{\tilde{V}}}(t) + \hat{\alpha}\hat{\tilde{V}}^{\frac{\delta+1}{2}}(t), \quad t \geqslant t^*,$$

where $\hat{\tilde{V}}(t) = \frac{\tilde{V}(t)}{2}$, $\hat{\alpha} = 2^{\frac{\delta+1}{2}}\alpha$.

Similarly, according to the proof method in Theorem 5.12 , the following conclusions can be easily derived.

Theorem 5.13 *The network* (5.21) *under the adaptive law* (5.17) *and controller* (5.22) *realizes FTISP if there exist* $W \in \mathbb{R}^{pN \times nN}$, $0 < W_1 \in \mathbb{R}^{nN \times nN}$ *and* $\mathbb{R}^{N \times N} \ni \Gamma = (\gamma_{ij})_{N \times N}$ *satisfying*

$$\begin{pmatrix} Q_1 & Z_1 \\ Z_1^T & W_1 - W^T(I_N \otimes M_2) - (I_N \otimes M_2^T)W \end{pmatrix} \leqslant 0, \tag{5.27}$$

where $Q_1 = \eta \otimes (-2B + CC^T + \vartheta) + d(F + \Gamma) \otimes H$, $Z_1 = \eta \otimes E - (I_N \otimes M_1^T)W$, *the definitions of* F *and* Γ *are the same as these in Theorem 5.12.*

Theorem 5.14 *The network* (5.21) *under the adaptive law* (5.17) *and controller* (5.22) *realizes FTOSP if there exist* $W \in \mathbb{R}^{pN \times nN}$, $0 < W_2 \in \mathbb{R}^{pN \times pN}$ *and* $\mathbb{R}^{N \times N} \ni \Gamma = (\gamma_{ij})_{N \times N}$ *satisfying*

$$\begin{pmatrix} Q_3 & Z_2 \\ Z_2^T & Q_4 \end{pmatrix} \leqslant 0, \tag{5.28}$$

where $Q_3 = \eta \otimes (-2B + CC^T + \vartheta) + d(F + \Gamma) \otimes H + (I_N \otimes M_1^T)W_2(I_N \otimes M_1)$, $Z_2 = \eta \otimes E - (I_N \otimes M_1^T)W + (I_N \otimes M_1^T)W_2(I_N \otimes M_2)$, $Q_4 = (I_N \otimes M_2^T)W_2 (I_N \otimes M_2) - W^T(I_N \otimes M_2) - (I_N \otimes M_2^T)W$, *the definitions of* F *and* Γ *are the same as these in Theorem 5.12.*

5.4.3 FTS Criteria

Theorem 5.15 *Suppose that* $V(t) : [0, +\infty) \to [0, +\infty)$ *is continuously differentiable and satisfies*

$$\upsilon_1(\|w(t)\|) \leqslant V(t),$$

where $\upsilon_1 : [0, +\infty) \to [0, +\infty)$ *is continuous and strictly monotonically increasing function with* $\upsilon_1(0) = 0$*. If network* (5.21) *under the adaptive law* (5.17) *and controller* (5.22) *realizes FTP (FTISP, FTOSP) with respect to* $V(t)$*, the network* (5.15) *under the adaptive law* (5.17) *and controller* (5.22) *realizes FTS.*

Proof Based on the proof method in Theorem 5.10, these results can be easily obtained.

According to Theorems 5.12 and 5.15, the following conclusion can be derived.

Corollary 5.16 *The network* (5.15) *under the adaptive law* (5.17) *and controller* (5.22) *realizes FTS if there exist* $W \in \mathbb{R}^{pN\times nN}$ *and* $\mathbb{R}^{N\times N} \ni \Gamma = (\gamma_{ij})_{N\times N}$ *satisfying*

$$\begin{pmatrix} Q_1 & Z_1 \\ Z_1^T & -W^T(I_N \otimes M_2) - (I_N \otimes M_2^T)W \end{pmatrix} \leqslant 0, \tag{5.29}$$

where $Q_1 = \eta \otimes (-2B + CC^T + \vartheta) + d(F + \Gamma) \otimes H$, $Z_1 = \eta \otimes E - (I_N \otimes M_1^T)W$*, the definitions of* F *and* Γ *are the same as these in Theorem 5.12.*

Remark 5.17 Recently, passivity has been investigated by some authors in CNNs [9, 10] since it is a valid approach to study synchronization. As is well known, in many nature phenomena and practical applications, CDNs require to realize synchronization within a finite time. Thus, it is meaningful to investigate the FTP of CNNs. Nevertheless, only a small group of authors have studied the FTP problem for CNNs [21]. Especially, the FTP and FTS of adaptive coupled neural networks with undirected and directed topologies have not yet been considered.

5.5 Numerical Examples

Example 5.18 The following undirected CNNs is discussed:

$$\begin{aligned}\dot{x}_i(t) = &-Bx_i(t) + Cg(x_i(t)) + 0.1\sum_{j=1}^{6} A_{ij}(t)Hx_j(t)\\ &+Eu_i(t) + v_i(t),\ i = 1, 2, \ldots, 6,\end{aligned} \tag{5.30}$$

where $g_s(\varepsilon) = \frac{1}{8}(|\varepsilon + 1| - |\varepsilon - 1|)$, $s = 1, 2, 3$, $B = \mathrm{diag}(0.7, 0.4, 0.3)$, $H = \mathrm{diag}(0.2, 0.3, 0.1)$,

$$C = \begin{pmatrix} 0.3 & 0.1 & 0.2 \\ 0.1 & 0.1 & 0.1 \\ 0.2 & 0.1 & 0.2 \end{pmatrix},$$

$$E = \begin{pmatrix} 0.4 & 0.1 & 0.1 \\ 0.1 & 0.3 & 0.1 \\ 0.1 & 0.1 & 0.2 \end{pmatrix},$$

$$A(0) = \begin{pmatrix} -0.8 & 0.2 & 0 & 0.2 & 0 & 0.4 \\ 0.2 & -0.5 & 0.1 & 0 & 0.2 & 0 \\ 0 & 0.1 & -0.4 & 0.3 & 0 & 0 \\ 0.2 & 0 & 0.3 & -0.9 & 0.4 & 0 \\ 0 & 0.2 & 0 & 0.4 & -0.9 & 0.3 \\ 0.4 & 0 & 0 & 0 & 0.3 & -0.7 \end{pmatrix}.$$

Obviously, function $g_s(\cdot)$ fulfills (A1) with $\vartheta_s = 0.25$.
Case 1: Take

$$M_1 = \begin{pmatrix} 0.3 & 0.5 & 0.7 \\ 0.4 & 0.6 & 0.5 \\ 0.3 & 0.7 & 0.4 \\ 0.4 & 0.3 & 0.5 \end{pmatrix},$$

$$M_2 = \begin{pmatrix} 0.4 & 0.2 & 0.3 \\ 0.2 & 0.2 & 0.3 \\ 0.5 & 0.7 & 0.6 \\ 0.7 & 0.8 & 0.9 \end{pmatrix}.$$

It can be found that

$$W = I_6 \otimes \begin{pmatrix} 4.4968 & -1.2034 & -2.3856 \\ -1.9634 & -2.3434 & 3.9901 \\ 1.0134 & 3.6734 & -3.8845 \\ -1.5200 & -1.2667 & 2.9978 \end{pmatrix}$$

satisfies (5.5). From the Theorem 5.6, the network (5.3) achieves FTP under the adaptive law (5.2) and controller (5.4). Figures 5.1 and 5.2 display the simulation results.

It can be found that

$$W = 10 I_6 \otimes \begin{pmatrix} 4.4968 & -1.2034 & -2.3856 \\ -1.9634 & -2.3434 & 3.9901 \\ 1.0134 & 3.6734 & -3.8845 \\ -1.5200 & -1.2667 & 2.9978 \end{pmatrix},$$

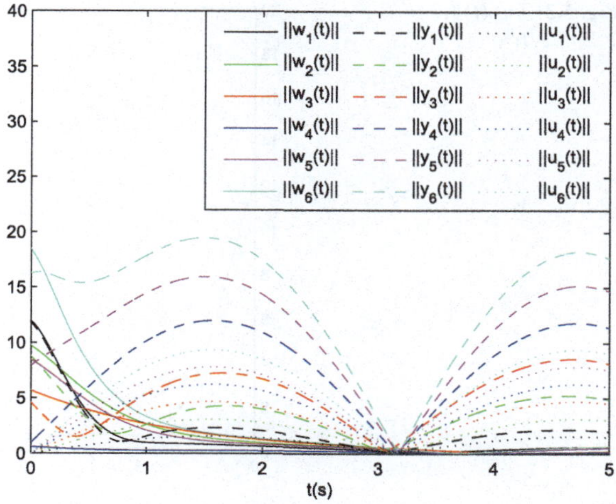

Fig. 5.1 $\|w_i(t)\|$, $\|y_i(t)\|$, $\|u_i(t)\|$, $i = 1, 2, \ldots, 6$

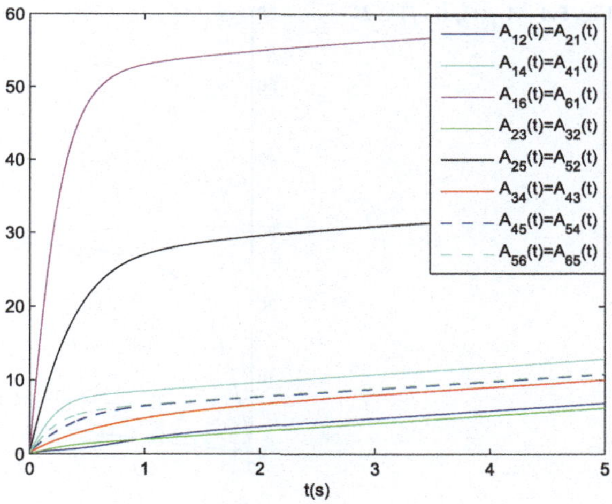

Fig. 5.2 $A_{ij}(t)$, $(i, j) \in \mathcal{B}$

$$W_1 = I_6 \otimes \begin{pmatrix} 8.4868 & -0.0000 & 0.0000 \\ -0.0000 & 8.4868 & 0.0000 \\ 0.0000 & 0.0000 & 8.4868 \end{pmatrix}$$

satisfy (5.12). By Theorem 5.7, the network (5.3) achieves FTISP under the adaptive law (5.2) and controller (5.4). Figures 5.1 and 5.2 display the simulation results.

Fig. 5.3 $\|w_i(t)\|, i = 1, 2, \ldots, 6$

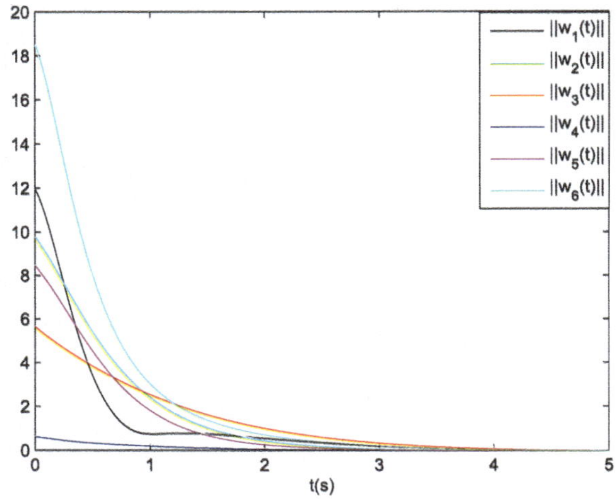

Fig. 5.4 $A_{ij}(t), (i, j) \in \mathcal{B}$

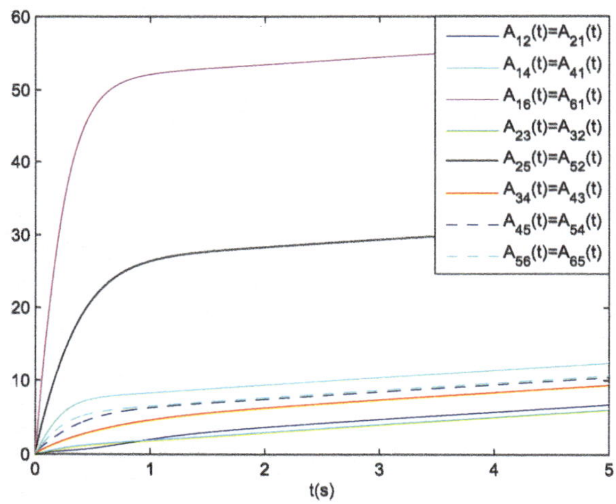

It can be found that

$$W = I_6 \otimes \begin{pmatrix} 4.4383 & -0.9486 & -1.9556 \\ -1.6460 & -1.9948 & 3.8954 \\ 1.3194 & 3.9163 & -3.0976 \\ -0.8498 & -0.5394 & 3.4521 \end{pmatrix},$$

$$W_2 = I_6 \otimes \begin{pmatrix} 1.5577 & -0.0000 & -0.0000 & -0.0000 \\ -0.0000 & 1.5577 & -0.0000 & -0.0000 \\ -0.0000 & -0.0000 & 1.5577 & -0.0000 \\ -0.0000 & -0.0000 & -0.0000 & 1.5577 \end{pmatrix}$$

satisfy (5.13). Based on Theorem 5.8, the network (5.3) achieves FTOSP under the adaptive law (5.2) and controller (5.4). Figures 5.1 and 5.2 display the simulation results.

Case 2: Apparently, the condition (5.14) also holds if we select

$$W = I_6 \otimes \begin{pmatrix} 4.4968 & -1.2034 & -2.3856 \\ -1.9634 & -2.3434 & 3.9901 \\ 1.0134 & 3.6734 & -3.8845 \\ -1.5200 & -1.2667 & 2.9978 \end{pmatrix}.$$

From Corollary 5.11, the network (5.1) achieves FTS under the adaptive law (5.2) and controller (5.4). Figures 5.3 and 5.4 display the simulation results.

Example 5.19 The following directed CNNs is discussed:

$$\begin{aligned} \dot{x}_i(t) = & -Bx_i(t) + Cg(x_i(t)) + 0.15 \sum_{j=1}^{6} A_{ij}(t)Hx_j(t) \\ & + Eu_i(t) + v_i(t), \ i = 1, 2, \ldots, 6, \end{aligned} \tag{5.31}$$

where $g_s(\varepsilon) = \frac{1}{8}\left(|\varepsilon + 1| - |\varepsilon - 1|\right)$, $s = 1, 2, 3$, $B = \text{diag}(1.4, 0.8, 0.6)$, $H = \text{diag}(0.3, 0.1, 0.1)$,

$$C = \begin{pmatrix} 0.2 & 0.3 & 0.1 \\ 0.2 & 0.1 & 0.1 \\ 0.1 & 0.1 & 0.2 \end{pmatrix},$$

$$E = \begin{pmatrix} 0.3 & 0.2 & 0.1 \\ 0.1 & 0.2 & 0.1 \\ 0.2 & 0.1 & 0.3 \end{pmatrix},$$

$$A(0) = \begin{pmatrix} -1 & 0.4 & 0 & 0.2 & 0.4 & 0 \\ 0.2 & -0.3 & 0.1 & 0 & 0 & 0 \\ 0 & 0.2 & -0.4 & 0.2 & 0 & 0 \\ 0.2 & 0 & 0.1 & -0.5 & 0.2 & 0 \\ 0.1 & 0 & 0 & 0.3 & -0.6 & 0.2 \\ 0 & 0 & 0 & 0 & 0.3 & -0.3 \end{pmatrix}.$$

Obviously, function $g_s(\cdot)$ fulfills (A1) with $\vartheta_s = 0.25$. Utilizing the function NULL in MATLAB, there exists the positive vector $\phi = (0.75, 1.5, 0.75, 1.5, 1.5, 1)^T \in \mathbb{R}^6$ satisfying $\phi^T A(0) = 0$. According to the definition of η, we can obtain $\eta = \text{diag}(\frac{3}{28}, \frac{3}{14}, \frac{3}{28}, \frac{3}{14}, \frac{3}{14}, \frac{1}{7})$.

Case 1: Take

$$M_1 = \begin{pmatrix} 0.3 & 0.5 & 0.7 \\ 0.4 & 0.3 & 0.6 \\ 0.2 & 0.4 & 0.6 \\ 0.4 & 0.3 & 0.5 \end{pmatrix}$$

$$M_2 = \begin{pmatrix} 0.5 & 0.3 & 0.2 \\ 0.4 & 0.3 & 0.2 \\ 0.4 & 0.3 & 0.6 \\ 0.7 & 0.6 & 0.9 \end{pmatrix}$$

It can be found that

$$W = I_6 \otimes \begin{pmatrix} 2.3752 & 0.8882 & -1.7253 \\ -1.4954 & -0.4027 & 0.0863 \\ -2.5355 & -1.4680 & 2.0371 \\ 1.4359 & 0.9733 & 0.0139 \end{pmatrix},$$

$$\Gamma = \begin{pmatrix} -9.2070 & 3.1526 & 0 & 3.0226 & 3.0318 & 0 \\ 3.1526 & -6.5735 & 3.4209 & 0 & 0 & 0 \\ 0 & 3.4209 & -6.7617 & 3.3408 & 0 & 0 \\ 3.0226 & 0 & 3.3408 & -9.0464 & 2.6829 & 0 \\ 3.0318 & 0 & 0 & 2.6829 & -9.1429 & 3.4282 \\ 0 & 0 & 0 & 0 & 3.4282 & -3.4282 \end{pmatrix}$$

satisfy (5.23). According to Theorem 5.12, the network (5.21) achieves FTP under the adaptive law (5.17) and controller (5.22). Figures 5.5 and 5.6 display the simulation results.

It can be found that

$$W = I_6 \otimes \begin{pmatrix} 5.0508 & 2.3438 & -4.4126 \\ -3.2736 & -1.6683 & -0.0278 \\ -5.6701 & -3.1501 & 5.0649 \\ 3.4016 & 2.3572 & 0.3623 \end{pmatrix},$$

$$W_1 = I_6 \otimes \begin{pmatrix} 1.0738 & 0.7429 & 0.1445 \\ 0.7429 & 0.5786 & 0.4489 \\ 0.1445 & 0.4489 & 2.4307 \end{pmatrix},$$

$$\Gamma = \begin{pmatrix} -16.8022 & 5.8903 & 0 & 5.2744 & 5.6375 & 0 \\ 5.8903 & -12.3236 & 6.4332 & 0 & 0 & 0 \\ 0 & 6.4332 & -12.9479 & 6.5147 & 0 & 0 \\ 5.2744 & 0 & 6.5147 & -16.5813 & 4.7923 & 0 \\ 5.6375 & 0 & 0 & 4.7923 & -17.6946 & 7.2648 \\ 0 & 0 & 0 & 0 & 7.2648 & -7.2648 \end{pmatrix}$$

satisfy (5.27). By Theorem 5.13, the network (5.21) achieves FTISP under the adaptive law (5.17) and controller (5.22). Figures 5.5 and 5.6 display the simulation results.

It can be found that

$$W = I_6 \otimes \begin{pmatrix} 2.0853 & 0.7176 & -1.5216 \\ -1.1905 & -0.2193 & -0.0670 \\ -2.3691 & -1.4248 & 1.5621 \\ 1.4036 & 1.0364 & 0.5844 \end{pmatrix},$$

$$W_2 = I_6 \otimes \begin{pmatrix} 1.0159 & -0.3918 & -0.4676 & -0.3497 \\ -0.3918 & 1.1695 & -0.3242 & -0.3481 \\ -0.4676 & -0.3242 & 1.1787 & -0.2827 \\ -0.3497 & -0.3481 & -0.2827 & 1.2655 \end{pmatrix}$$

$$\Gamma = \begin{pmatrix} -7.0686 & 2.4192 & 0 & 2.3242 & 2.3252 & 0 \\ 2.4192 & -5.0298 & 2.6106 & 0 & 0 & 0 \\ 0 & 2.6106 & -5.1636 & 2.5530 & 0 & 0 \\ 2.3242 & 0 & 2.5530 & -6.9473 & 2.0701 & 0 \\ 2.3252 & 0 & 0 & 2.0701 & -6.9856 & 2.5903 \\ 0 & 0 & 0 & 0 & 2.5903 & -2.5903 \end{pmatrix}$$

satisfy (5.28). According to Theorem 5.14, the network (5.21) achieves FTOSP under the adaptive law (5.17) and controller (5.22). Figures 5.5 and 5.6 display the simulation results.

Case 2: Apparently, the condition (5.29) also holds if we select

$$W = I_6 \otimes \begin{pmatrix} 2.3752 & 0.8882 & -1.7253 \\ -1.4954 & -0.4027 & 0.0863 \\ -2.5355 & -1.4680 & 2.0371 \\ 1.4359 & 0.9733 & 0.0139 \end{pmatrix},$$

$$\Gamma = \begin{pmatrix} -9.2070 & 3.1526 & 0 & 3.0226 & 3.0318 & 0 \\ 3.1526 & -6.5735 & 3.4209 & 0 & 0 & 0 \\ 0 & 3.4209 & -6.7617 & 3.3408 & 0 & 0 \\ 3.0226 & 0 & 3.3408 & -9.0464 & 2.6829 & 0 \\ 3.0318 & 0 & 0 & 2.6829 & -9.1429 & 3.4282 \\ 0 & 0 & 0 & 0 & 3.4282 & -3.4282 \end{pmatrix}.$$

From Corollary 5.16, the network (5.15) achieves FTS under the adaptive law (5.17) and controller (5.22). Figures 5.7 and 5.8 display the simulation results.

Figures. 5.1 and 5.5 describe the trends of $\|w_i(t)\|$, $\|y_i(t)\|$ and $\|u_i(t)\|$ of the CNNs under the adaptive laws and controllers. Figs. 5.2, 5.4, 5.6 and 5.8 display the change tendencies of the adaptive coupling weights. The change processes of $\|w_i(t)\|, i = 1, 2, \ldots, 6$, are shown in Figs. 5.3 and 5.7. It can be seen from Figs. 5.3 and 5.7 that $w_i(t)$ can realize FTS.

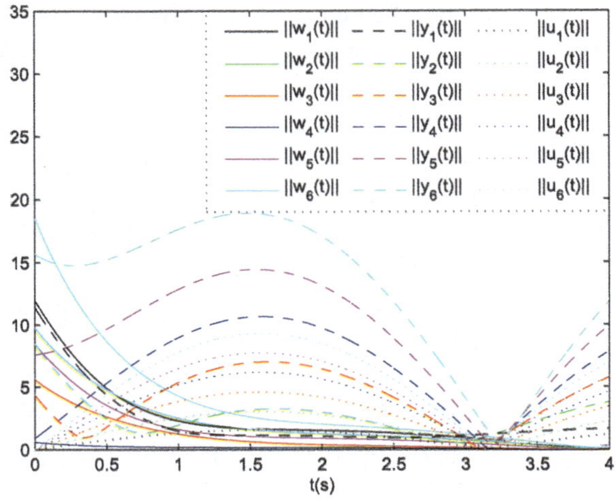

Fig. 5.5 $\|w_i(t)\|$, $\|y_i(t)\|$, $\|u_i(t)\|$, $i = 1, 2, \ldots, 6$

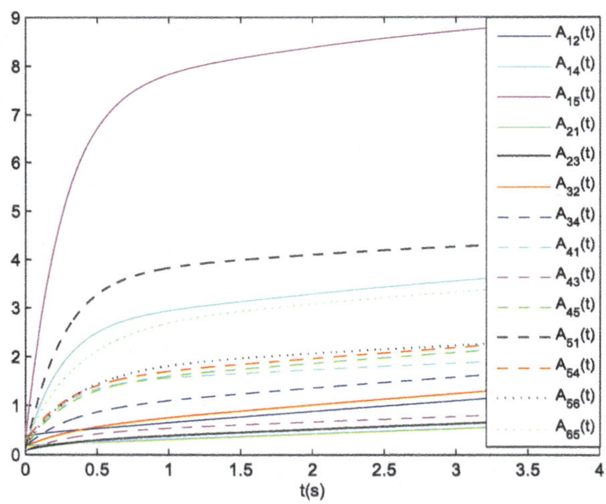

Fig. 5.6 $A_{ij}(t)$, $(i, j) \in \mathcal{B}$

Remark 5.20 During the last few decades, the synchronization [1–8] and passivity [9, 10] for CNNs with fixed coupling weights have been discussed. As is well known, the given coupling weights are generally larger than the needed values for ensuring the synchronization and passivity of CNNs. Thus, two adaptive control schemes are developed such that the coupling weights can achieve the appropriate values (see Figs. 5.2, 5.4, 5.6 and 5.8). Practically, the coupling weights do not need to be adjusted when $t > t^*$ (see Theorems 5.6 and 5.12). Thus, we take $\dot{A}_{ij}(t) = 0$ when $t > t^*$.

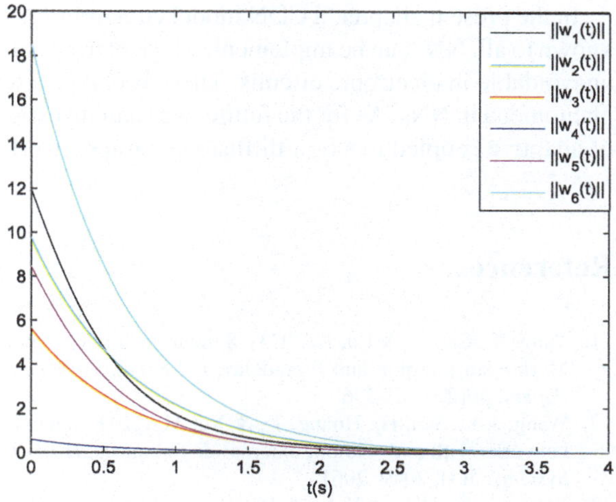

Fig. 5.7 $\|w_i(t)\|, i = 1, 2, \ldots, 6$

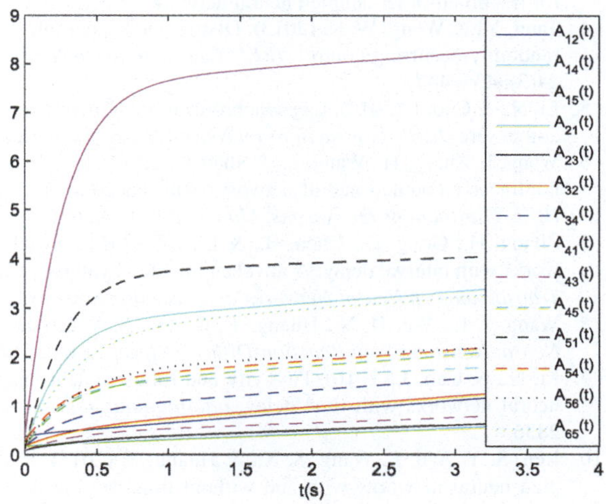

Fig. 5.8 $A_{ij}(t), (i, j) \in \mathcal{B}$

5.6 Conclusion

The FTP for adaptive coupled neural networks with undirected and directed topologies has been considered. On one hand, several sufficient conditions for ensuring the FTP of directed and undirected CNNs have been derived by designing appropriate adaptive laws and controllers. On the other hand, we also have discussed the relationship between FTP and FTS of the adaptive coupled neural networks with undirected and directed topologies. Finally, the effectiveness of obtained criteria has been verified by exploiting two numerical examples.

In the present chapter, a CDN model containing N identical NNs is studied. As is known to all, NNs can be implemented by electronic circuits, and diffusion effects are unavoidable in electronic circuits. Therefore, it is essential to consider the diffusion phenomena in NNs. As for the future, we shall investigate the FTP and FTS problems of adaptive coupled reaction-diffusion neural networks.

References

1. Yang, X., Cao, J., & Lu, J. (2013). Synchronization of randomly coupled neural networks with Markovian jumping and time-delay. *IEEE Transactions on Circuits and Systems I: Regular Papers*, *60*(2), 363–376.
2. Wang, J. L., Wu, H., Huang, T., & Xu, M. (2018). Output synchronization in coupled neural networks with and without external disturbances. *IEEE Transactions on Control of Network Systems*, *5*(4), 2049–2061.
3. Wang, S. X., Huang, Y. L., & Ren, S. Y. (2017). Synchronization and robust synchronization for fractional-order coupled neural networks. *IEEE Access*, *5*, 12439–12448.
4. Tang, Y., & Wong, W. K. (2013). Distributed synchronization of coupled neural networks via randomly occurring control. *IEEE Transactions on Neural Networks and Learning Systems*, *24*(3), 435–447.
5. Li, N., & Cao, J. (2016). Lag synchronization of memristor-based coupled neural networks via ω-measure. *IEEE Transactions on Neural Networks and Learning Systems*, *27*(3), 686–697.
6. Wang, J., Zhang, H., Wang, Z., & Shan, Q. (2017). Local synchronization criteria of Markovian nonlinearly coupled neural networks with uncertain and partially unknown transition rates. *IEEE Transactions on Systems, Man, and Cybernetics: Systems*, *47*(8), 1953–1964.
7. Zhang, H., Gong, D., Chen, B., & Liu, Z. (2013). Synchronization for coupled neural networks with interval delay: A novel augmented Lyapunov-Krasovskii functional method. *IEEE Transactions on Neural Networks and Learning Systems*, *24*(1), 58–70.
8. Wang, J. L., Wu, H. N., Huang, T., & Ren, S. Y. (2018). *Analysis and Control of Coupled Neural Networks With Reaction-Diffusion Terms*. Berlin: Springer.
9. Li, N., & Cao, J. (2016). Passivity and robust synchronisation of switched interval coupled neural networks with time delay. *International Journal of Systems Science*, *47*(12), 2827–2836.
10. Ren, S. Y., Wu, J., Wang, S. X., & Huang, Y. (2018). Passivity and pinning control of coupled neural networks with and without time-varying delay. *Transactions of the Institute of Measurement and Control*, *40*(9), 2708–2717.
11. Wu, Y., Cao, J., Li, Q., Alsaedi, A., & Alsaadi, F. E. (2017). Finite-time synchronization of uncertain coupled switched neural networks under asynchronous switching. *Neural Networks*, *85*, 128–139.
12. Wang, J., Zhang, H., Wang, Z., & Gao, D. W. (2017). Finite-time synchronization of coupled hierarchical hybrid neural networks with time-varying delays. *IEEE Transactions on Cybernetics*, *47*(10), 2995–3004.
13. Yang, X., Ca, J., Song, Q., Xu, C., & Feng, J. (2017). Finite-time synchronization of coupled Markovian discontinuous neural networks with mixed delays. *Circuits, Systems, and Signal Processing*, *36*(5), 1860–1889.
14. Liu, X., Su, H., & Chen, M. Z. Q. (2016). A switching approach to designing finite-time synchronization controllers of coupled neural networks. *IEEE Transactions on Neural Networks and Learning Systems*, *27*(2), 471–482.
15. Qiu, S. H., Huang, Y. L., Wang, J. L., Ren, S. Y., Liu, D. F. (2017). Finite-time synchronization of coupled cohen-grossberg neural networks with time-varying delays. In *2017 36th Chinese Control Conference (CCC)* (pp. 3966–3971).

16. Tan, M., Li, X., Liu, Y. (2019). Finite-time stability and synchronization of the coupled switched neural networks with nodes of different dimensions. *Neural Processing Letters, 49*(1), 285–303.
17. Shen, J., & Cao, J. (2011). Finite-time synchronization of coupled neural networks via discontinuous controllers. *Cognitive Neurodynamics, 5*(4), 373–385.
18. Liu, X., Cao, J., Yu, W., & Song, Q. (2016). Nonsmooth finite-time synchronization of switched coupled neural networks. *IEEE Transactions on Cybernetics, 46*(10), 2360–2371.
19. Zhang, J., Li, C., Huang, T. (2017). Finite-time synchronization of coupled switched impulsive neural networks. In *2017 13th IEEE International Conference on Control Automation (ICCA)* (pp. 577–581).
20. Yang, C., & Huang, L. (2017). Finite-time synchronization of coupled time-delayed neural networks with discontinuous activations. *Neurocomputing, 249*, 64–71.
21. Wang, J. L., Xu, M., Wu, H. N., & Huang, T. (2018). Finite-time passivity of coupled neural networks with multiple weights. *IEEE Transactions on Network Science and Engineering, 5*(3), 184–197.
22. Wang, J. L., Xu, M., Wu, H. N., & Huang, T. (2019). Passivity analysis and pinning control of multi-weighted complex dynamical networks. *IEEE Transactions on Network Science and Engineering, 6*(1), 60–73.
23. Zhang, X. X., Wang, J. L., Huang, Y. L., & Ren, S. Y. (2018). Analysis and pinning control for passivity of multi-weighted complex dynamical networks with fixed and switching topologies. *Neurocomputing, 275*, 958–968.
24. Xu, Q., Zhuang, S., Zeng, Y., & Xiao, J. (2017). Decentralized adaptive strategies for synchronization of fractional-order complex networks. *IEEE/CAA Journal of Automatica Sinica, 4*(3), 543–550.
25. Wang, J. L., Wu, H. N., Huang, T., Ren, S. Y., & Wu, J. (2017). Passivity of directed and undirected complex dynamical networks with adaptive coupling weights. *IEEE Transactions on Neural Networks and Learning Systems, 28*(8), 1827–1839.
26. Wang, J. L., Wu, H. N., Huang, T., & Ren, S. Y. (2015). Passivity and synchronization of linearly coupled reaction-diffusion neural networks with adaptive coupling. *IEEE Transactions on Cybernetics, 45*(9), 1942–1952.
27. Jin, X. Z., He, T., Xia, J. W., Wang, D., & Guan, W. (2019). Adaptive general pinned synchronization of a class of disturbed complex networks. *Communications in Nonlinear Science and Numerical Simulation, 67*, 658–669.

Chapter 6
FTP of CNNs with Multiple Weights

6.1 Introduction

Recently, NN has become a very hot topic in both theory and practice because of its wide spread applications in associative memories, image processing, pattern recognition, fixed-point computations, and so on. Therefore, it is very important and meaningful to investigate the NNs. Up to now, the dynamical behaviors of NNs have been extensively studied, and many important results have been reported for various ordinary differential [1–5] and partial differential [6–10] equation models.

In particular, the passivity of NNs has been one of the most active research areas [11–20]. In [12], Li et al. analyzed the passivity of uncertain continuous-time NNs with mixed time-varying delays. The authors [15] took into account the passivity problem for NNs with leakage delay and time-varying transmission delay. In [17], Lian and Wang considered the passivity of switched NNs with time-varying delays. The passivity analysis problem for delayed discrete-time NNs with randomly occurring quantization effects was addressed in [18]. Li et al. [19] dealt with the problem of robust passivity for delayed NN with norm-bounded parameter uncertainties. In [20], Wang et al. analyzed the passivity and stability of RDNNs with Dirichlet boundary conditions.

Many NNs can result in a complex network by mutually coupling, which is called CNNs. More recently, CNNs have received much attention due to their promising potential applications in many fields, such as brain science, chaos generators design, secure communication, and harmonic oscillation generation. As is known to us, such applications heavily depend on the dynamical behaviors of CNNs, especially the passivity of CNNs [21–25]. In [21], Li and Cao concerned with robust passivity of switched CNNs with uncertain parameters. In [22], a general array model of CRDNNs with delay coupling was studied, in which input vector has the same dimension as the output vector. Wang et al. [23] discussed the passivity of CNNs with reaction-diffusion terms and adaptive coupling. In [24], the authors analyzed the dissipativity and passivity of CNNs with reaction-diffusion terms and different dimensions of

The original version of this book was revised: The belated corrections have been incorporated. The corrections to this book are available at https://doi.org/10.1007/978-981-33-4287-3_10

J.-L. Wang et al., *Passivity of Complex Dynamical Networks*,
https://doi.org/10.1007/978-981-33-4287-3_6

input and output. Wang et al. [25] investigated passivity problem for CRDNNs with undirected topology and directed topology by updating the coupling weights, in which the input and output vectors have different dimensions. Unfortunately, in these existing works about passivity, the time-delay in CNNs is single [21–25]. Practically, in many circumstances, it is more reasonable to consider the network model with multiple coupling delays [26]. To our knowledge, very few results on passivity of CNNs with multiple coupling delays have been reported.

As we all know, passivity theory provides a powerful tool to analyze synchronization of CDNs [22, 23, 25, 27–32]. In [28], Liu and Zhao studied the output synchronization of CDNs with nonidentical nodes by exploiting the output quasi-passivity property. Under the assumptions that each agent is input-output passive and the topology structure is strongly connected, several output synchronization criteria for multi-agent systems were established by selecting appropriate control inputs in [30]. In [31], Wang et al. respectively discussed the relationship between output synchronization and output strict passivity of CDNs with fixed and adaptive coupling strength. Furthermore, some authors [22, 23, 25] studied the synchronization of passive CRDNNs with fixed and adaptive coupling weights, respectively. However, in these existing works [22, 23, 25, 27–32], synchronization is defined over the infinite time interval. But, in many practical applications, synchronization is realized over a finite time interval may be more reasonable. Therefore, it is more meaningful and valuable to study FTS of CDNs [33–37]. In [33], Zhang et al. considered FTS of a multi-layer nonlinear coupled complex network model by utilizing intermittent feedback control method. Wu et al. [36] investigated FTS of uncertain coupled switched neural networks by exploiting average dwell time technique and Lyapunov functional method. In [37], some FTS criteria were presented for coupled hierarchical hybrid neural networks. Obviously, it is very important to define the FTP for studying the FTS of CDNs. As far as we know, very few researchers have discussed FTP of CDNs in recent years. Especially, the FTP and the relationship between FTP and FTS of the CNNs have not yet been considered.

Motivated by the above discussions, we respectively investigate FTP of MWCNNs with and without coupling delays. The main contribution of this chapter lies in three aspects. First, three concepts of FTP are presented, which extend the traditional passivity definitions. Second, several FTP criteria are developed for MWCNNs by designing appropriate controller and constructing suitable Lyapunov functional, and some sufficient conditions about FTS are derived for finite-time passive MWCNNs. Third, we also consider FTP of MWCNNs with coupling delays, and the relationship between FTP and FTS.

6.2 Preliminaries

Definition 6.1 A system with input $u(t) \in \mathbb{R}^n$ and output $y(t) \in \mathbb{R}^n$ is finite-time passive if

$$u^T(t)y(t) \geqslant \dot{V}(t) + \beta V^{\alpha}(t)$$

for some $\alpha \in (0, 1)$ and $\mathbb{R} \ni \beta > 0$, where V is a nonnegative function.

Definition 6.2 A system with input $u(t) \in \mathbb{R}^n$ and output $y(t) \in \mathbb{R}^n$ is finite-time input strictly passive if

$$u^T(t)y(t) - r_1 u^T(t)u(t) \geqslant \dot{V}(t) + \beta V^{\alpha}(t)$$

for some $\alpha \in (0, 1)$, $\mathbb{R} \ni \beta > 0$ and $\mathbb{R} \ni r_1 > 0$, where V is a nonnegative function.

Definition 6.3 A system with input $u(t) \in \mathbb{R}^n$ and output $y(t) \in \mathbb{R}^n$ is finite-time output strictly passive if

$$u^T(t)y(t) - r_2 y^T(t)y(t) \geqslant \dot{V}(t) + \beta V^{\alpha}(t)$$

for some $\alpha \in (0, 1)$, $\mathbb{R} \ni \beta > 0$ and $\mathbb{R} \ni r_2 > 0$, where V is a nonnegative function.

Remark 6.4 As we all know, passivity is a powerful tool to analyze infinite-time synchronization of CDNs [22, 23, 25, 27–32]. But, in many practical applications, synchronization is realized over a finite time interval may be more reasonable. Therefore, it is very important to define the FTP for studying the FTS of CDNs. Nevertheless, very few researchers have discussed FTP and the relationship between FTP and FTS of CDNs. In this chapter, we present three FTP definitions (see Definitions 6.1, 6.2, and 6.3), which extend some existing passivity concepts [21–25, 27–32].

Remark 6.5 Nonnegative function $V(\cdot)$, called the storage function, represents the energy stored inside the system. Based on storage function $V(\cdot)$ and supply rate $w(u, y)$, FTP, FTISP and FTOSP are defined (see Definitions 6.1, 6.2, and 6.3) through selecting different supply rates $w(u(t), y(t)) = u^T(t)y(t)$, $u^T(t)y(t) - r_1 u^T(t)u(t)$ and $u^T(t)y(t) - r_2 y^T(t)y(t)$, which denote the energy dissipated inside the system $V(t_2) - V(t_1)$ is less than the energy supplied from external source $\int_{t_1}^{t_2} w(u(t), y(t))dt$.

6.3 FTP of MWCNNs

6.3.1 Network Model

The network model considered in this section is described by:

$$\begin{aligned}
\dot{x}_i(t) = & -Ax_i(t) + Dg(x_i(t)) + J + c_1 \sum_{j=1}^{N} G_{ij}^1 \Gamma^1 x_j(t) \\
& + c_2 \sum_{j=1}^{N} G_{ij}^2 \Gamma^2 x_j(t) + \cdots + c_m \sum_{j=1}^{N} G_{ij}^m \Gamma^m x_j(t) \\
& + u_i(t) + v_i(t), \ i = 1, 2, \ldots, N,
\end{aligned} \tag{6.1}$$

where $x_i(t) = (x_{i1}(t), x_{i2}(t), \ldots, x_{in}(t))^T \in \mathbb{R}^n$ represents the state vector of the ith node; $u_i(t) \in \mathbb{R}^n$ denotes the input vector; $v_i(t) \in \mathbb{R}^n$ is the control input; $A = \text{diag}(a_1, a_2, \ldots, a_n) \in \mathbb{R}^{n\times n} > 0$, $D \in \mathbb{R}^{n\times n}$ is a constant matrix; $g(x_i(t)) = (g_1(x_{i1}(t)), g_2(x_{i2}(t)), \ldots, g_n(x_{in}(t)))^T \in \mathbb{R}^n$ and $J = (J_1, J_2, \ldots, J_n)^T \in \mathbb{R}^n$; $\mathbb{R} \ni c_s > 0$ $(s = 1, 2, \ldots, m)$ is coupling strength for the sth coupling form; $\Gamma^s \in \mathbb{R}^{n\times n} > 0$ $(s = 1, 2, \ldots, m)$ denotes the inner coupling matrix for the sth coupling form; $\mathbb{R}^{N\times N} \ni G^s = (G^s_{ij})_{N\times N}$ $(s = 1, 2, \ldots, m)$ represents coupling weights in the sth coupling form, where $\mathbb{R} \ni G^s_{ij}$ is defined as follows: if there exists a connection between node i and node j, then $\mathbb{R} \ni G^s_{ij} = G^s_{ji} > 0$; otherwise, $\mathbb{R} \ni G^s_{ij} = G^s_{ji} = 0$ $(i \neq j)$; and the diagonal elements of matrix G^s are defined by

$$G^s_{ii} = -\sum_{\substack{j=1\\ j\neq i}}^{N} G^s_{ij}, i = 1, 2, \ldots, N.$$

To obtain our results, the following useful assumption is introduced:

Throughout this chapter, the function $g_k(\cdot)(k = 1, 2, \ldots, n)$ satisfies the Lipschitz condition, that is, there exists positive constant ρ_k such that

$$|g_k(\varepsilon_1) - g_k(\varepsilon_2)| \leqslant \rho_k|\varepsilon_1 - \varepsilon_2|$$

for any $\varepsilon_1, \varepsilon_2 \in \mathbb{R}$. Take $\Theta = \text{diag}(\rho_1^2, \rho_2^2, \ldots, \rho_n^2) \in \mathbb{R}^{n\times n}$.

Remark 6.6 In recent years, the passivity of CNNs has drawn much attention, and several passivity criteria have been established for different kinds of CNNs [21–25]. Nevertheless, in these existing works about passivity, the CNNs models with single weight were investigated. Practically, in many circumstances, it is more reasonable to consider the CNNs with multiple weights [26]. To our knowledge, very few researchers have studied the passivity of CNNs with multiple weights.

6.3.2 FTP

Let $\bar{x}(t) = (\bar{x}_1(t), \bar{x}_2(t), \ldots, \bar{x}_n(t))^T = \frac{1}{N}\sum_{i=1}^{N} x_i(t)$, one gets

$$\begin{aligned}\dot{\bar{x}}(t) = &-\frac{A}{N}\sum_{i=1}^{N} x_i(t) + \frac{D}{N}\sum_{i=1}^{N} g(x_i(t)) + J\\ &+\frac{1}{N}\sum_{j=1}^{N} c_1\left(\sum_{i=1}^{N} G^1_{ij}\right)\Gamma^1 x_j(t)\\ &+\frac{1}{N}\sum_{j=1}^{N} c_2\left(\sum_{i=1}^{N} G^2_{ij}\right)\Gamma^2 x_j(t)\end{aligned}$$

$$
\begin{aligned}
&+\cdots+\frac{1}{N}\sum_{j=1}^{N}c_m\left(\sum_{i=1}^{N}G_{ij}^{m}\right)\Gamma^{m}x_j(t)\\
&+\frac{1}{N}\sum_{i=1}^{N}u_i(t)+\frac{1}{N}\sum_{i=1}^{N}v_i(t)\\
=&-A\bar{x}(t)+\frac{D}{N}\sum_{i=1}^{N}g(x_i(t))+J\\
&+\frac{1}{N}\sum_{i=1}^{N}u_i(t)+\frac{1}{N}\sum_{i=1}^{N}v_i(t). \qquad (6.2)
\end{aligned}
$$

Take $z_i(t) = x_i(t) - \bar{x}(t)$. By (6.1) and (6.2), we can get

$$
\begin{aligned}
\dot{z}_i(t) =& -Az_i(t) + Dg(x_i(t)) - \frac{D}{N}\sum_{k=1}^{N}g(x_k(t))\\
&+\sum_{s=1}^{m}\sum_{j=1}^{N}c_s G_{ij}^{s}\Gamma^{s}z_j(t)+u_i(t)+v_i(t)\\
&-\frac{1}{N}\sum_{k=1}^{N}v_k(t)-\frac{1}{N}\sum_{k=1}^{N}u_k(t), \qquad (6.3)
\end{aligned}
$$

where $i = 1, 2, \ldots, N$.

The output vector $y_i(t) \in \mathbb{R}^n$ of network (6.3) is defined as follows:

$$
y_i(t) = B_1 z_i(t) + B_2 u_i(t),
$$

where $B_1, B_2 \in \mathbb{R}^{n\times n}$.

Design the following controller for network (6.1):

$$
\begin{aligned}
v_i(t) =& -K_i(x_i(t)-\bar{x}(t))\\
&-\beta M^{\frac{\alpha-1}{2}}\operatorname{sign}(x_i(t)-\bar{x}(t))|x_i(t)-\bar{x}(t)|^{\alpha}, \qquad (6.4)
\end{aligned}
$$

where $K_i \in \mathbb{R}^{n\times n}$, $0 < \alpha < 1$, $\operatorname{sign}(x_i(t)-\bar{x}(t)) = \operatorname{diag}(\operatorname{sign}(x_{i1}(t)-\bar{x}_1(t)), \operatorname{sign}(x_{i2}(t)-\bar{x}_2(t)), \ldots, \operatorname{sign}(x_{in}(t)-\bar{x}_n(t)))$, $|x_i(t)-\bar{x}(t)|^{\alpha} = (|x_{i1}(t)-\bar{x}_1(t)|^{\alpha}, |x_{i2}(t)-\bar{x}_2(t)|^{\alpha}, \ldots, |x_{in}(t)-\bar{x}_n(t)|^{\alpha})^T$, $\mathbb{R} \ni \beta > 0$, $0 < M = \operatorname{diag}(M_1, M_2, \ldots, M_n) \in \mathbb{R}^{n\times n}$, $M^{\frac{\alpha-1}{2}} = \operatorname{diag}(M_1^{\frac{\alpha-1}{2}}, M_2^{\frac{\alpha-1}{2}}, \ldots, M_n^{\frac{\alpha-1}{2}})$.

Theorem 6.7 *The network* (6.3) *is finite-time passive under the controller* (6.4) *if there exist matrices* $0 < M = \operatorname{diag}(M_1, M_2, \ldots, M_n) \in \mathbb{R}^{n\times n}$, $K = \operatorname{diag}(K_1, K_2, \ldots, K_N) \in \mathbb{R}^{nN\times nN}$ *such that*

$$\begin{pmatrix} Q_1 & E_1 \\ E_1^T & -I_N \otimes \frac{B_2+B_2^T}{2} \end{pmatrix} \leqslant 0, \tag{6.5}$$

where $Q_1 = I_N \otimes (-MA - AM + MDD^T M + \Theta) - (I_N \otimes M)K - K^T(I_N \otimes M) + \sum_{s=1}^m c_s G^s \otimes (M\Gamma^s + \Gamma^s M)$, $E_1 = I_N \otimes M - \frac{I_N \otimes B_1^T}{2}$.

Proof Choose the following Lyapunov functional for network (6.3):

$$V_1(t) = \sum_{i=1}^N z_i^T(t) M z_i(t), \tag{6.6}$$

one has

$$\begin{aligned}
\dot{V}_1(t) &= 2\sum_{i=1}^N z_i^T(t) M \dot{z}_i(t) \\
&= 2\sum_{i=1}^N z_i^T(t) M \Big[-Az_i(t) + Dg(x_i(t)) \\
&\quad - Dg(\bar{x}(t)) + Dg(\bar{x}(t)) - \frac{D}{N}\sum_{k=1}^N g(x_k(t)) \\
&\quad + \sum_{s=1}^m \sum_{j=1}^N c_s G_{ij}^s \Gamma^s z_j(t) + u_i(t) - \frac{1}{N}\sum_{k=1}^N u_k(t) \\
&\quad - K_i z_i(t) - \beta M^{\frac{\alpha-1}{2}} \operatorname{sign}(z_i(t)) |z_i(t)|^{\alpha} \\
&\quad - \frac{1}{N}\sum_{k=1}^N v_k(t) \Big].
\end{aligned} \tag{6.7}$$

Obviously,

$$2z_i^T(t) MD\left(g(x_i(t)) - g(\bar{x}(t))\right) \leqslant z_i^T(t)\left(MDD^T M + \Theta\right) z_i(t). \tag{6.8}$$

On the other hand,

$$\begin{aligned}
\sum_{i=1}^N z_i(t) &= \sum_{i=1}^N (x_i(t) - \bar{x}(t)) \\
&= \sum_{i=1}^N \left(x_i(t) - \frac{1}{N}\sum_{k=1}^N x_k(t) \right) \\
&= \sum_{i=1}^N x_i(t) - \sum_{k=1}^N x_k(t) \\
&= 0.
\end{aligned} \tag{6.9}$$

By (6.9), one gets

$$\sum_{i=1}^{N} z_i^T(t)MD\left(g(\bar{x}(t)) - \frac{1}{N}\sum_{k=1}^{N} g(x_k(t))\right) = 0, \tag{6.10}$$

$$\sum_{i=1}^{N} z_i^T(t)M\left(\frac{1}{N}\sum_{k=1}^{N} u_k(t) + \frac{1}{N}\sum_{k=1}^{N} v_k(t)\right) = 0. \tag{6.11}$$

From (6.7)–(6.11), we can obtain

$$\begin{aligned}
\dot{V}_1(t) \leqslant & \sum_{i=1}^{N} z_i^T(t)(-MA - AM + MDD^T M + \Theta)z_i(t) \\
& +2\sum_{s=1}^{m}\sum_{i=1}^{N}\sum_{j=1}^{N} c_s G_{ij}^s z_i^T(t)M\Gamma^s z_j(t) \\
& +2\sum_{i=1}^{N} z_i^T(t)Mu_i(t) - 2\sum_{i=1}^{N} z_i^T(t)MK_i z_i(t) \\
& -2\beta\sum_{i=1}^{N} z_i^T(t)M^{\frac{\alpha+1}{2}}\operatorname{sign}(z_i(t))|z_i(t)|^{\alpha} \\
\leqslant & z^T(t)\Big[I_N \otimes (-MA - AM + MDD^T M + \Theta) \\
& -(I_N \otimes M)K - K^T(I_N \otimes M) \\
& +\sum_{s=1}^{m} c_s G^s \otimes (M\Gamma^s + \Gamma^s M)\Big]z(t) \\
& +2z^T(t)(I_N \otimes M)u(t) \\
& -2\beta\sum_{i=1}^{N} z_i^T(t)M^{\frac{\alpha+1}{2}}\operatorname{sign}(z_i(t))|z_i(t)|^{\alpha},
\end{aligned} \tag{6.12}$$

where $z(t) = (z_1^T(t), z_2^T(t), \ldots, z_N^T(t))^T$, $u(t) = (u_1^T(t), u_2^T(t), \ldots, u_N^T(t))^T$.

Obviously,

$$\begin{aligned}
& \sum_{i=1}^{N} z_i^T(t)M^{\frac{\alpha+1}{2}}\operatorname{sign}(z_i(t))|z_i(t)|^{\alpha} \\
= & \sum_{i=1}^{N}\sum_{j=1}^{n} M_j^{\frac{\alpha+1}{2}}|z_{ij}(t)|^{\alpha+1}
\end{aligned}$$

$$
\begin{aligned}
&\geqslant \sum_{i=1}^{N}\left(\sum_{j=1}^{n} M_j z_{ij}^2(t)\right)^{\frac{\alpha+1}{2}} \\
&= \sum_{i=1}^{N}\left(z_i^T(t) M z_i(t)\right)^{\frac{\alpha+1}{2}}.
\end{aligned} \tag{6.13}
$$

From (6.12) and (6.13), we can get

$$
\begin{aligned}
&\dot{V}_1(t) - u^T(t)y(t) \\
&= \dot{V}_1(t) - u^T(t)(I_N \otimes B_1)z(t) - u^T(t)(I_N \otimes B_2)u(t) \\
&\leqslant -2\beta \sum_{i=1}^{N}\left(z_i^T(t) M z_i(t)\right)^{\frac{\alpha+1}{2}} \\
&\quad + \xi^T(t)\begin{pmatrix} Q_1 & E_1 \\ E_1^T & -I_N \otimes \frac{B_2+B_2^T}{2} \end{pmatrix}\xi(t) \\
&\leqslant -2\beta \sum_{i=1}^{N}\left(z_i^T(t) M z_i(t)\right)^{\frac{\alpha+1}{2}} \\
&\leqslant -2\beta \left(\sum_{i=1}^{N} z_i^T(t) M z_i(t)\right)^{\frac{\alpha+1}{2}} \\
&= -2\beta V_1^{\frac{\alpha+1}{2}}(t),
\end{aligned}
$$

where $\xi(t) = (z^T(t), u^T(t))^T$ and $y(t) = (y_1^T(t), y_2^T(t), \ldots, y_N^T(t))^T$.
Then, we can obtain

$$
u^T(t)y(t) \geqslant \dot{V}_1(t) + 2\beta V_1^{\frac{\alpha+1}{2}}(t).
$$

Therefore, the network (6.3) is finite-time passive under the controller (6.4).

Theorem 6.8 *The network* (6.3) *is finite-time input strictly passive under the controller* (6.4) *if there exist matrices* $0 < M = \operatorname{diag}(M_1, M_2, \ldots, M_n) \in \mathbb{R}^{n\times n}$, $K = \operatorname{diag}(K_1, K_2, \ldots, K_N) \in \mathbb{R}^{nN\times nN}$ *and a positive real number* r_1 *such that*

$$
\begin{pmatrix} Q_1 & E_1 \\ E_1^T & r_1 I_{nN} - I_N \otimes \frac{B_2+B_2^T}{2} \end{pmatrix} \leqslant 0, \tag{6.14}
$$

where Q_1, E_1 *have the same meanings as in Theorem 6.7.*

Proof For network (6.3), we choose the same $V_1(t)$ as (6.6). By (6.12), one obtains

$$\begin{aligned}
& \dot{V}_1(t) - u^T(t)y(t) + r_1 u^T(t)u(t) \\
\leqslant\ & z^T(t)\Big[I_N \otimes (-MA - AM + MDD^TM + \Theta) \\
& -(I_N \otimes M)K - K^T(I_N \otimes M) \\
& + \sum_{s=1}^{m} c_s G^s \otimes (M\Gamma^s + \Gamma^s M)\Big]z(t) \\
& + 2z^T(t)(I_N \otimes M)u(t) + r_1 u^T(t)u(t) \\
& - 2\beta \sum_{i=1}^{N} z_i^T(t) M^{\frac{\alpha+1}{2}} \mathrm{sign}(z_i(t))|z_i(t)|^{\alpha} \\
& - u^T(t)(I_N \otimes B_1)z(t) - u^T(t)(I_N \otimes B_2)u(t) \\
=\ & -2\beta \sum_{i=1}^{N} z_i^T(t) M^{\frac{\alpha+1}{2}} \mathrm{sign}(z_i(t))|z_i(t)|^{\alpha} \\
& + \xi^T(t) \begin{pmatrix} Q_1 & E_1 \\ E_1^T & r_1 I_{nN} - I_N \otimes \frac{B_2 + B_2^T}{2} \end{pmatrix} \xi(t) \\
\leqslant\ & -2\beta \left(\sum_{i=1}^{N} z_i^T(t) M z_i(t) \right)^{\frac{\alpha+1}{2}} \\
=\ & -2\beta V_1^{\frac{\alpha+1}{2}}(t).
\end{aligned}$$

Then, one gets

$$u^T(t)y(t) - r_1 u^T(t)u(t) \geqslant \dot{V}_1(t) + 2\beta V_1^{\frac{\alpha+1}{2}}(t).$$

Therefore, the network (6.3) is finite-time input strictly passive under the controller (6.4).

Theorem 6.9 *The network* (6.3) *is finite-time output strictly passive under the controller* (6.4) *if there exists matrices* $0 < M = \mathrm{diag}(M_1, M_2, \ldots, M_n) \in \mathbb{R}^{n \times n}$, $K = \mathrm{diag}(K_1, K_2, \ldots, K_N) \in \mathbb{R}^{nN \times nN}$ *and a positive real number* r_2 *such that*

$$\begin{pmatrix} Q_2 & E_2 \\ E_2^T & Q_3 \end{pmatrix} \leqslant 0, \tag{6.15}$$

where $Q_2 = Q_1 + r_2(I_N \otimes B_1^T)(I_N \otimes B_1)$, $E_2 = E_1 + r_2(I_N \otimes B_1^T)(I_N \otimes B_2)$, $Q_3 = r_2(I_N \otimes B_2^T)(I_N \otimes B_2) - I_N \otimes \frac{B_2 + B_2^T}{2}$, Q_1, E_1 *have the same meanings as in Theorem 6.7.*

Proof Select the same $V_1(t)$ as (6.6) for network (6.3). By (6.12), we obtain

$$
\begin{aligned}
&\dot{V}_1(t) - u^T(t)y(t) + r_2 y^T(t)y(t) \\
\leqslant\ & z^T(t)\Big[I_N \otimes (-MA - AM + MDD^T M + \Theta) \\
& -(I_N \otimes M)K - K^T(I_N \otimes M) \\
& + \sum_{s=1}^{m} c_s G^s \otimes (M\Gamma^s + \Gamma^s M)\Big] z(t) \\
& + 2z^T(t)(I_N \otimes M)u(t) - u^T(t)(I_N \otimes B_1)z(t) \\
& + r_2 z^T(t)(I_N \otimes B_1^T)(I_N \otimes B_1)z(t) \\
& + r_2 z^T(t)(I_N \otimes B_1^T)(I_N \otimes B_2)u(t) \\
& + r_2 u^T(t)(I_N \otimes B_2^T)(I_N \otimes B_1)z(t) \\
& + r_2 u^T(t)(I_N \otimes B_2^T)(I_N \otimes B_2)u(t) \\
& - u^T(t)(I_N \otimes B_2)u(t) \\
& - 2\beta \sum_{i=1}^{N} z_i^T(t) M^{\frac{\alpha+1}{2}} \operatorname{sign}(z_i(t))|z_i(t)|^{\alpha} \\
=\ & -2\beta \sum_{i=1}^{N} z_i^T(t) M^{\frac{\alpha+1}{2}} \operatorname{sign}(z_i(t))|z_i(t)|^{\alpha} \\
& + \xi^T(t) \begin{pmatrix} Q_2 & E_2 \\ E_2^T & Q_3 \end{pmatrix} \xi(t) \\
\leqslant\ & -2\beta \left(\sum_{i=1}^{N} z_i^T(t) M z_i(t) \right)^{\frac{\alpha+1}{2}} \\
=\ & -2\beta V_1^{\frac{\alpha+1}{2}}(t).
\end{aligned}
$$

Then, one has

$$
u^T(t)y(t) - r_2 y^T(t)y(t) \geqslant \dot{V}_1(t) + 2\beta V_1^{\frac{\alpha+1}{2}}(t).
$$

Therefore, the network (6.3) is finite-time output strictly passive under the controller (6.4).

6.3.3 FTS of Finite-Time Passive MWCNNs

In what follows, we introduce the definition of FTS for MWCNNs (6.1).
Definition 3.1. The MWCNNs (6.1) is finite-time synchronized if there exists a constant $T > 0$ such that

$$\lim_{t \to T^-} \left\| x_i(t) - \frac{1}{N} \sum_{i=1}^{N} x_i(t) \right\| = 0,$$

$$\left\| x_i(t) - \frac{1}{N} \sum_{i=1}^{N} x_i(t) \right\| = 0$$

for $t \geqslant T, i = 1, 2, \ldots, N$, where $u_i(t) = 0, i = 1, 2, \ldots, N$.

Theorem 6.10 *Suppose that $\hat{V}(t) : [0, +\infty) \to [0, +\infty)$ is continuously differentiable and satisfies the following condition:*

$$\upsilon_1(\|z(t)\|) \leqslant \hat{V}(t), \tag{6.16}$$

where $\upsilon_1 : [0, +\infty) \to [0, +\infty)$ is continuous and strictly monotonically increasing function, $\upsilon_1(s)$ is positive for $s > 0$ with $\upsilon_1(0) = 0$. If network (6.3) is finite-time passive (finite-time input strictly passive, finite-time output strictly passive) with respect to $\hat{V}(t)$, MWCNNs (6.1) is finite-time synchronized under the controller (6.4).

Proof If network (6.3) is finite-time passive with respect to $\hat{V}(t)$ under the controller (6.4), there exist $\alpha \in (0, 1)$ and $\mathbb{R} \ni \beta > 0$ such that

$$u^T(t)y(t) \geqslant \dot{\hat{V}}(t) + \beta \hat{V}^{\alpha}(t).$$

Letting $u(t) = 0$, one gets

$$\dot{\hat{V}}(t) \leqslant -\beta \hat{V}^{\alpha}(t).$$

By Lemma 1.6, we can obtain $\hat{V}(t) = 0$ for $t \geqslant T_1$, where $T_1 = \frac{\hat{V}^{1-\alpha}(0)}{\beta(1-\alpha)}$. On the other hand, since

$$\upsilon_1(\|z(t)\|) \leqslant \hat{V}(t),$$

one has

$$\upsilon_1(\|z(t)\|) \leqslant \hat{V}(t) = 0$$

for $t \geqslant T_1$. Then, we can conclude that $\|z(t)\| = 0, t \geqslant T_1$. Namely, MWCNNs (6.1) is finite-time synchronized under the controller (6.4).

Similarly, we can easily prove that MWCNNs (6.1) is finite-time synchronized under the controller (6.4) if network (6.3) is finite-time input strictly passive or finite-time output strictly passive.

According to Theorems 6.10 and 6.7, we can derive the following conclusion.

Corollary 6.11 *If there exist matrices* $0 < M = \mathrm{diag}(M_1, M_2, \ldots, M_n) \in \mathbb{R}^{n\times n}$, $K = \mathrm{diag}(K_1, K_2, \ldots, K_N) \in \mathbb{R}^{nN\times nN}$, $B_1 \in \mathbb{R}^{n\times n}$, *and* $B_2 \in \mathbb{R}^{n\times n}$ *such that*

$$\begin{pmatrix} Q_1 & E_1 \\ E_1^T & -I_N \otimes \frac{B_2+B_2^T}{2} \end{pmatrix} \leqslant 0,$$

where $Q_1 = I_N \otimes (-MA - AM + MDD^TM + \Theta) - (I_N \otimes M)K - K^T(I_N \otimes M) + \sum_{s=1}^{m} c_s G^s \otimes (M\Gamma^s + \Gamma^s M)$, $E_1 = I_N \otimes M - \frac{I_N \otimes B_1^T}{2}$, *then network* (6.1) *is finite-time synchronized under the controller* (6.4).

6.4 FTP of MWCNNs with Coupling Delays

6.4.1 Network Model

The network model considered in this section is described by:

$$\begin{aligned}\dot{x}_i(t) = {} & -Ax_i(t) + Dg(x_i(t)) + J + c_1 \sum_{j=1}^{N} G_{ij}^1 \Gamma^1 x_j(t - \tau_1) \\ & + c_2 \sum_{j=1}^{N} G_{ij}^2 \Gamma^2 x_j(t - \tau_2) + \cdots + c_m \sum_{j=1}^{N} G_{ij}^m \Gamma^m x_j(t - \tau_m) \\ & + u_i(t) + v_i(t), \end{aligned} \tag{6.17}$$

where $i = 1, 2, \ldots, N$, $x_i(t)$, $g(\cdot)$, $u_i(t)$, $v_i(t)$, A, D, J, c_s, G_{ij}^s, Γ^s, $s = 1, 2, \ldots, m$ have the same meanings as in Section 6.3, $\tau_s (s = 1, 2, \ldots, m)$ is the time-delay.

6.4.2 FTP

Define $\bar{x}(t) = \frac{1}{N}\sum_{i=1}^{N} x_i(t)$. From (6.17), one obtains

$$\begin{aligned}\dot{\bar{x}}(t) = {} & -\frac{A}{N}\sum_{i=1}^{N} x_i(t) + \frac{D}{N}\sum_{i=1}^{N} g(x_i(t)) + J \\ & + \frac{1}{N}\sum_{s=1}^{m} c_s \sum_{j=1}^{N} \left(\sum_{i=1}^{N} G_{ij}^s\right) \Gamma^s x_j(t - \tau_s) \\ & + \frac{1}{N}\sum_{i=1}^{N} u_i(t) + \frac{1}{N}\sum_{i=1}^{N} v_i(t)\end{aligned}$$

$$= -A\bar{x}(t) + \frac{D}{N}\sum_{i=1}^{N} g(x_i(t)) + J + \frac{1}{N}\sum_{i=1}^{N} u_i(t)$$
$$+\frac{1}{N}\sum_{i=1}^{N} v_i(t). \tag{6.18}$$

Let $z_i(t) = x_i(t) - \bar{x}(t)$. By (6.17) and (6.18), we can get

$$\dot{z}_i(t) = -Az_i(t) + Dg(x_i(t)) - \frac{D}{N}\sum_{k=1}^{N} g(x_k(t))$$
$$+\sum_{s=1}^{m}\sum_{j=1}^{N} c_s G_{ij}^s \Gamma^s z_j(t-\tau_s) + u_i(t) + v_i(t)$$
$$-\frac{1}{N}\sum_{k=1}^{N} u_k(t) - \frac{1}{N}\sum_{k=1}^{N} v_k(t), \tag{6.19}$$

where $i = 1, 2, \ldots, N$.

The output vector $y_i(t) \in \mathbb{R}^n$ of network (6.19) is defined as follows:

$$y_i(t) = B_1 z_i(t) + B_2 u_i(t),$$

where $B_1, B_2 \in \mathbb{R}^{n\times n}$.

Design the following controller for network (6.17):

$$v_i(t) = -\beta M^{-1}\sum_{s=1}^{m}\left(c_s \int_{t-\tau_s}^{t} (x_i(h) - \bar{x}(h))^T P_i^s (x_i(h)\right.$$
$$\left.-\bar{x}(h))dh\right)^{\frac{\alpha+1}{2}} \frac{x_i(t) - \bar{x}(t)}{\|x_i(t) - \bar{x}(t)\|^2}$$
$$-\beta M^{\frac{\alpha-1}{2}} \operatorname{sign}(x_i(t) - \bar{x}(t))|x_i(t) - \bar{x}(t)|^{\alpha}$$
$$-K_i(x_i(t) - \bar{x}(t)), \tag{6.20}$$

where $0 < P_i^s \in \mathbb{R}^{n\times n}$, $P_s = \operatorname{diag}(P_1^s, P_2^s, \ldots, P_N^s)$, $s = 1, 2, \ldots, m$, K_i, α, M, β, $M^{\frac{\alpha-1}{2}}$, $\operatorname{sign}(x_i(t) - \bar{x}(t))$, $|x_i(t) - \bar{x}(t)|^{\alpha}$ have the same meanings as in (6.4).

Theorem 6.12 *The network* (6.19) *is finite-time passive under the controller* (6.20) *if there exist matrices* $0 < M = \operatorname{diag}(M_1, M_2, \ldots, M_n) \in \mathbb{R}^{n\times n}$, $K = \operatorname{diag}(K_1, K_2, \ldots, K_N) \in \mathbb{R}^{nN\times nN}$, $0 < P_s = \operatorname{diag}(P_1^s, P_2^s, \ldots, P_N^s) \in \mathbb{R}^{nN\times nN}$, $s = 1, 2, \ldots, m$ *such that*

$$\begin{pmatrix} W_1 & \Xi_1 \\ \Xi_1^T & -I_N \otimes \frac{B_2+B_2^T}{2} \end{pmatrix} \leqslant 0, \tag{6.21}$$

where $W_1 = I_N \otimes (-MA - AM + MDD^TM + \Theta) - (I_N \otimes M)K - K^T(I_N \otimes M) + \sum_{s=1}^{m} c_s P_s + \sum_{s=1}^{m} c_s [G^s \otimes (M\Gamma^s)] P_s^{-1} [G^s \otimes (\Gamma^s M)]$, $\Xi_1 = I_N \otimes M - \frac{I_N \otimes B_1^T}{2}$.

Proof Choose the following Lyapunov functional for network (6.19):

$$V_2(t) = \sum_{i=1}^{N} z_i^T(t) M z_i(t) + \sum_{s=1}^{m} c_s \int_{t-\tau_s}^{t} z^T(h) P_s z(h) dh, \tag{6.22}$$

where $z(t) = (z_1^T(t), z_2^T(t), \ldots, z_N^T(t))^T$.

The derivative of $V_2(t)$ is given by

$$\begin{aligned}
\dot{V}_2(t) &= 2\sum_{i=1}^{N} z_i^T(t) M \dot{z}_i(t) + \sum_{s=1}^{m} c_s z^T(t) P_s z(t) \\
&\quad - \sum_{s=1}^{m} c_s z^T(t-\tau_s) P_s z(t-\tau_s) \\
&= 2\sum_{i=1}^{N} z_i^T(t) M \Big[- A z_i(t) + D g(x_i(t)) \\
&\quad - D g(\bar{x}(t)) + D g(\bar{x}(t)) - \frac{D}{N} \sum_{k=1}^{N} g(x_k(t)) \\
&\quad + u_i(t) - \frac{1}{N} \sum_{k=1}^{N} u_k(t) - \frac{1}{N} \sum_{k=1}^{N} v_k(t) \\
&\quad + \sum_{s=1}^{m} \sum_{j=1}^{N} c_s G_{ij}^s \Gamma^s z_j(t-\tau_s) - K_i z_i(t) \Big] \\
&\quad - 2\beta \sum_{i=1}^{N} \sum_{s=1}^{m} \left(c_s \int_{t-\tau_s}^{t} z_i^T(h) P_i^s z_i(h) dh \right)^{\frac{\alpha+1}{2}} \\
&\quad - 2\beta \sum_{i=1}^{N} z_i^T(t) M^{\frac{\alpha+1}{2}} \operatorname{sign}(z_i(t)) |z_i(t)|^{\alpha} \\
&\quad - \sum_{s=1}^{m} c_s z^T(t-\tau_s) P_s z(t-\tau_s) \\
&\quad + \sum_{s=1}^{m} c_s z^T(t) P_s z(t)
\end{aligned}$$

$$
\begin{aligned}
&\leqslant z^T(t)\Big[I_N \otimes (-MA - AM + MDD^T M \\
&\quad +\Theta) - (I_N \otimes M)K - K^T(I_N \otimes M)\Big]z(t) \\
&\quad +2\sum_{s=1}^{m} c_s z^T(t)[G^s \otimes (M\Gamma^s)]z(t-\tau_s) \\
&\quad -2\beta \sum_{s=1}^{m}\left(c_s \int_{t-\tau_s}^{t} z^T(h)P_s z(h)dh\right)^{\frac{\alpha+1}{2}} \\
&\quad -2\beta\left(\sum_{i=1}^{N} z_i^T(t)Mz_i(t)\right)^{\frac{\alpha+1}{2}} \\
&\quad -\sum_{s=1}^{m} c_s z^T(t-\tau_s)P_s z(t-\tau_s) \\
&\quad +\sum_{s=1}^{m} c_s z^T(t)P_s z(t) + 2z^T(t)(I_N \otimes M)u(t).
\end{aligned}
$$

It is not difficult to acquire

$$
\begin{aligned}
&2\sum_{s=1}^{m} c_s z^T(t)[G^s \otimes (M\Gamma^s)]z(t-\tau_s) \\
\leqslant &\sum_{s=1}^{m} c_s z^T(t)[G^s \otimes (M\Gamma^s)]P_s^{-1}[G^s \otimes (\Gamma^s M)]z(t) \\
&+\sum_{s=1}^{m} c_s z^T(t-\tau_s)P_s z(t-\tau_s).
\end{aligned} \tag{6.23}
$$

From (6.23), one has

$$
\begin{aligned}
\dot{V}_2(t) \leqslant\ & z^T(t)\Big\{I_N \otimes (-MA - AM + MDD^T M + \Theta) \\
&-(I_N \otimes M)K - K^T(I_N \otimes M) + \sum_{s=1}^{m} c_s P_s \\
&+\sum_{s=1}^{m} c_s[G^s \otimes (M\Gamma^s)]P_s^{-1}[G^s \otimes (\Gamma^s M)]\Big\}z(t) \\
&+2z^T(t)(I_N \otimes M)u(t) \\
&-2\beta \sum_{s=1}^{m}\left(c_s \int_{t-\tau_s}^{t} z^T(h)P_s z(h)dh\right)^{\frac{\alpha+1}{2}}
\end{aligned}
$$

$$-2\beta\Big(\sum_{i=1}^{N} z_i^T(t)Mz_i(t)\Big)^{\frac{\alpha+1}{2}}. \tag{6.24}$$

Thus,

$$\begin{aligned}
&\dot{V}_2(t) - u^T(t)y(t) \\
&= \dot{V}_2(t) - u^T(t)(I_N \otimes B_1)z(t) - u^T(t)(I_N \otimes B_2)u(t) \\
&\leqslant \xi^T(t)\begin{pmatrix} W_1 & \Xi_1 \\ \Xi_1^T & -I_N \otimes \frac{B_2+B_2^T}{2} \end{pmatrix}\xi(t) \\
&\quad -2\beta\sum_{s=1}^{m}\Big(c_s\int_{t-\tau_s}^{t} z^T(h)P_s z(h)dh\Big)^{\frac{\alpha+1}{2}} \\
&\quad -2\beta\Big(\sum_{i=1}^{N} z_i^T(t)Mz_i(t)\Big)^{\frac{\alpha+1}{2}} \\
&\leqslant -2\beta\sum_{s=1}^{m}\Big(c_s\int_{t-\tau_s}^{t} z^T(h)P_s z(h)dh\Big)^{\frac{\alpha+1}{2}} \\
&\quad -2\beta\Big(\sum_{i=1}^{N} z_i^T(t)Mz_i(t)\Big)^{\frac{\alpha+1}{2}} \\
&\leqslant -2\beta\Big(\sum_{s=1}^{m} c_s\int_{t-\tau_s}^{t} z^T(h)P_s z(h)dh + \sum_{i=1}^{N} z_i^T(t)Mz_i(t)\Big)^{\frac{\alpha+1}{2}} \\
&= -2\beta V_2^{\frac{\alpha+1}{2}}(t),
\end{aligned}$$

where $u(t) = (u_1^T(t), u_2^T(t), \ldots, u_N^T(t))^T$, $y(t) = (y_1^T(t), y_2^T(t), \ldots, y_N^T(t))^T$, $\xi(t) = (z^T(t), u^T(t))^T$.

Then, we have

$$u^T(t)y(t) \geqslant \dot{V}_2(t) + 2\beta V_2^{\frac{\alpha+1}{2}}(t).$$

Consequently, the network (6.19) is finite-time passive under the controller (6.20).

Theorem 6.13 *The network* (6.19) *is finite-time input strictly passive under the controller* (6.20) *if there exist matrices* $0 < M = \text{diag}(M_1, M_2, \ldots, M_n) \in \mathbb{R}^{n\times n}$, $K = \text{diag}(K_1, K_2, \ldots, K_N) \in \mathbb{R}^{nN\times nN}$, $0 < P_s = \text{diag}(P_1^s, P_2^s, \ldots, P_N^s) \in \mathbb{R}^{nN\times nN}$, $s = 1, 2, \ldots, m$, *and a positive real number* r_3 *such that*

$$\begin{pmatrix} W_1 & \Xi_1 \\ \Xi_1^T & r_3 I_{nN} - I_N \otimes \frac{B_2+B_2^T}{2} \end{pmatrix} \leqslant 0,$$

where W_1, Ξ_1 *have the same meanings as in Theorem 6.12.*

Proof Select the same $V_2(t)$ as (6.22) for network (6.19). By (6.24), we get

$$\begin{aligned}
&\dot{V}_2(t) - u^T(t)y(t) + r_3 u^T(t)u(t) \\
\leqslant\ & \xi^T(t) \begin{pmatrix} W_1 & \Xi_1 \\ \Xi_1^T & r_3 I_{nN} - I_N \otimes \frac{B_2+B_2^T}{2} \end{pmatrix} \xi(t) \\
& -2\beta \sum_{s=1}^{m} \left(c_s \int_{t-\tau_s}^{t} z^T(h) P_s z(h) dh \right)^{\frac{\alpha+1}{2}} \\
& -2\beta \left(\sum_{i=1}^{N} z_i^T(t) M z_i(t) \right)^{\frac{\alpha+1}{2}} \\
\leqslant\ & -2\beta \left(\sum_{s=1}^{m} c_s \int_{t-\tau_s}^{t} z^T(h) P_s z(h) dh + \sum_{i=1}^{N} z_i^T(t) M z_i(t) \right)^{\frac{\alpha+1}{2}} \\
=\ & -2\beta V_2^{\frac{\alpha+1}{2}}(t).
\end{aligned}$$

Then, one has

$$u^T(t)y(t) - r_3 u^T(t)u(t) \geqslant \dot{V}_2(t) + 2\beta V_2^{\frac{\alpha+1}{2}}(t).$$

Therefore, the network (6.19) is finite-time input strictly passive under the controller (6.20).

Theorem 6.14 *The network* (6.19) *is finite-time output strictly passive under the controller* (6.20) *if there exist matrices* $0 < M = \mathrm{diag}(M_1, M_2, \ldots, M_n) \in \mathbb{R}^{n \times n}$, $K = \mathrm{diag}(K_1, K_2, \ldots, K_N) \in \mathbb{R}^{nN \times nN}$, $0 < P_s = \mathrm{diag}(P_1^s, P_2^s, \ldots, P_N^s) \in \mathbb{R}^{nN \times nN}$, $s = 1, 2, \ldots, m$, *and a positive real number* r_4 *such that*

$$\begin{pmatrix} W_2 & \Xi_2 \\ \Xi_2^T & W_3 \end{pmatrix} \leqslant 0,$$

where $W_2 = W_1 + r_4(I_N \otimes B_1^T)(I_N \otimes B_1)$, $\Xi_2 = \Xi_1 + r_4(I_N \otimes B_1^T)(I_N \otimes B_2)$, $W_3 = r_4(I_N \otimes B_2^T)(I_N \otimes B_2) - I_N \otimes \frac{B_2+B_2^T}{2}$, W_1, Ξ_1 *have the same meanings as in Theorem 6.12.*

Proof Construct the same $V_2(t)$ as (6.22) for network (6.19). By (6.24), one obtains

$$\dot{V}_2(t) - u^T(t)y(t) + r_4 y^T(t)y(t)$$

$$
\begin{aligned}
&\leqslant \xi^T(t)\begin{pmatrix} W_2 & \Xi_2 \\ \Xi_2^T & W_3 \end{pmatrix}\xi(t) \\
&\quad -2\beta\sum_{s=1}^{m}\Big(c_s\int\limits_{t-\tau_s}^{t} z^T(h)P_s z(h)dh\Big)^{\frac{\alpha+1}{2}} \\
&\quad -2\beta\Big(\sum_{i=1}^{N} z_i^T(t)Mz_i(t)\Big)^{\frac{\alpha+1}{2}} \\
&\leqslant -2\beta\Big(\sum_{s=1}^{m} c_s\int\limits_{t-\tau_s}^{t} z^T(h)P_s z(h)dh + \sum_{i=1}^{N} z_i^T(t)Mz_i(t)\Big)^{\frac{\alpha+1}{2}} \\
&= -2\beta V_2^{\frac{\alpha+1}{2}}(t).
\end{aligned}
$$

Then, we have

$$
u^T(t)y(t) - r_4 y^T(t)y(t) \geqslant \dot{V}_2(t) + 2\beta V_2^{\frac{\alpha+1}{2}}(t).
$$

Therefore, the network (6.19) is finite-time output strictly passive under the controller (6.20).

6.4.3 FTS of Finite-Time Passive MWCNNs

Theorem 6.15 *Suppose that $\bar{V}(t) : [0, +\infty) \to [0, +\infty)$ is continuously differentiable and satisfies the following condition:*

$$
\upsilon_2(\|z(t)\|) \leqslant \bar{V}(t),
$$

where $\upsilon_2 : [0, +\infty) \to [0, +\infty)$ is continuous and strictly monotonically increasing function, $\upsilon_2(s)$ is positive for $s > 0$ with $\upsilon_2(0) = 0$. If network (6.19) is finite-time passive (finite-time input strictly passive, finite-time output strictly passive) with respect to $\bar{V}(t)$, MWCNNs (6.17) is finite-time synchronized under controller (6.20).

Proof If network (6.19) is finite-time passive with respect to $\bar{V}(t)$ under controller (6.20), there exist $\alpha \in (0, 1)$ and $\mathbb{R} \ni \beta > 0$ such that

$$
u^T(t)y(t) \geqslant \dot{\bar{V}}(t) + \beta\bar{V}^{\alpha}(t).
$$

Letting $u(t) = 0$, we have

$$\dot{\bar{V}}(t) \leqslant -\beta \bar{V}^{\alpha}(t).$$

By Lemma 1.6, we can obtain $\bar{V}(t) = 0$ for $t \geqslant T_2$, where $T_2 = \frac{\bar{V}^{1-\alpha}(0)}{\beta(1-\alpha)}$. On the other hand, since

$$\upsilon_2(\|z(t)\|) \leqslant \bar{V}(t),$$

one has

$$\upsilon_2(\|z(t)\|) \leqslant \bar{V}(t) = 0$$

for $t \geqslant T_2$. Then, we can conclude that $\|z_i(t)\| = 0, t \geqslant T_2$. Namely, MWCNNs (6.17) is finite-time synchronized under the controller (6.20).

Similarly, we can easily prove that MWCNNs (6.17) is finite-time synchronized under the controller (6.20) if network (6.19) is finite-time input strictly passive or finite-time output strictly passive.

According to Theorems 6.7 and 6.10, we can derive the following conclusion.

Corollary 6.16 *If there exist matrices* $0 < M = \text{diag}(M_1, M_2, \dots, M_n) \in \mathbb{R}^{n\times n}$, $K=\text{diag}(K_1, K_2, \dots, K_N) \in \mathbb{R}^{nN\times nN}, 0 < P_s = \text{diag}(P_1^s, P_2^s, \dots, P_N^s) \in \mathbb{R}^{nN\times nN}$, $s = 1, 2, \dots, m$, $B_1 \in \mathbb{R}^{n\times n}$, *and* $B_2 \in \mathbb{R}^{n\times n}$ *such that*

$$\begin{pmatrix} W_1 & \Xi_1 \\ \Xi_1^T & -I_N \otimes \frac{B_2+B_2^T}{2} \end{pmatrix} \leqslant 0,$$

where W_1, Ξ_1 *have the same meanings as in Theorem 6.12, then MWCNNs* (6.17) *is finite-time synchronized under the controller* (6.20).

Remark 6.17 In the last few years, synchronization [22, 23, 25] of CNNs has been extensively investigated due to the fact that it has been found fruitful applications in various fields such as secure communication, brain science, and so on. Obviously, in many circumstances, synchronization is realized over a finite time interval may be more reasonable. Therefore, it is very important to study the FTS of CNNs. Nevertheless, very few researchers have discussed the FTS problem for CNNs. In this chapter, some sufficient conditions to guarantee the FTS of CNNs with constant and delayed couplings are presented based on the FTP.

6.5 Numerical Examples

Example 6.18 Consider the following MWCNNs:

$$\begin{aligned}\dot{x}_i(t) = &-Ax_i(t) + Dg(x_i(t)) + J + c_1 \sum_{j=1}^{6} G^1_{ij}\Gamma^1 x_j(t) \\ &+c_2 \sum_{j=1}^{6} G^2_{ij}\Gamma^2 x_j(t) + c_3 \sum_{j=1}^{6} G^3_{ij}\Gamma^3 x_j(t) \\ &+u_i(t) + v_i(t), i = 1, 2, \ldots, 6, \end{aligned} \tag{6.25}$$

where $g_k(\varphi) = \frac{1}{4}(|\varphi+1| - |\varphi-1|)$, $k = 1, 2, 3$, A=diag$(0.5, 0.4, 0.3)$, $c_1 = 0.1$, $c_2 = 0.3$, $c_3 = 0.2$, $\Gamma^1 = \mathrm{diag}(0.8, 0.6, 0.5)$, $\Gamma^2 = \mathrm{diag}(0.7, 0.5, 0.6)$, $\Gamma^3 = \mathrm{diag}$ $(0.8, 0.6, 0.5)$, J=$(0, 0, 0)^T$, and the matrices D, G^1, G^2, G^3 are chosen as, respectively,

$$D = \begin{pmatrix} 0.3 & 0.2 & 0.2 \\ 0.2 & 0.2 & 0.3 \\ 0.2 & 0.3 & 0.4 \end{pmatrix},$$

$$G^1 = \begin{pmatrix} -0.5 & 0.1 & 0 & 0.1 & 0.2 & 0.1 \\ 0.1 & -0.5 & 0.1 & 0.2 & 0.1 & 0 \\ 0 & 0.1 & -0.6 & 0 & 0.3 & 0.2 \\ 0.1 & 0.2 & 0 & -0.6 & 0.2 & 0.1 \\ 0.2 & 0.1 & 0.3 & 0.2 & -0.9 & 0.1 \\ 0.1 & 0 & 0.2 & 0.1 & 0.1 & -0.5 \end{pmatrix},$$

$$G^2 = \begin{pmatrix} -0.7 & 0.3 & 0 & 0.2 & 0.1 & 0.1 \\ 0.3 & -0.7 & 0.2 & 0.1 & 0.1 & 0 \\ 0 & 0.2 & -0.9 & 0 & 0.4 & 0.3 \\ 0.2 & 0.1 & 0 & -0.6 & 0.2 & 0.1 \\ 0.1 & 0.1 & 0.4 & 0.2 & -0.9 & 0.1 \\ 0.1 & 0 & 0.3 & 0.1 & 0.1 & -0.6 \end{pmatrix},$$

$$G^3 = \begin{pmatrix} -0.6 & 0.2 & 0 & 0.2 & 0.1 & 0.1 \\ 0.2 & -0.8 & 0.3 & 0.1 & 0.2 & 0 \\ 0 & 0.3 & -0.9 & 0 & 0.4 & 0.2 \\ 0.2 & 0.1 & 0 & -0.7 & 0.3 & 0.1 \\ 0.1 & 0.2 & 0.4 & 0.3 & -1.1 & 0.1 \\ 0.1 & 0 & 0.2 & 0.1 & 0.1 & -0.5 \end{pmatrix}.$$

Obviously, function $g_k(\cdot)$ satisfies Lipschitz condition with $\rho_k = 0.5$. Choose $B_2 = \mathrm{diag}(0.4, 0.5, 0.6)$ and

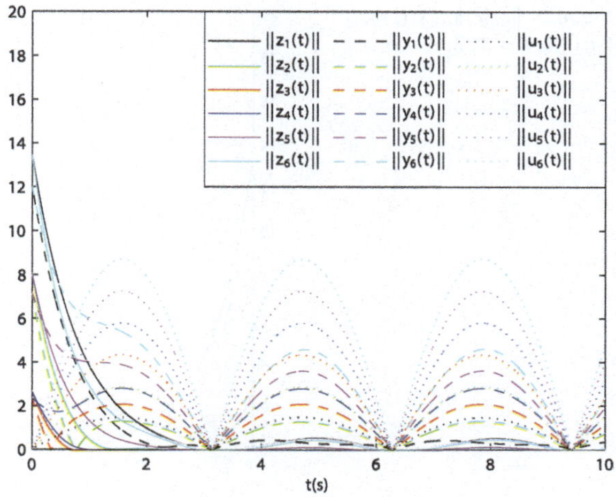

Fig. 6.1 $\|z_i(t)\|$, $\|y_i(t)\|$, $\|u_i(t)\|$, $i = 1, 2, \ldots, 6$

$$B_1 = \begin{pmatrix} 0.4 & 0.2 & 0.3 \\ 0.5 & 0.3 & 0.3 \\ 0.2 & 0.2 & 0.4 \end{pmatrix}.$$

Case 1: Take $K = \mathrm{diag}(0.2I_3, 0.7I_3, 0.3I_3, 0.4I_3, 0.5I_3, 0.6I_3)$. Then, we can easily find the following matrix $M = \mathrm{diag}(0.3756, 0.3688, 0.4236)$ satisfying the condition of Theorem 6.7. From Theorem 6.7, the network is finite-time passive under the controller (6.4). The simulation results are shown in Fig. 6.1.

Case 2: Take $K = \mathrm{diag}(0.3I_3, 0.5I_3, 0.2I_3, 0.7I_3, 0.4I_3, 0.8I_3)$. Then, we can easily find the following parameters $r_1 = 0.0200$, $M = \mathrm{diag}(0.4876, 0.4050, 0.5445)$ satisfying the condition of Theorem 6.8. Therefore, the network is finite-time input strictly passive under the controller (6.4). The simulation results are shown in Fig. 6.2.

Case 3: Take $K = \mathrm{diag}(0.7I_3, 0.5I_3, 0.3I_3, 0.2I_3, 0.9I_3, 0.6I_3)$. Then, we can easily find the following parameters $r_2 = 0.0383$, $M = \mathrm{diag}(0.5129, 0.4403, 0.5559)$ satisfying the condition of Theorem 6.9. By Theorem 6.9, the network is finite-time output strictly passive under the controller (6.4). The simulation results are shown in Fig. 6.3.

Case 4: Take $K = \mathrm{diag}(0.2I_3, 0.7I_3, 0.3I_3, 0.4I_3, 0.5I_3, 0.6I_3)$. Then, we can easily find the following matrix $M = \mathrm{diag}(0.3756, 0.3688, 0.4236)$ satisfying the condition of Corollary 6.11. According to Corollary 6.11, the network (6.25) is finite-time synchronized under the controller (6.4). The simulation results are shown in Fig. 6.4.

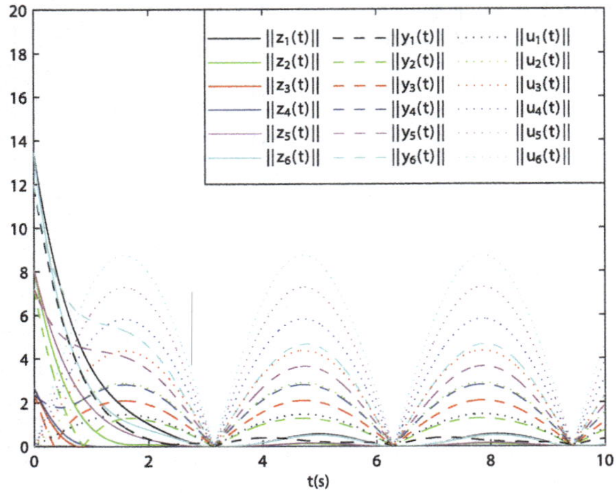

Fig. 6.2 $\|z_i(t)\|$, $\|y_i(t)\|$, $\|u_i(t)\|$, $i = 1, 2, \ldots, 6$

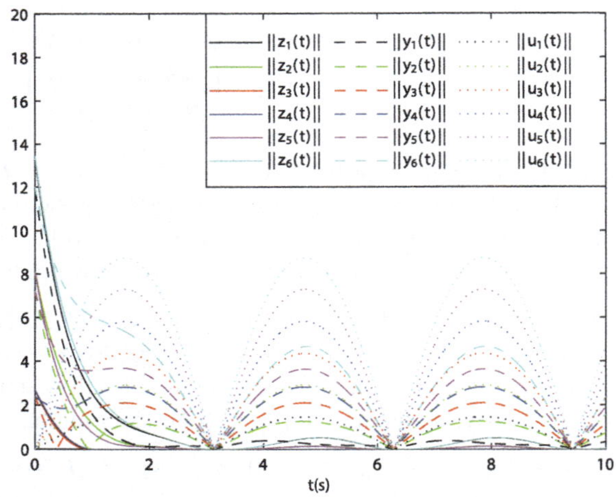

Fig. 6.3 $\|z_i(t)\|$, $\|y_i(t)\|$, $\|u_i(t)\|$, $i = 1, 2, \ldots, 6$

Example 6.19 Consider the following MWCNNs with coupling delays:

$$\begin{aligned}\dot{x}_i(t) = &-Ax_i(t) + Dg(x_i(t)) + J + c_1 \sum_{j=1}^{6} G_{ij}^1 \Gamma^1 x_j(t - \tau_1) \\ &+ c_2 \sum_{j=1}^{6} G_{ij}^2 \Gamma^2 x_j(t - \tau_2) + u_i(t) \\ &+ c_3 \sum_{j=1}^{6} G_{ij}^3 \Gamma^3 x_j(t - \tau_3) + v_i(t), \qquad (6.26)\end{aligned}$$

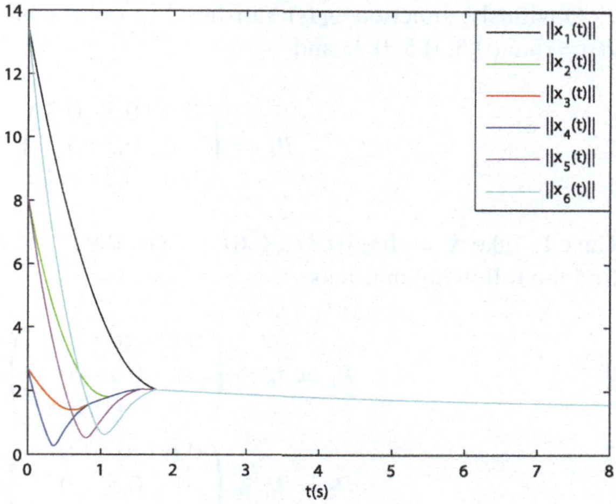

Fig. 6.4 $\|x_i(t)\|, i = 1, 2, \ldots, 6$

where i=1, 2, ..., 6, $g_k(\varphi) = \frac{1}{4}(|\varphi + 1| - |\varphi - 1|)$, k=1, 2, 3, $c_1 = 0.2$, c_2=0.1, $c_3 = 0.3$, $A = \text{diag}(0.8, 0.8, 0.6)$, Γ^1=diag(0.5, 0.4, 0.6), Γ^2=diag(0.6, 0.4, 0.6), $\Gamma^3 = \text{diag}(0.8, 0.6, 0.5)$, $J = (0, 0, 0)^T$, $\tau_1 = 0.2$, $\tau_2 = 0.3$, $\tau_3 = 0.4$ and the matrices D, G^1, G^2, G^3 are chosen as, respectively,

$$D = \begin{pmatrix} 0.3 & 0.2 & 0.2 \\ 0.2 & 0.2 & 0.3 \\ 0.2 & 0.3 & 0.4 \end{pmatrix},$$

$$G^1 = \begin{pmatrix} -0.6 & 0.2 & 0 & 0.1 & 0.2 & 0.1 \\ 0.2 & -0.5 & 0.1 & 0.1 & 0.1 & 0 \\ 0 & 0.1 & -0.7 & 0 & 0.4 & 0.2 \\ 0.1 & 0.1 & 0 & -0.5 & 0.2 & 0.1 \\ 0.2 & 0.1 & 0.4 & 0.2 & -1.0 & 0.1 \\ 0.1 & 0 & 0.2 & 0.1 & 0.1 & -0.5 \end{pmatrix},$$

$$G^2 = \begin{pmatrix} -0.7 & 0.3 & 0 & 0.2 & 0.1 & 0.1 \\ 0.3 & -0.7 & 0.2 & 0.1 & 0.1 & 0 \\ 0 & 0.2 & -0.9 & 0 & 0.4 & 0.3 \\ 0.2 & 0.1 & 0 & -0.6 & 0.2 & 0.1 \\ 0.1 & 0.1 & 0.4 & 0.2 & -0.9 & 0.1 \\ 0.1 & 0 & 0.3 & 0.1 & 0.1 & -0.6 \end{pmatrix},$$

$$G^3 = \begin{pmatrix} -0.6 & 0.2 & 0 & 0.2 & 0.1 & 0.1 \\ 0.2 & -0.8 & 0.3 & 0.1 & 0.2 & 0 \\ 0 & 0.3 & -0.9 & 0 & 0.4 & 0.2 \\ 0.2 & 0.1 & 0 & -0.7 & 0.3 & 0.1 \\ 0.1 & 0.2 & 0.4 & 0.3 & -1.1 & 0.1 \\ 0.1 & 0 & 0.2 & 0.1 & 0.1 & -0.5 \end{pmatrix}.$$

Obviously, function $g_k(\cdot)$ satisfies Lipschitz condition with $\rho_k = 0.5$. Choose $B_2 = \text{diag}(0.5, 0.5, 0.4)$ and

$$B_1 = \begin{pmatrix} 0.4 & 0.5 & 0.3 \\ 0.2 & 0.3 & 0.3 \\ 0.3 & 0.2 & 0.4 \end{pmatrix}.$$

Case 1: Take $K = \text{diag}(0.7I_3, 0.6I_3, 0.8I_3, 0.9I_3, 0.8I_3, 0.7I_3)$. Then, we can easily find the following matrices:

$$P_1 = I_6 \otimes \begin{pmatrix} 0.6 & 0 & 0 \\ 0 & 0.7 & 0 \\ 0 & 0 & 0.6 \end{pmatrix},$$

$$P_2 = I_6 \otimes \begin{pmatrix} 0.6 & 0 & 0 \\ 0 & 0.5 & 0 \\ 0 & 0 & 0.2 \end{pmatrix},$$

$$P_3 = I_6 \otimes \begin{pmatrix} 0.5 & 0 & 0 \\ 0 & 0.4 & 0 \\ 0 & 0 & 0.5 \end{pmatrix},$$

$$M = \begin{pmatrix} 0.6343 & 0 & 0 \\ 0 & 0.6797 & 0 \\ 0 & 0 & 0.5380 \end{pmatrix}$$

satisfying the condition of Theorem 6.12. From Theorem 6.12, the network is finite-time passive under the controller (6.20). The simulation results are shown in Fig. 6.5.

Case 2: Take $K = \text{diag}(0.5I_3, 0.6I_3, 0.7I_3, 0.8I_3, 0.6I_3, 0.7I_3)$. Then, we can easily find the following matrices:

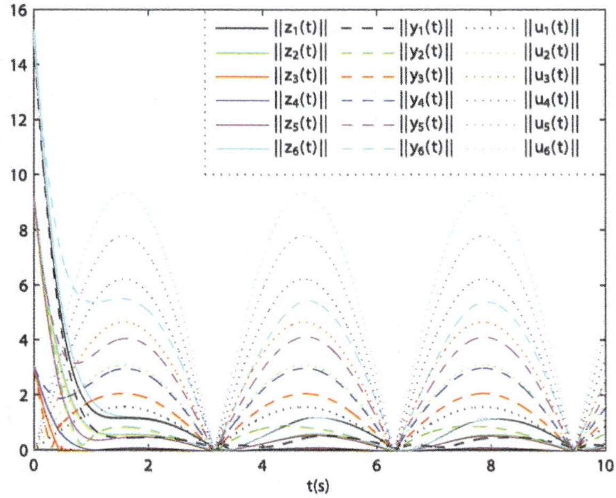

Fig. 6.5 $\|z_i(t)\|$, $\|y_i(t)\|$, $\|u_i(t)\|$, $i = 1, 2, \ldots, 6$

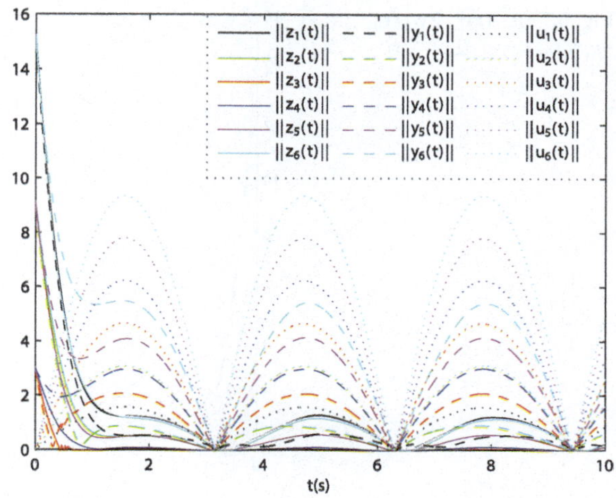

Fig. 6.6 $\|z_i(t)\|$, $\|y_i(t)\|$, $\|u_i(t)\|$, $i = 1, 2, \ldots, 6$

$$P_1 = I_6 \otimes \begin{pmatrix} 0.5 & 0 & 0 \\ 0 & 0.4 & 0 \\ 0 & 0 & 0.6 \end{pmatrix},$$

$$P_2 = I_6 \otimes \begin{pmatrix} 0.3 & 0 & 0 \\ 0 & 0.5 & 0 \\ 0 & 0 & 0.3 \end{pmatrix},$$

$$P_3 = I_6 \otimes \begin{pmatrix} 0.4 & 0 & 0 \\ 0 & 0.6 & 0 \\ 0 & 0 & 0.5 \end{pmatrix},$$

$$M = \begin{pmatrix} 0.5194 & 0 & 0 \\ 0 & 0.6651 & 0 \\ 0 & 0 & 0.4950 \end{pmatrix},$$

and $r_3 = 0.0119$ satisfying the condition of Theorem 6.13. Therefore, the network is finite-time input strictly passive under the controller (6.20). The simulation results are shown in Fig. 6.6.

Case 3: Take $K = \mathrm{diag}(0.6I_3, 0.8I_3, 0.7I_3, 0.9I_3, 0.7I_3, 0.8I_3)$. Then, we can easily find the following matrices:

$$P_1 = I_6 \otimes \begin{pmatrix} 0.6 & 0 & 0 \\ 0 & 0.8 & 0 \\ 0 & 0 & 0.7 \end{pmatrix},$$

$$P_2 = I_6 \otimes \begin{pmatrix} 0.7 & 0 & 0 \\ 0 & 0.5 & 0 \\ 0 & 0 & 0.6 \end{pmatrix},$$

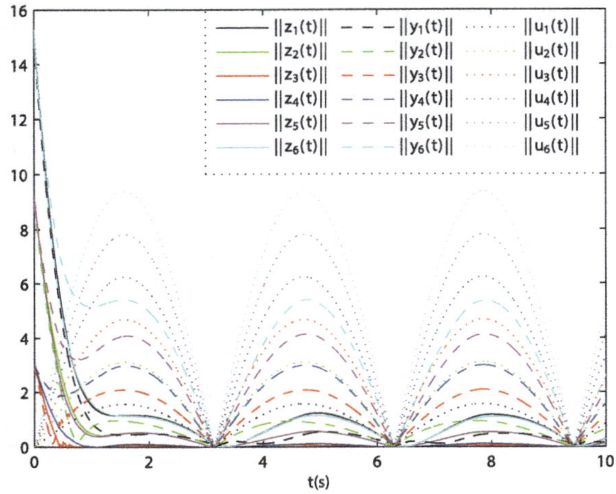

Fig. 6.7 $\|z_i(t)\|$, $\|y_i(t)\|$, $\|u_i(t)\|$, $i = 1, 2, \ldots, 6$

$$P_3 = I_6 \otimes \begin{pmatrix} 0.8 & 0 & 0 \\ 0 & 0.7 & 0 \\ 0 & 0 & 0.7 \end{pmatrix},$$

$$M = \begin{pmatrix} 0.6618 & 0 & 0 \\ 0 & 0.7312 & 0 \\ 0 & 0 & 0.5635 \end{pmatrix},$$

and $r_4 = 0.0306$ satisfying the condition of Theorem 6.14. By Theorem 6.14, the network is finite-time output strictly passive under the controller (6.20). The simulation results are shown in Fig. 6.7.

Case 4: Take $K = \text{diag}(0.7I_3, 0.6I_3, 0.8I_3, 0.9I_3, 0.8I_3, 0.7I_3)$. Then, we can easily find the following matrices:

$$P_1 = I_6 \otimes \begin{pmatrix} 0.6 & 0 & 0 \\ 0 & 0.7 & 0 \\ 0 & 0 & 0.6 \end{pmatrix},$$

$$P_2 = I_6 \otimes \begin{pmatrix} 0.6 & 0 & 0 \\ 0 & 0.5 & 0 \\ 0 & 0 & 0.2 \end{pmatrix},$$

$$P_3 = I_6 \otimes \begin{pmatrix} 0.5 & 0 & 0 \\ 0 & 0.4 & 0 \\ 0 & 0 & 0.5 \end{pmatrix},$$

$$M = \begin{pmatrix} 0.6343 & 0 & 0 \\ 0 & 0.6797 & 0 \\ 0 & 0 & 0.5380 \end{pmatrix}$$

Fig. 6.8 $\|x_i(t)\|, i = 1, 2, \ldots, 6$

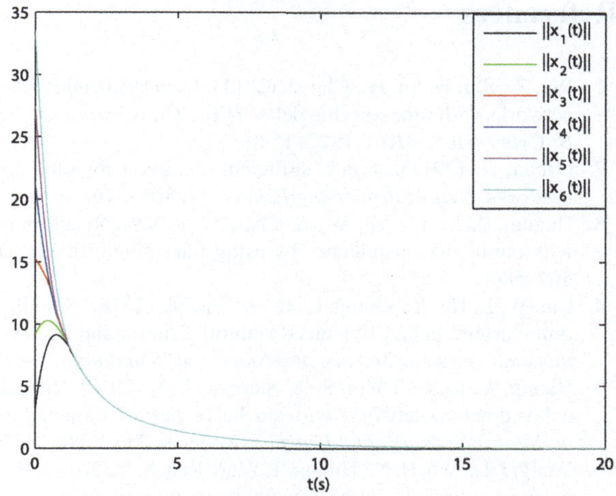

satisfying the condition of Corollary 6.16. According to Corollary 6.16, the network (6.26) is finite-time synchronized under the controller (6.20). The simulation results are shown in Fig. 6.8.

6.6 Conclusion

In this chapter, two kinds of MWCNNs with and without coupling delays have been proposed. On the one hand, we have analyzed the FTP of the proposed network models, and some sufficient conditions have been established. On the other hand, several FTS criteria for finite-time passive MWCNNs with and without coupling delays have been presented. Two numerical examples have also been given to show the effectiveness of the obtained FTP and FTS criteria.

In recent years, studies have been carried out on the consensus or synchronization control problem of multiple autonomous mobile agents, which not only can help us understand some cooperative phenomena in biology (such as fish schooling, bird flocking, and ant swarming) but also solve many coordination problems in engineering applications. Considering that the passivity properties of systems can keep the systems internally stable, the authors[28–30, 32] studied infinite-time consensus or infinite-time synchronization of multi-agent systems by employing the passivity theory. According to Theorems 6.10 and 6.15, finite-time passive MWCNNs is finite-time synchronized. It will provide the foundation for designing the appropriate control inputs for finite-time consensus or FTS in multi-agent systems. Obviously, this is an important and interesting topic and will become our future investigative direction.

References

1. Wu, Z., Shi, P., Su, H., Chu, J. (2011). Delay-dependent stability analysis for switched neural networks with time-varying delay. *IEEE Transactions on Systems, Man, and Cybernetics, Part B: Cybernetics*, *41*(6), 1522–1530.
2. Ozcan, N. (2011). A new sufficient condition for global robust stability of delayed neural networks. *Neural Processing Letters*, *34*, 305–316.
3. Huang, T., Li, C., Yu, W., & Chen, G. (2009). Synchronization of delayed chaotic systems with parameter mismatches by using intermittent linear state feedback. *Nonlinearity*, *22*(3), 569–584.
4. Lin, W. J., He, Y., Zhang, C. K., & Wu, M. (2018). Stability analysis of neural networks with time-varying delay: Enhanced stability criteria and conservatism comparisons. *Communications in Nonlinear Science and Numerical Simulation*, *54*(1), 118–135.
5. Huang, T., Li, C., Duan, S., & Starzyk, J. A. (2012). Robust exponential stability of uncertain delayed neural networks with stochastic perturbation and impulse effects. *IEEE Transactions on Neural Networks and Learning Systems*, *23*(6), 866–875.
6. Wang, J. L., Wu, H. N., Huang, T. W., & Ren, S. Y. (2016). Pinning control strategies for synchronization of linearly coupled neural networks with reaction-diffusion terms. *IEEE Transactions on Neural Networks and Learning Systems*, *27*(4), 749–761.
7. Wang, J., & Wu, H. (2014). Synchronization and adaptive control of an array of linearly coupled reaction-diffusion neural networks with hybrid coupling. *IEEE Transactions on Cybernetics*, *44*(8), 1350–1361.
8. Wang, K., Teng, Z., & Jiang, H. (2012). Adaptive synchronization in an array of linearly coupled neural networks with reaction-diffusion terms and time delays. *Communications in Nonlinear Science and Numerical Simulation*, *17*(10), 3866–3875.
9. Liu, L., Chen, W. H., & Lu, X. (2017). Aperiodically intermittent $\mathcal{H}_\infty$ synchronization for a class of reaction-diffusion neural networks. *Neurocomputing*, *222*(26), 105–115.
10. Tu, Z., Ding, N., Li, L., Feng, Y., Zou, L., & Zhang, W. (2017). Adaptive synchronization of memristive neural networks with time-varying delays and reaction-diffusion term. *Applied Mathematics and Computation*, *311*(15), 118–128.
11. Lou, X., & Cui, B. (2007). Passivity analysis of integro-differential neural networks with time-varying delays. *Neurocomputing*, *70*(4), 1071–1078.
12. Li, H., Lam, J., & Cheng, K. C. (2012). Passivity criteria for continuous-time neural networks with mixed time-varying delays. *Applied Mathematics and Computation*, *218*(22), 11062–11074.
13. Song, Q., Liang, J., & Wang, Z. (2009). Passivity analysis of discrete-time stochastic neural networks with time-varying delays. *Neurocomputing*, *72*(7–9), 1782–1788.
14. Yao, J., Wang, H. O., Guan, Z. H., & Xu, W. (2009). Passive stability and synchronization of complex spatio-temporal switching networks with time delays. *Automatica*, *45*(7), 1721–1728.
15. Chen, Y., Fu, Z., Liu, Y., & Alsaadi, F. E. (2017). Further results on passivity analysis of delayed neural networks with leakage delay. *Neurocomputing*, *224*(8), 135–141.
16. Zeng, H., He, Y., Wu, M., & Xiao, H. (2014). Improved conditions for passivity of neural networks with a time-varying delay. *IEEE Transactions on Cybernetics*, *44*(6), 785–792.
17. Lian, J., & Wang, J. (2015). Passivity of switched recurrent neural networks with time-varying delays. *IEEE Transactions on Neural Networks and Learning Systems*, *26*(2), 357–366.
18. Zhang, J., Ma, L., & Liu, Y. (2016). Passivity analysis for discrete-time neural networks with mixed time-delays and randomly occurring quantization effects. *Neurocomputing*, *216*, 657–665.
19. Li, Y., Zhong, S., Cheng, J., Shi, K., & Ren, J. (2016) New passivity criteria for uncertain neural networks with time-varying delay. *Neurocomputing*, *171*(1), 1003–1012.
20. Wang, J., Wu, H., & Guo, L. (2011). Passivity and stability analysis of reaction-diffusion neural networks with Dirichlet boundary conditions. *IEEE Transactions on Neural Networks*, *22*(12), 2105–2116.

21. Li, N., & Cao, J. (2016). Passivity and robust synchronisation of switched interval coupled neural networks with time delay. *International Journal of Systems Science*, *47*(12), 2827–2836.
22. Wang, J. L., Wu, H. N., & Huang, T. (2015). Passivity-based synchronization of a class of complex dynamical networks with time-varying delay. *Automatica*, *56*, 105–112.
23. Wang, J. L., Wu, H. N., Huang, T., & Ren, S. Y. (2015). Passivity and synchronization of linearly coupled reaction-diffusion neural networks with adaptive coupling. *IEEE Transactions on Cybernetics*, *45*(9), 1942–1952.
24. Wang, J., Wu, H., Huang, T., Ren, S., & Wu, J. (2017). Passivity analysis of coupled reaction-diffusion neural networks with Dirichlet boundary conditions. *IEEE Transactions on Systems, Man, and Cybernetics: Systems*, *47*(8), 2148–2159.
25. Wang, J. L., Wu, H. N., Huang, T., Ren, S. Y., & Wu, J. (2017). Passivity of directed and undirected complex dynamical networks with adaptive coupling weights. *IEEE Transactions on Neural Networks and Learning Systems*, *28*(8), 1827–1839.
26. Zhao, Y. P., He, P., Nik, H. S., Ren, J. (2015). Robust adaptive synchronization of uncertain complex networks with multiple time-varying coupled delays. *Complexity*, *20*(6), 62–73.
27. Steur, E., & Nijmeijer, H. (2011). Synchronization in networks of diffusively time-delay coupled (semi-)passive systems. *IEEE Transactions on Circuits and Systems I: Regular Papers*, *58*(6), 1358–1371.
28. Liu, Y., & Zhao, J. (2012). Generalized output synchronization of dynamical networks using output quasi-passivity. *IEEE Transactions on Circuits and Systems I: Regular Papers*, *59*(6), 1290–1298.
29. Yu, H., & Antsaklis, P. J. (2014). Output synchronization of networked passive systems with event-driven communication. *IEEE Transactions on Automatic Control*, *59*(3), 750–756.
30. Chopra, N. (2012). Output synchronization on strongly connected graphs. *IEEE Transactions on Automatic Control*, *57*(11), 2896–2901.
31. Wang, J., Wu, H., Huang, T., Ren, S., & Wu, J. (2018). Passivity and output synchronization of complex dynamical networks with fixed and adaptive coupling strength. *IEEE Transactions on Neural Networks and Learning Systems*, *29*(2), 364–376.
32. Chopra, N., Spong, M. W. (2006). *Passivity-Based Control of Multi-Agent Systems*. Berlin: Springer.
33. Zhang, D., Shen, Y., & Mei, J. (2017). Finite-time synchronization of multi-layer nonlinear coupled complex networks via intermittent feedback control. *Neurocomputing*, *225*(15), 129–138.
34. Zhao, H., Li, L., Peng, H., Xiao, J., Yang, Y., Zheng, M., Li, S. (2017). Finite-time synchronization for multi-link complex networks via discontinuous control. *Optik 138*, 440–454.
35. Yang, C., & Huang, L. (2017). Finite-time synchronization of coupled time-delayed neural networks with discontinuous activations. *Neurocomputing*, *249*, 64–71.
36. Wu, Y., Cao, J., Li, Q., Alsaedi, A., & Alsaadi, F. E. (2017). Finite-time synchronization of uncertain coupled switched neural networks under asynchronous switching. *Neural Networks*, *85*, 128–139.
37. Wang, J., Zhang, H., Wang, Z., & Gao, D. W. (2017). Finite-time synchronization of coupled hierarchical hybrid neural networks with time-varying delays. *IEEE Transactions on Cybernetics*, *47*(10), 2995–3004.

Chapter 7
Passivity and Stability Analysis of RDNNs with Dirichlet Boundary Conditions

7.1 Introduction

Recently, DNNs have received considerable attention due to the wide applications in various areas such as image processing, signal processing, associative memory, pattern classification, optimization, and moving object speed detection. As we know, such applications heavily depend on the dynamical behaviors (e.g., stability, instability, bifurcation, periodic oscillatory behavior, and chaos), especially the asymptotic stability, of DNNs. Hence, study of the dynamical behaviors is a necessary step for practical design of NNs. In particular, the stability of DNNs has been one of the most active research area.

Stability problems are often linked to the theory of dissipative systems, which postulate that the energy dissipated inside a dynamic system is less than that supplied from external source. Passivity is part of a broader and a general theory of dissipativeness [1, 2]. The main point of passivity theory is that the passive properties of systems can keep the systems internally stable. The passivity theory was firstly proposed in the circuit analysis [3], and since then has found successful applications in diverse areas such as stability [4, 5], complexity [6], signal processing [7], chaos control and synchronization [8, 9], and fuzzy control [10]. In recent years, as a powerful tool, passivity has played an important role in group coordination, synchronization, network control, process control, energy management and so on [11–21]. These are the main reasons why passivity theory has become a very hot topic across many fields, and much investigative attention has been focused on this topic.

Although research on passivity has attracted so much attention, little of that had been devoted to the passivity properties of DNNs until Li and Liao [22] obtained the conditions for passivity of DNNs. Recently, many authors have studied the passivity of DNNs [23–33]. Several sufficient conditions on passivity were derived for various NNs such as time-invariant [27], time-varying [28, 31], discrete [25, 26, 30], uncertain [23, 24, 29, 30, 33], and stochastic [25, 30, 32] network models.

The original version of this book was revised: The belated corrections have been incorporated. The corrections to this book are available at https://doi.org/10.1007/978-981-33-4287-3_10

J.-L. Wang et al., *Passivity of Complex Dynamical Networks*,
https://doi.org/10.1007/978-981-33-4287-3_7

Unfortunately, in most existing works on the passivity of DNNs, the diffusion effects have not been considered. Strictly speaking, the diffusion phenomena could not be ignored in NNs and electric circuits once electrons transport in a nonuniform electromagnetic field [34–36]. Moreover, the input and output variables are also dependent on the time and space in many practical situations. To our knowledge, for this case, the passivity theory has not yet been established. Thus, it is important and interesting to study the passivity of RDNNs, in which the input and output variables are varied with the time and space variables. Chen and Lee [37] first developed input strict passivity to study both input-output and internal stability problems for feedback connections of discrete-time linear descriptor systems. By applying the input strict passivity and Lyapunov theory, Chang et al. [21] derived the relaxed stability conditions for continuous-time affine Takagi-Sugeno fuzzy models. In order to better study the dynamical behaviors of RDNNs, the input strict passivity of RDNNs is also discussed in this paper.

In NNs, besides the diffusion phenomena, there might also be some uncertainties due to the existence of modeling errors, external disturbance, and parameter fluctuation, which might lead to undesirable dynamic network behaviors such as oscillation and instability. Therefore, it is important to study the robust stability and robust passivity of RDNNs against these uncertainties. To our knowledge, there are very few works on the robust stability of the RDNNs [38–40], and the robust passivity of RDNNs has not yet been established.

Motivated by the above discussions, the objective of this chapter is to investigate the passivity and stability of RDNNs. The main contributions of this chapter is stated as follows:

(i) We first give the passivity definition for the case where input and output variables are varied with the time and space variables. Several sufficient conditions are established to guarantee the passivity of the RDNNs.
(ii) To our knowledge, in most existing works on the stability of RDNNs [34, 41], they always assume that the input variable is only dependent on the time. In this chapter, we remove this restriction and consider the global exponential stability of the RDNNs. Moreover, the existence and uniqueness of the periodic solution are also discussed.
(iii) A class of interval RDNNs (IRDNNs), which contain uncertain parameters whose values are unknown but bounded, is considered. By utilizing the Lyapunov functional method combined with the inequality techniques, several sufficient conditions are presented, ensuring the robust passivity and robust global exponential stability of IRDNNs.

7.2 Networks Model and Preliminaries

In this chapter, we consider the following RDNN with time-varying delays:

$$\begin{cases} \frac{\partial w_i(x,t)}{\partial t} = \sum\limits_{k=1}^{m} \frac{\partial}{\partial x_k}\left(a_{ik}\frac{\partial w_i(x,t)}{\partial x_k}\right) - c_i w_i(x,t) + \sum\limits_{j=1}^{n} d_{ij} f_j(w_j(x,t)) \\ \qquad +u_i(x,t) + \sum\limits_{j=1}^{n} e_{ij} f_j(w_j(x,t-\tau_j(t))), \\ y_i(x,t) = h_i w_i(x,t) + g_i u_i(x,t), \end{cases} \tag{7.1}$$

where $i = 1, 2, \ldots, n$, n is the number of neurons in the networks; $x = (x_1, x_2, \ldots, x_\sigma)^T \in \Omega \subset \mathbb{R}^\sigma$ and $\Omega = \{x = (x_1, x_2, \ldots, x_\sigma)^T \mid |x_k| < l_k, k = 1, 2, \ldots, \sigma\}$ is a bounded compact set with smooth boundary $\partial\Omega$ and $mes\Omega > 0$ in space $\mathbb{R}^\sigma$; $w_i(x,t) \in \mathbb{R}$ is the state of ith neuron at time t and in space x; $\mathbb{R} \ni a_{ik} > 0$ represents the transmission diffusion coefficient along the ith neuron; $f_j(\cdot)$ denotes the activation function of jth neuron; $\mathbb{R} \ni c_i > 0$ represents the rate with which the ith neuron will reset its potential to the resting state when disconnected from the networks and external inputs in space x; $u_i(x,t) \in \mathbb{R}$ and $y_i(x,t) \in \mathbb{R}$ denote input and output of ith neuron at time t and in space x, respectively; $d_{ij} \in \mathbb{R}$ and $e_{ij} \in \mathbb{R}$ stand for the weights of neuron interconnections; h_i and g_i are known constants; $\tau_j(t) \in \mathbb{R}$ is the transmission delay and satisfies $0 \leqslant \tau_j(t) \leqslant \tau$.

Throughout this chapter, we make the following assumptions:

(A1) The neuron activation function $f_j(\cdot)(j = 1, 2, \ldots, n)$ satisfies

$$0 \leqslant \frac{f_j(\xi_1) - f_j(\xi_2)}{\xi_1 - \xi_2} \leqslant F_j$$

for any $\xi_1, \xi_2 \in \mathbb{R}$, $\xi_1 \neq \xi_2$, where $F_j > 0$. Further, $f_j(0) = 0$.

(A2) $\dot{\tau}_j(t) \leqslant \delta < 1 (\delta \geqslant 0)$, $j = 1, 2, \ldots, n$.

The initial value and boundary value conditions associated with the RDNN (7.1) are given in form

$$w_i(x,t) = \varphi_i(x,t), \ (x,t) \in \Omega \times [-\tau, 0], \tag{7.2}$$
$$w_i(x,t) = 0, \ (x,t) \in \partial\Omega \times [-\tau, +\infty), \tag{7.3}$$

where $i = 1, 2, \ldots, n$, $\varphi_i(x,t)$ is bounded and continuous on $\Omega \times [-\tau, 0]$. Let $w(x,t,\Psi) = (w_1(x,t,\Psi), \ w_2(x,t,\Psi), \ldots, w_n(x,t,\Psi))^T$ be the state trajectory of RDNN (7.1) with initial condition $\Psi(x,t) = (\varphi_1(x,t), \varphi_2(x,t), \ldots, \varphi_n(x,t))^T$.

To begin with, we introduce some definitions.

Definition 7.1 System (7.1) is said to be globally exponentially stable if there exist constants $\epsilon > 0$ and $M \geqslant 1$ such that for any two solutions $w(x, t, \Phi)$, $w(x, t, \Psi)$ of RDNN (7.1) with initial functions Φ, Ψ, respectively, it holds that

$$\|w(\cdot, t, \Psi) - w(\cdot, t, \Phi)\|_2 \leqslant M\|\Psi(\cdot, 0) - \Phi(\cdot, 0)\|_\tau e^{-\epsilon t}$$

for all $t \geqslant 0$. If such $w(x, t, \Phi)$ is an equilibrium solution (or periodic solution), then this equilibrium solution (or periodic solution) is said to be globally exponentially stable.

Definition 7.2 A system with input $u(x, t)$ and output $y(x, t)$ where $u(x, t)$, $y(x, t)$ $\in \mathbb{R}^p$ is said to be passive if there is a $\beta \in \mathbb{R}$ such that

$$\int_0^{t_p}\int_\Omega y^T(x, s)u(x, s)dxds \geqslant -\beta^2$$

for all $t_p \geqslant 0$, where Ω is a bounded compact set. If in addition, there are constants $\gamma_1 \geqslant 0$ and $\gamma_2 \geqslant 0$ such that

$$\begin{aligned}\int_0^{t_p}\int_\Omega y^T(x, s)u(x, s)dxds \geqslant -\beta^2 + \gamma_1 \int_0^{t_p}\int_\Omega u^T(x, s)u(x, s)dxds \\ +\gamma_2 \int_0^{t_p}\int_\Omega y^T(x, s)y(x, s)dxds\end{aligned}$$

for all $t_p \geqslant 0$, then the system is input strictly passive if $\gamma_1 > 0$ and output strictly passive if $\gamma_2 > 0$.

Remark 7.3 Based on passivity property, [42, 43] addressed the passive stability, control, and synchronization of complex networks, in which the input and output variables are only dependent on the time. However, the input and output variables in many systems are varied with the time and space variables. To our knowledge, for this case, the passivity theory has not yet been established. As a natural extension of the definition of passivity in [21, 42–45], we propose the Definition 7.2.

In the existing works, there is another widely accepted definition of passivity. Correspondingly, we introduce the following concept of passivity.

Definition 7.4 A system with input $u(x, t)$ and output $y(x, t)$ where $u(x, t)$, $y(x, t)$ $\in \mathbb{R}^p$ is said to be passive if there are constants $\gamma \geqslant 0$ and β such that

$$2\int_0^{t_p}\int_\Omega y^T(x,s)u(x,s)dxds \geqslant -\beta^2 - \gamma\int_0^{t_p}\int_\Omega u^T(x,s)u(x,s)dxds$$

for all $t_p \geqslant 0$, where Ω is a bounded compact set.

Remark 7.5 It is obvious that the Definition 7.4 extends the definition of passivity in [22, 23, 27–29].

7.3 Passivity and Stability of RDNNs

7.3.1 Passivity Analysis

Theorem 7.6 *System* (7.1) *is input strictly passive in the sense of Definition 7.2, if there exist constants* $\lambda_i > 0$ *and* $\gamma > 0$ *such that*

$$\begin{pmatrix} \hat{M}_i & \lambda_i - h_i \\ \lambda_i - h_i & \gamma - 2g_i \end{pmatrix} \leqslant 0, \tag{7.4}$$

where $i = 1, 2, \ldots, n,\ \hat{M}_i = -2\lambda_i\sum_{k=1}^{\sigma}\frac{a_{ik}}{l_k^2} - 2c_i\lambda_i + \lambda_i\sum_{j=1}^{n}(|d_{ij}| + |e_{ij}|)F_j + \sum_{j=1}^{n}\lambda_j\left(|d_{ji}| + \frac{|e_{ji}|}{1-\delta}\right)F_i$.

Proof First, construct a Lyapunov functional for (7.1) as follows:

$$V(t) = \sum_{i=1}^{n}\lambda_i\Bigg\{\int_\Omega w_i^2(x,t)dx + \frac{1}{1-\delta}\sum_{j=1}^{n}\left[|e_{ij}|F_j\int_{t-\tau_j(t)}^{t}\int_\Omega w_j^2(x,s)dxds\right]\Bigg\}. \tag{7.5}$$

Then the upper right derivative $D^+V(t)$ of $V(t)$ is given by

$$\begin{aligned} D^+V(t) = \sum_{i=1}^{n}\lambda_i\Bigg\{&2\int_\Omega w_i(x,t)\frac{\partial w_i(x,t)}{\partial t}dx \\ &+\frac{1}{1-\delta}\sum_{j=1}^{n}|e_{ij}|F_j\int_\Omega w_j^2(x,t)dx \\ &-\frac{1}{1-\delta}\sum_{j=1}^{n}\left[|e_{ij}|(1-\dot{\tau}_j(t))F_j\int_\Omega w_j^2(x,t-\tau_j(t))dx\right]\Bigg\} \end{aligned}$$

$$
\begin{aligned}
&\leqslant \sum_{i=1}^{n} \lambda_i \left\{ 2 \int_{\Omega} w_i(x,t) \sum_{k=1}^{\sigma} \frac{\partial}{\partial x_k} \left(a_{ik} \frac{\partial w_i(x,t)}{\partial x_k} \right) dx \right. \\
&\quad -2c_i \int_{\Omega} w_i^2(x,t)dx + 2 \int_{\Omega} w_i(x,t)u_i(x,t)dx \\
&\quad +2 \int_{\Omega} w_i(x,t) \sum_{j=1}^{n} d_{ij} f_j(w_j(x,t))dx \\
&\quad +2 \int_{\Omega} w_i(x,t) \sum_{j=1}^{n} e_{ij} f_j(w_j(x,t-\tau_j(t)))dx \\
&\quad +\frac{1}{1-\delta} \sum_{j=1}^{n} |e_{ij}| F_j \int_{\Omega} w_j^2(x,t)dx \\
&\quad \left. - \sum_{j=1}^{n} \left[|e_{ij}| F_j \int_{\Omega} w_j^2(x,t-\tau_j(t))dx \right] \right\} \\
&\leqslant \sum_{i=1}^{n} \lambda_i \left\{ 2 \int_{\Omega} w_i(x,t) \sum_{k=1}^{\sigma} \frac{\partial}{\partial x_k} \left(a_{ik} \frac{\partial w_i(x,t)}{\partial x_k} \right) dx \right. \\
&\quad -2c_i \int_{\Omega} w_i^2(x,t)dx + 2 \int_{\Omega} w_i(x,t)u_i(x,t)dx \\
&\quad +2 \sum_{j=1}^{n} |d_{ij}| F_j \sqrt{\int_{\Omega} w_i^2(x,t)dx} \sqrt{\int_{\Omega} w_j^2(x,t)dx} \\
&\quad +2 \sum_{j=1}^{n} |e_{ij}| F_j \sqrt{\int_{\Omega} w_i^2(x,t)dx} \sqrt{\int_{\Omega} w_j^2(x,t-\tau_j(t))dx} \\
&\quad +\frac{1}{1-\delta} \sum_{j=1}^{n} |e_{ij}| F_j \int_{\Omega} w_j^2(x,t)dx \\
&\quad \left. - \sum_{j=1}^{n} \left[|e_{ij}| F_j \int_{\Omega} w_j^2(x,t-\tau_j(t))dx \right] \right\} \\
&\leqslant \sum_{i=1}^{n} \lambda_i \left\{ 2 \int_{\Omega} w_i(x,t) \sum_{k=1}^{\sigma} \frac{\partial}{\partial x_k} \left(a_{ik} \frac{\partial w_i(x,t)}{\partial x_k} \right) dx \right. \\
&\quad -2c_i \int_{\Omega} w_i^2(x,t)dx + 2 \int_{\Omega} w_i(x,t)u_i(x,t)dx
\end{aligned}
$$

$$+\sum_{j=1}^{n}(|d_{ij}|+|e_{ij}|)F_j\int_{\Omega} w_i^2(x,t)dx$$
$$+\sum_{j=1}^{n}\left(|d_{ij}|+\frac{|e_{ij}|}{1-\delta}\right)F_j\int_{\Omega} w_j^2(x,t)dx\Bigg\}.$$

From Green's formula and the Dirichlet boundary condition, we have

$$\int_{\Omega} w_i(x,t)\sum_{k=1}^{\sigma}\frac{\partial}{\partial x_k}\left(a_{ik}\frac{\partial w_i(x,t)}{\partial x_k}\right)dx=-\sum_{k=1}^{\sigma}\int_{\Omega} a_{ik}\left(\frac{\partial w_i(x,t)}{\partial x_k}\right)^2 dx.$$

According to Lemma 1.5, we can obtain

$$\sum_{k=1}^{\sigma}\int_{\Omega} a_{ik}\left(\frac{\partial w_i(x,t)}{\partial x_k}\right)^2 dx\geqslant\sum_{k=1}^{\sigma}\int_{\Omega}\frac{a_{ik}}{l_k^2}w_i^2(x,t)dx.$$

Therefore,

$$D^+V(t)-2\sum_{i=1}^{n}\int_{\Omega} y_i(x,t)u_i(x,t)dx+\gamma\sum_{i=1}^{n}\int_{\Omega} u_i^2(x,t)dx$$
$$\leqslant\sum_{i=1}^{n}\Bigg\{\left(-2\lambda_i\sum_{k=1}^{\sigma}\frac{a_{ik}}{l_k^2}-2c_i\lambda_i\right)\int_{\Omega} w_i^2(x,t)dx$$
$$+2(\lambda_i-h_i)\int_{\Omega} w_i(x,t)u_i(x,t)dx+(\gamma-2g_i)\int_{\Omega} u_i^2(x,t)dx$$
$$+\lambda_i\sum_{j=1}^{n}(|d_{ij}|+|e_{ij}|)F_j\int_{\Omega} w_i^2(x,t)dx$$
$$+\lambda_i\sum_{j=1}^{n}\left(|d_{ij}|+\frac{|e_{ij}|}{1-\delta}\right)F_j\int_{\Omega} w_j^2(x,t)dx\Bigg\}$$
$$=\sum_{i=1}^{n}\Bigg\{\Bigg[-2\lambda_i\sum_{k=1}^{\sigma}\frac{a_{ik}}{l_k^2}-2c_i\lambda_i+\lambda_i\sum_{j=1}^{n}(|d_{ij}|+|e_{ij}|)F_j$$
$$+\sum_{j=1}^{n}\lambda_j\left(|d_{ji}|+\frac{|e_{ji}|}{1-\delta}\right)F_i\Bigg]\int_{\Omega} w_i^2(x,t)dx$$
$$+2(\lambda_i-h_i)\int_{\Omega} w_i(x,t)u_i(x,t)dx$$

$$+(\gamma - 2g_i) \int_{\Omega} u_i^2(x,t)dx \Big\}$$

$$= \sum_{i=1}^{n} \int_{\Omega} \begin{pmatrix} w_i(x,t) \\ u_i(x,t) \end{pmatrix}^T \begin{pmatrix} \hat{M}_i & \lambda_i - h_i \\ \lambda_i - h_i & \gamma - 2g_i \end{pmatrix} \begin{pmatrix} w_i(x,t) \\ u_i(x,t) \end{pmatrix} dx.$$

It follows from (7.4) that

$$D^+V(t) - 2\sum_{i=1}^{n} \int_{\Omega} y_i(x,t)u_i(x,t)dx + \gamma \sum_{i=1}^{n} \int_{\Omega} u_i^2(x,t)dx \leqslant 0. \tag{7.6}$$

By integrating (7.6) with respect to t over the time period 0 to t_p, we can get

$$2\sum_{i=1}^{n} \int_0^{t_p} \int_{\Omega} y_i(x,t)u_i(x,t)dxdt \geqslant V(t_p) - V(0) + \gamma \sum_{i=1}^{n} \int_0^{t_p} \int_{\Omega} u_i^2(x,t)dxdt.$$

From the definition of $V(t)$, we have $V(t_p) \geqslant 0$ and $V(0) \geqslant 0$. Thus,

$$\sum_{i=1}^{n} \int_0^{t_p} \int_{\Omega} y_i(x,t)u_i(x,t)dxdt \geqslant -\beta^2 + \frac{\gamma}{2} \sum_{i=1}^{n} \int_0^{t_p} \int_{\Omega} u_i^2(x,t)dxdt$$

for all $t_p \geqslant 0$, where $\beta = \sqrt{\frac{V(0)}{2}}$. The proof is completed.

Remark 7.7 In Theorem 7.6, one input strict passivity condition is obtained for the RDNN (7.1). In a similar manner to the proof of Theorem 7.6, sufficient condition ensuring the output strict passivity can also be derived for the RDNN (7.1).

By a minor modification of the proof of Theorem 7.6, we can easily get the following.

Theorem 7.8 *System* (7.1) *is passive in the sense of Definition 7.4 if there exist constants* $\lambda_i > 0$ *and* $\gamma \geqslant 0$ *such that*

$$\begin{pmatrix} \hat{M}_i & \lambda_i - h_i \\ \lambda_i - h_i & -\gamma - 2g_i \end{pmatrix} \leqslant 0, \tag{7.7}$$

where $i = 1, 2, \ldots, n$, $\hat{M}_i = -2\lambda_i \sum_{k=1}^{\sigma} \frac{a_{ik}}{l_k^2} - 2c_i\lambda_i + \lambda_i \sum_{j=1}^{n}(|d_{ij}| + |e_{ij}|)F_j + \sum_{j=1}^{n} \lambda_j \left(|d_{ji}| + \frac{|e_{ji}|}{1-\delta} \right) F_i$.

Remark 7.9 Li and Liao [22], Zhang et al. [27], and Balasubramaniam et al. [33], investigated the passivity of NNs with time-varying delay, in which the input and output variables are only dependent on the time. In these works, the following passivity definition was considered:

A system with input $u(t)$ and output $y(t)$ where $u(t), y(t) \in \mathbb{R}^p$, is called passive if there exists a scalar $\gamma \geqslant 0$ such that

$$2\int_0^{t_p} y^T(s)u(s)ds \geqslant -\gamma \int_0^{t_p} u^T(s)u(s)ds \tag{7.8}$$

for all $t_p \geqslant 0$ and for all solution $x(t, 0)$. Moreover, the diffusion effects were not considered in [22, 27, 33]. Therefore, our results can be viewed as the extension of existing ones.

7.3.2 Stability Analysis

Theorem 7.10 *System* (7.1) *is globally exponentially stable if there exist constants* $\lambda_i > 0$ *and* $\epsilon > 0$ *such that*

$$\begin{aligned} &2\lambda_i\epsilon - 2\lambda_i c_i + \lambda_i \sum_{j=1}^{n}\Big(|d_{ij}| + |e_{ij}|\Big)F_j + \sum_{j=1}^{n}\lambda_j\Big(|d_{ji}| \\ &+\frac{e^{2\epsilon\tau}|e_{ji}|}{1-\delta}\Big)F_i - 2\lambda_i\sum_{k=1}^{\sigma}\frac{a_{ik}}{l_k^2} \leqslant 0, \end{aligned} \tag{7.9}$$

where $i = 1, 2, \ldots, n$.

Proof Define $z(x,t) = w(x,t,\Psi) - w(x,t,\Phi)$, $\Psi_z(x,t) = \Psi(x,t) - \Phi(x,t)$, then the dynamics of the difference vector $z(x,t) = (z_1(x,t), z_2(x,t), \ldots, z_n(x,t))^T$ is governed by the following equation:

$$\begin{aligned} \frac{\partial z_i(x,t)}{\partial t} &= \sum_{k=1}^{\sigma}\frac{\partial}{\partial x_k}\left(a_{ik}\frac{\partial z_i(x,t)}{\partial x_k}\right) - c_i z_i(x,t) \\ &+\sum_{j=1}^{n} d_{ij}[f_j(w_j(x,t,\Psi)) - f_j(w_j(x,t,\Phi))] \\ &+\sum_{j=1}^{n} e_{ij}[f_j(w_j(x,t-\tau_j(t),\Psi)) \\ &-f_j(w_j(x,t-\tau_j(t),\Phi))], \end{aligned} \tag{7.10}$$

where $i = 1, 2, \ldots, n$. Define the following Lyapunov functional for (7.10):

$$V(t) = \sum_{i=1}^{n} \lambda_i \left\{ e^{2\epsilon t} \int_{\Omega} z_i^2(x,t)dx + \frac{e^{2\epsilon\tau}}{1-\delta} \sum_{j=1}^{n} \left(|e_{ij}| F_j \int_{t-\tau_j(t)}^{t} \int_{\Omega} z_j^2(x,s) e^{2\epsilon s} dx ds \right) \right\}. \tag{7.11}$$

Then the upper right derivative $D^+V(t)$ of $V(t)$ is given by

$$\begin{aligned}
D^+V(t) = & \sum_{i=1}^{n} \lambda_i e^{2\epsilon t} \Big\{ 2\epsilon \int_{\Omega} z_i^2(x,t)dx + \int_{\Omega} 2z_i(x,t) \frac{\partial z_i(x,t)}{\partial t} dx \\
& + \frac{e^{2\epsilon\tau}}{1-\delta} \sum_{j=1}^{n} |e_{ij}| F_j \int_{\Omega} z_j^2(x,t)dx \\
& - \frac{e^{2\epsilon\tau}}{1-\delta} \sum_{j=1}^{n} (1-\dot{\tau}_j(t)) |e_{ij}| F_j e^{-2\epsilon\tau_j(t)} \int_{\Omega} z_j^2(x,t-\tau_j(t))dx \Big\} \\
\leqslant & \sum_{i=1}^{n} \lambda_i e^{2\epsilon t} \Big\{ (2\epsilon - 2c_i) \int_{\Omega} z_i^2(x,t)dx \\
& + \int_{\Omega} 2z_i(x,t) \sum_{k=1}^{\sigma} \frac{\partial}{\partial x_k} \left(a_{ik} \frac{\partial z_i(x,t)}{\partial x_k} \right) dx \\
& + \int_{\Omega} 2z_i(x,t) \sum_{j=1}^{n} d_{ij} [f_j(w_j(x,t,\Psi)) - f_j(w_j(x,t,\Phi))] dx \\
& + \int_{\Omega} 2z_i(x,t) \sum_{j=1}^{n} e_{ij} [f_j(w_j(x,t-\tau_j(t),\Psi)) \\
& - f_j(w_j(x,t-\tau_j(t),\Phi))] dx + \frac{e^{2\epsilon\tau}}{1-\delta} \sum_{j=1}^{n} |e_{ij}| F_j \int_{\Omega} z_j^2(x,t)dx \\
& - \sum_{j=1}^{n} |e_{ij}| F_j \int_{\Omega} z_j^2(x,t-\tau_j(t))dx \Big\} \\
\leqslant & e^{2\epsilon t} \sum_{i=1}^{n} \Big\{ \Big[2\lambda_i \epsilon - 2\lambda_i c_i + \lambda_i \sum_{j=1}^{n} (|d_{ij}| + |e_{ij}|) F_j \\
& + \sum_{j=1}^{n} \lambda_j \left(|d_{ji}| + \frac{e^{2\epsilon\tau} |e_{ji}|}{1-\delta} \right) F_i - 2\lambda_i \sum_{k=1}^{\sigma} \frac{a_{ik}}{l_k^2} \Big] \int_{\Omega} z_i^2(x,t)dx \Big\}.
\end{aligned}$$

It follows from (7.9) that

$$D^+V(t) \leqslant 0.$$

So $V(t) \leqslant V(0),\ t \leqslant 0$. Since

$$V(0) \leqslant \sum_{i=1}^{n} \lambda_i \left\{ \int_{\Omega} z_i^2(x,0)dx + \frac{e^{2\epsilon\tau}}{1-\delta} \sum_{j=1}^{n} \left[|e_{ij}|F_j \int_{-\tau_j(0)}^{0} \int_{\Omega} z_j^2(x,s)dxds \right] \right\}$$
$$\leqslant \left[\max_{i=1,2,\ldots,n} \{\lambda_i\} + \max_{i=1,2,\ldots,n} \left\{ \frac{\tau e^{2\epsilon\tau} F_i}{1-\delta} \sum_{j=1}^{n} \lambda_j |e_{ji}| \right\} \right] \|\Psi_z(\cdot,0)\|_{\tau}^2,$$

and

$$V(t) \geqslant \min_{i=1,2,\ldots,n} \{\lambda_i\} e^{2\epsilon t} \|z(\cdot,t)\|_2^2.$$

Let $\lambda_- = \min_{i=1,2,\ldots,n}\{\lambda_i\}$, $\lambda^+ = \max_{i=1,2,\ldots,n}\{\lambda_i\}$, and

$$M = \sqrt{\frac{\lambda^+ + \max_{i=1,2,\ldots,n} \left\{ \frac{\tau e^{2\epsilon\tau} F_i}{1-\delta} \sum_{j=1}^{n} \lambda_j |e_{ji}| \right\}}{\lambda_-}}.$$

Then $M \geqslant 1$, and we can obtain

$$\|z(\cdot,t)\|_2 \leqslant M e^{-\epsilon t} \|\Psi_z(\cdot,0)\|_{\tau}.$$

Namely,

$$\|w(\cdot,t,\Psi) - w(\cdot,t,\Phi)\|_2 \leqslant M \|\Psi(\cdot,0) - \Phi(\cdot,0)\|_{\tau} e^{-\epsilon t}.$$

This completes the proof of Theorem 7.10.

Remark 7.11 Practically, Theorem 7.10 not only can judge the global exponential stability of RDNN (7.1) but also can guarantee the existence and uniqueness of the periodic solution in some circumstances.

In the following, we shall consider the existence and uniqueness of periodic solution of RDNN (7.1), in which $u_i(x,t)$ and $\tau_i(t)$ are periodic continuous functions with period ϖ, that is, $u_i(x,t+\varpi) = u_i(x,t)$, $\tau_i(t+\varpi) = \tau_i(t)$, $i = 1, 2, \ldots, n$.

Corollary 7.12 *System* (7.1) *has a unique* ϖ*-periodic solution if there exist constants* $\lambda_i > 0$ *and* $\epsilon > 0$ *such that*

$$2\lambda_i\epsilon - 2\lambda_i c_i + \lambda_i \sum_{j=1}^{n}(|d_{ij}| + |e_{ij}|)F_j + \sum_{j=1}^{n}\lambda_j\Bigg(|d_{ji}| + \frac{e^{2\epsilon\tau}|e_{ji}|}{1-\delta}\Bigg)F_i - 2\lambda_i \sum_{k=1}^{\sigma}\frac{a_{ik}}{l_k^2} \leqslant 0, \tag{7.12}$$

where $i = 1, 2, \ldots, n$.

Proof Define

$$w_t(\Psi) = w(x, t+s, \Psi),\ s \in [-\tau, 0],\ t \geqslant 0,$$
$$w_t(\Phi) = w(x, t+s, \Phi),\ s \in [-\tau, 0],\ t \geqslant 0.$$

Obviously, from Theorem 7.10, we can derive

$$\|w(\cdot, t, \Psi) - w(\cdot, t, \Phi)\|_2 \leqslant M\|\Psi(\cdot, 0) - \Phi(\cdot, 0)\|_\tau e^{-\epsilon t} \tag{7.13}$$

for $t \geqslant 0$, where $M \geqslant 1$ is a constant. We can choose a positive integer N such that

$$Me^{-\epsilon(N\omega-\tau)} \leqslant \frac{1}{3}. \tag{7.14}$$

Now define a Poincaré mapping $P : C(\Omega \times [-\tau, 0], \mathbb{R}^n) \to C(\Omega \times [-\tau, 0], \mathbb{R}^n)$ by

$$\Psi \to P(\Psi) = w_\varpi(\Psi).$$

Since the periodicity of system (7.1), one has $P^N(\Psi) = w_{N\varpi}(\Psi)$. From (7.13) and (7.14), one obtains

$$\|P^N(\Psi) - P^N(\Phi)\|_2 \leqslant \frac{1}{3}\|\Psi(\cdot, 0) - \Phi(\cdot, 0)\|_\tau.$$

This implies that P^N is a contraction mapping, so there exists one unique fixed point $\Psi^* \in C(\Omega \times [-\tau, 0], \mathbb{R}^n)$ such that

$$P^N(\Psi^*) = \Psi^*.$$

Since $P^N(P(\Psi^*)) = P(P^N(\Psi^*)) = P(\Psi^*)$, then $P(\Psi^*)$ is also a fixed point of P^N, then

$$P(\Psi^*) = \Psi^*.$$

Let $w(x, t, \Psi^*)$ be the solution of system (7.1) with initial conditions Ψ^*, then $w(x, t+\omega, \Psi^*)$ is also a solution of system (7.1). Obviously,

$$w(x, t+\omega, \Psi^*) = w(x, t, w_\omega(x, \Psi^*)) = w(x, t, \Psi^*)$$

for all $t \geqslant 0$.

This means that $w(x, t, \Psi^*)$ is exactly one ω-periodic solution of system (7.1). This completes the proof of Corollary 7.12.

Remark 7.13 From (7.13), it is obvious that all other solutions of system (7.1) converge exponentially to the periodic solution $w(x, t, \Psi^*)$ as $t \to +\infty$. This means that the periodic solution $w(x, t, \Psi^*)$ is globally exponentially stable.

Remark 7.14 In recent years, some researchers have investigated the stability and periodicity of RDNNs [34, 41, 46]. But, in these papers, they always assume that the input variable is only dependent on the time. However, in many practical situations, the input variable is varied with the time and space variables. Therefore, it is essential to consider the input variable varying both temporally and spatially in NNs. By constructing suitable Lyapunov functionals and utilizing some inequality techniques, Lu [41] analyzed the global exponential stability and periodicity of a class of reaction-diffusion recurrent neural networks with constant delays and Dirichlet boundary conditions. However, absolute constant delay may be scarce, and delays are frequently varied with time. In this chapter, we remove these restrictions, and use similar methods to obtain criteria for the global exponential stability and periodicity of the RDNNs with time-varying delays and Dirichlet boundary conditions.

Remark 7.15 Recently, many authors have studied the passivity of the traditional NNs without reaction-diffusion, which are described by ordinary differential equations. However, strictly speaking, diffusion effects can not be avoided in the NNs when electrons are moving in asymmetric electromagnetic field, thus we must consider the diffusion effects in NNs. The main difficulty for passivity analysis of RDNNs comes from the reaction-diffusion terms, which can not be dealt with by those techniques used in traditional NNs. By employing Cauchy-Schwarz inequality, Green's formula, and Lemma 1.5, several new passivity criteria are proposed in this chapter, which are dependent on the reaction-diffusion terms.

7.4 Robust Passivity and Robust Stability of IRDNNs

It is well known that parameters acquired in NNs are inaccurate in the design and hardware implementation. Moreover, parameter fluctuation in NN implementation is also unavoidable in some circumstances. This may lead to parameter deviations. In fact, these deviations are bounded. Hence, the quantities a_{ik}, c_i, d_{ij}, e_{ij}, h_i, and g_i may be intervalized as follows:

$$
f\begin{cases}
A_I := \{A = (a_{ik})_{n\times\sigma} : \underline{A} \leqslant A \leqslant \overline{A}, i.e., \\
\quad \underline{a}_{ik} \leqslant a_{ik} \leqslant \overline{a}_{ik},\ i = 1, 2, \ldots, n, \\
\quad k = 1, 2, \ldots, \sigma, \forall A \in A_I\}, \\
C_I := \{C = \mathrm{diag}(c_i) : \underline{C} \leqslant C \leqslant \overline{C}, i.e., \\
\quad \underline{c}_i \leqslant c_i \leqslant \overline{c}_i, i = 1, 2, \ldots, n, \\
\quad \forall C \in C_I\}, \\
D_I := \{D = (d_{ij})_{n\times n} : \underline{D} \leqslant D \leqslant \overline{D}, i.e., \\
\quad \underline{d}_{ij} \leqslant d_{ij} \leqslant \overline{d}_{ij}, i, j = 1, 2, \ldots, n, \\
\quad \forall D \in D_I\}, \\
E_I := \{E = (e_{ij})_{n\times n} : \underline{E} \leqslant E \leqslant \overline{E}, i.e., \\
\quad \underline{e}_{ij} \leqslant e_{ij} \leqslant \overline{e}_{ij}, i, j = 1, 2, \ldots, n, \\
\quad \forall E \in E_I\}, \\
H_I := \{H = \mathrm{diag}(h_i) : \underline{H} \leqslant H \leqslant \overline{H}, i.e., \\
\quad \underline{h}_i \leqslant h_i \leqslant \overline{h}_i, i = 1, 2, \ldots, n, \\
\quad \forall H \in H_I\}, \\
G_I := \{G = \mathrm{diag}(g_i) : \underline{G} \leqslant G \leqslant \overline{G}, i.e., \\
\quad \underline{g}_i \leqslant g_i \leqslant \overline{g}_i, i = 1, 2, \ldots, n, \\
\quad \forall G \in G_I\}.
\end{cases} \tag{7.15}
$$

7.4.1 Robust Passivity Analysis

Definition 7.16 System (7.1) with the parameter ranges defined by (7.15) is called input robustly passive, if for all $A \in A_I, C \in C_I, D \in D_I, E \in E_I, H \in H_I, G \in G_I$, there exist constants $\beta \in \mathbb{R}$ and $\gamma > 0$ such that

$$
\int_0^{t_p}\int_\Omega y^T(x,s)u(x,s)dxds \geqslant -\beta^2 + \gamma \int_0^{t_p}\int_\Omega u^T(x,s)u(x,s)dxds
$$

for all $t_p \geqslant 0$, where $u(x,t) = (u_1(x,t), u_2(x,t), \ldots, u_n(x,t))^T$, $y(x,t) = (y_1(x,t), y_2(x,t), \ldots, y_n(x,t))^T$.

Definition 7.17 System (7.1) with the parameter ranges defined by (7.15) is called robustly passive, if for all $A \in A_I, C \in C_I, D \in D_I, E \in E_I, H \in H_I, G \in G_I$, there exist constants $\beta \in \mathbb{R}$ and $\gamma \geqslant 0$ such that

$$
2\int_0^{t_p}\int_\Omega y^T(x,s)u(x,s)dxds \geqslant -\beta^2 - \gamma \int_0^{t_p}\int_\Omega u^T(x,s)u(x,s)dxds
$$

for all $t_p \geqslant 0$, where $u(x,t) = (u_1(x,t), u_2(x,t), \ldots, u_n(x,t))^T$, $y(x,t) = (y_1(x,t), y_2(x,t), \ldots, y_n(x,t))^T$.

Theorem 7.18 *System* (7.1) *with the parameter ranges defined by* (7.15) *is input robustly passive in the sense of Definition 7.16, if there exist constants* $\lambda_i > 0$ *and* $\gamma > 0$ *such that* $\gamma - 2\underline{g}_i < 0$ *and*

$$-2\lambda_i \sum_{k=1}^{\sigma} \frac{\underline{a}_{ik}}{l_k^2} - 2\underline{c}_i \lambda_i + \lambda_i \sum_{j=1}^{n} (d_{ij}^* + e_{ij}^*) F_j + \sum_{j=1}^{n} \lambda_j \left(d_{ji}^* + \frac{e_{ji}^*}{1-\delta} \right) F_i - \frac{\hat{m}_i^2}{\gamma - 2\underline{g}_i} \leqslant 0, \tag{7.16}$$

where $i = 1, 2, \ldots, n$, $d_{ij}^* = max\{|\underline{d}_{ij}|, |\overline{d}_{ij}|\}$, $e_{ij}^* = max\{|\underline{e}_{ij}|, |\overline{e}_{ij}|\}$, $\hat{m}_i = max\{|\lambda_i - \underline{h}_i|, |\lambda_i - \overline{h}_i|\}$.

Proof Firstly, construct a Lyapunov functional for RDNN (7.1) as follows:

$$V(t) = \sum_{i=1}^{n} \lambda_i \left\{ \int_{\Omega} w_i^2(x,t)dx + \frac{1}{1-\delta} \sum_{j=1}^{n} \left(e_{ij}^* F_j \int_{t-\tau_j(t)}^{t} \int_{\Omega} w_j^2(x,s)dxds \right) \right\}. \tag{7.17}$$

Then the upper right derivative $D^+V(t)$ of $V(t)$ is given by

$$\begin{aligned}
D^+V(t) = & \sum_{i=1}^{n} \lambda_i \left\{ 2\int_{\Omega} w_i(x,t) \frac{\partial w_i(x,t)}{\partial t} dx + \frac{1}{1-\delta} \sum_{j=1}^{n} e_{ij}^* F_j \int_{\Omega} w_j^2(x,t)dx \right. \\
& \left. - \frac{1}{1-\delta} \sum_{j=1}^{n} \left[e_{ij}^* F_j (1 - \dot{\tau}_j(t)) \int_{\Omega} w_j^2(x, t-\tau_j(t))dx \right] \right\} \\
\leqslant & \sum_{i=1}^{n} \lambda_i \left\{ 2\int_{\Omega} w_i(x,t) \sum_{k=1}^{\sigma} \frac{\partial}{\partial x_k} \left(a_{ik} \frac{\partial w_i(x,t)}{\partial x_k} \right) dx \right. \\
& -2\underline{c}_i \int_{\Omega} w_i^2(x,t)dx + 2\int_{\Omega} w_i(x,t) u_i(x,t)dx \\
& +2\int_{\Omega} w_i(x,t) \sum_{j=1}^{n} d_{ij} f_j(w_j(x,t))dx \\
& +2\int_{\Omega} w_i(x,t) \sum_{j=1}^{n} e_{ij} f_j(w_j(x, t-\tau_j(t)))dx
\end{aligned}$$

$$
\begin{aligned}
&+\frac{1}{1-\delta}\sum_{j=1}^{n} e_{ij}^{*}F_j\int\limits_{\Omega} w_j^2(x,t)dx \\
&-\sum_{j=1}^{n}\left(e_{ij}^{*}F_j\int\limits_{\Omega} w_j^2(x,t-\tau_j(t))dx\right)\Bigg\} \\
\leqslant&\sum_{i=1}^{n}\lambda_i\Bigg\{2\int\limits_{\Omega} w_i(x,t)\sum_{k=1}^{\sigma}\frac{\partial}{\partial x_k}\left(a_{ik}\frac{\partial w_i(x,t)}{\partial x_k}\right)dx \\
&-2\underline{c}_i\int\limits_{\Omega} w_i^2(x,t)dx+2\int\limits_{\Omega} w_i(x,t)u_i(x,t)dx \\
&+\sum_{j=1}^{n}(d_{ij}^{*}+e_{ij}^{*})F_j\int\limits_{\Omega} w_i^2(x,t)dx \\
&+\sum_{j=1}^{n}\left(d_{ij}^{*}+\frac{e_{ij}^{*}}{1-\delta}\right)F_j\int\limits_{\Omega} w_j^2(x,t)dx\Bigg\} \\
\leqslant&\sum_{i=1}^{n}\lambda_i\Bigg\{-2\sum_{k=1}^{\sigma}\int\limits_{\Omega}\frac{a_{ik}}{l_k^2}w_i^2(x,t)dx \\
&-2\underline{c}_i\int\limits_{\Omega} w_i^2(x,t)dx+2\int\limits_{\Omega} w_i(x,t)u_i(x,t)dx \\
&+\sum_{j=1}^{n}(d_{ij}^{*}+e_{ij}^{*})F_j\int\limits_{\Omega} w_i^2(x,t)dx \\
&+\sum_{j=1}^{n}\left(d_{ij}^{*}+\frac{e_{ij}^{*}}{1-\delta}\right)F_j\int\limits_{\Omega} w_j^2(x,t)dx\Bigg\}.
\end{aligned}
$$

Therefore,

$$
\begin{aligned}
&D^{+}V(t)-2\sum_{i=1}^{n}\int\limits_{\Omega} y_i(x,t)u_i(x,t)dx+\gamma\sum_{i=1}^{n}\int\limits_{\Omega} u_i^2(x,t)dx \\
\leqslant&\sum_{i=1}^{n}\Bigg\{\left(-2\lambda_i\sum_{k=1}^{\sigma}\frac{a_{ik}}{l_k^2}-2\underline{c}_i\lambda_i\right)\int\limits_{\Omega} w_i^2(x,t)dx \\
&+2m_i\int\limits_{\Omega} w_i(x,t)u_i(x,t)dx+q_i\int\limits_{\Omega} u_i^2(x,t)dx \\
&+\lambda_i\sum_{j=1}^{n}(d_{ij}^{*}+e_{ij}^{*})F_j\int\limits_{\Omega} w_i^2(x,t)dx
\end{aligned}
$$

$$+\lambda_i \sum_{j=1}^{n} (d_{ij}^* + \frac{e_{ij}^*}{1-\delta}) F_j \int_{\Omega} w_j^2(x,t)dx \Big\}$$

$$= \sum_{i=1}^{n} \Big\{ \Big[-2\lambda_i \sum_{k=1}^{\sigma} \frac{a_{ik}}{l_k^2} - 2\underline{c}_i \lambda_i + \lambda_i \sum_{j=1}^{n} (d_{ij}^* + e_{ij}^*) F_j$$

$$+ \sum_{j=1}^{n} \lambda_j \Big(d_{ji}^* + \frac{e_{ji}^*}{1-\delta} \Big) F_i \Big] \int_{\Omega} w_i^2(x,t)dx$$

$$+2m_i \int_{\Omega} w_i(x,t)u_i(x,t)dx + q_i \int_{\Omega} u_i^2(x,t)dx \Big\}$$

$$= \sum_{i=1}^{n} \int_{\Omega} \begin{pmatrix} w_i(x,t) \\ u_i(x,t) \end{pmatrix}^T \begin{pmatrix} \hat{M}_i & m_i \\ m_i & q_i \end{pmatrix} \begin{pmatrix} w_i(x,t) \\ u_i(x,t) \end{pmatrix} dx,$$

where $\hat{M}_i = -2\lambda_i \sum_{k=1}^{\sigma} \frac{a_{ik}}{l_k^2} - 2\underline{c}_i \lambda_i + \lambda_i \sum_{j=1}^{n} (d_{ij}^* + e_{ij}^*) F_j + \sum_{j=1}^{n} \lambda_j \Big(d_{ji}^* + \frac{e_{ji}^*}{1-\delta} \Big) F_i$, $m_i = \lambda_i - h_i$, $q_i = \gamma - 2g_i$. Since $q_i \leqslant \gamma - 2\underline{g}_i < 0$, it is obvious

$$\begin{pmatrix} \hat{M}_i & m_i \\ m_i & q_i \end{pmatrix} \leqslant 0$$

is equivalent to

$$\hat{M}_i - m_i^2 q_i^{-1} \leqslant 0.$$

On the other hand,

$$\hat{M}_i - m_i^2 q_i^{-1} \leqslant \hat{M}_i - \frac{\hat{m}_i^2}{\gamma - 2\underline{g}_i}.$$

It follows from (7.16) that

$$D^+V(t) + \gamma \sum_{i=1}^{n} \int_{\Omega} u_i^2(x,t)dx - 2\sum_{i=1}^{n} \int_{\Omega} y_i(x,t)u_i(x,t)dx \leqslant 0. \quad (7.18)$$

By integrating (7.18) with respect to t over the time period 0 to t_p, we have

$$2\sum_{i=1}^{n} \int_0^{t_p} \int_{\Omega} y_i(x,t)u_i(x,t)dxdt \geqslant V(t_p) - V(0) + \gamma \sum_{i=1}^{n} \int_0^{t_p} \int_{\Omega} u_i^2(x,t)dxdt.$$

From the definition of $V(t)$, we have $V(t_p) \geqslant 0$ and $V(0) \geqslant 0$. Thus,

$$\sum_{i=1}^{n} \int_{0}^{t_p} \int_{\Omega} y_i(x,t) u_i(x,t) dx dt \geqslant -\beta^2 + \frac{\gamma}{2} \sum_{i=1}^{n} \int_{0}^{t_p} \int_{\Omega} u_i^2(x,t) dx dt$$

for all $t_p \geqslant 0$, where $\beta = \sqrt{\frac{V(0)}{2}}$. The proof is completed.

By a minor modification of the proof of Theorem 7.18, we can easily get the following.

Theorem 7.19 *System* (7.1) *with the parameter ranges defined by* (7.15) *is robustly passive in the sense of Definition 7.17, if there exist constants* $\lambda_i > 0$ *and* $\gamma \geqslant 0$ *such that* $-\gamma - 2\underline{g}_i < 0$ *and*

$$\begin{aligned} &-2\lambda_i \sum_{k=1}^{\sigma} \frac{\underline{a}_{ik}}{l_k^2} - 2\underline{c}_i \lambda_i + \lambda_i \sum_{j=1}^{n} (d_{ij}^* + e_{ij}^*) F_j \\ &+ \sum_{j=1}^{n} \lambda_j \left(d_{ji}^* + \frac{e_{ji}^*}{1-\delta} \right) F_i - \frac{\hat{m}_i^2}{-\gamma - 2\underline{g}_i} \leqslant 0, \end{aligned} \tag{7.19}$$

where $i = 1, 2, \ldots, n$, $d_{ij}^* = max\{|\underline{d}_{ij}|, |\overline{d}_{ij}|\}$, $e_{ij}^* = max\{|\underline{e}_{ij}|, |\overline{e}_{ij}|\}$, $\hat{m}_i = max \{|\lambda_i - \underline{h}_i|, |\lambda_i - \overline{h}_i|\}$.

7.4.2 Robust Stability Analysis

In the following, we shall discuss the robust stability of RDNN (7.1) with the parameter ranges defined by (7.1). Next, we introduce the definition of robust global exponential stability.

Definition 7.20 System (7.1) with the parameter ranges defined by (7.15) is robustly globally exponentially stable if the RDNN (7.1) is globally exponentially stable for all $A \in A_I, C \in C_I, D \in D_I,$ *and* $E \in E_I$.

Theorem 7.21 *System* (7.1) *with the parameter ranges defined by* (7.15) *is robustly globally exponentially stable if there exist constants* $\lambda_i > 0$ *and* $\epsilon > 0$ *such that*

$$\begin{aligned} &\lambda_i \left[2\epsilon - 2\sum_{k=1}^{\sigma} \frac{\underline{a}_{ik}}{l_k^2} - 2\underline{c}_i + \sum_{j=1}^{n} (d_{ij}^* + e_{ij}^*) F_j \right] \\ &+ \sum_{j=1}^{n} \lambda_j \left(d_{ji}^* + \frac{e^{2\epsilon\tau} e_{ji}^*}{1-\delta} \right) F_i \leqslant 0, \end{aligned} \tag{7.20}$$

where $i, j = 1, 2, \ldots, n, d_{ij}^* = max\{|\underline{d}_{ij}|, |\overline{d}_{ij}|\}, e_{ij}^* = max\{|\underline{e}_{ij}|, |\overline{e}_{ij}|\}$.

Proof Define the following Lyapunov functional for the system (7.10).

$$V(t) = \sum_{i=1}^{n} \lambda_i \Bigg\{ e^{2\epsilon t} \int_{\Omega} z_i^2(x,t)dx + \frac{e^{2\epsilon\tau}}{1-\delta} \sum_{j=1}^{n} \Bigg[e_{ij}^* F_j \int_{t-\tau_j(t)}^{t} \int_{\Omega} z_j^2(x,s) e^{2\epsilon s} dxds \Bigg] \Bigg\}. \tag{7.21}$$

Then, following similar arguments as in the proof of Theorem 7.10, we can obtain the desired result immediately.

Remark 7.22 In some circumstances, the Theorem 7.21 can also guarantee the existence and uniqueness of the equilibrium solution.

Let $u_i(x,t) = 0, i = 1, 2, \ldots, n$, we can get from (7.1) that

$$\begin{aligned} \frac{\partial w_i(x,t)}{\partial t} = & \sum_{k=1}^{\sigma} \frac{\partial}{\partial x_k} \left(a_{ik} \frac{\partial w_i(x,t)}{\partial x_k} \right) - c_i w_i(x,t) \\ & + \sum_{j=1}^{n} d_{ij} f_j(w_j(x,t)) \\ & + \sum_{j=1}^{n} e_{ij} f_j(w_j(x, t-\tau_j(t))), \end{aligned} \tag{7.22}$$

where $i = 1, 2, \ldots, n$.

Corollary 7.23 *System* (7.22) *with the parameter ranges defined by* (7.15) *has a unique equilibrium solution if there exist constants* $\lambda_i > 0, i = 1, 2, \ldots, n$, *such that*

$$\lambda_i \Bigg[-2 \sum_{k=1}^{\sigma} \frac{\underline{a}_{ik}}{l_k^2} - 2\underline{c}_i + \sum_{j=1}^{n} (d_{ij}^* + e_{ij}^*) F_j \Bigg] + \sum_{j=1}^{n} \lambda_j (d_{ji}^* + e_{ji}^*) F_i < 0, \tag{7.23}$$

where $i = 1, 2, \ldots, n, d_{ij}^* = max\{|\underline{d}_{ij}|, |\overline{d}_{ij}|\}, e_{ij}^* = max\{|\underline{e}_{ij}|, |\overline{e}_{ij}|\}$.

Proof In the following, we complete the proof in two steps. In Step 1, we prove the existence of the equilibrium solution; in Step 2, we prove the uniqueness of the equilibrium solution.

Step 1: According to (A1), we can easily obtain that 0 is an equilibrium solution of (7.22). The existence of the equilibrium solution is proven.

Step 2: We shall prove its uniqueness. Without loss of generality, let $v^*(x) = (v_1^*(x), v_2^*(x), \ldots, v_n^*(x))^T$ and $s^*(x) = (s_1^*(x), s_2^*(x), \ldots, s_n^*(x))^T$ be any two equilibrium solutions of system (7.22). Then, we have

$$
\begin{aligned}
0 = &\sum_{k=1}^{\sigma} \frac{\partial}{\partial x_k}\left(a_{ik}\frac{\partial(v_i^*(x) - s_i^*(x))}{\partial x_k}\right) - c_i(v_i^*(x) - s_i^*(x)) \\
&+ \sum_{j=1}^{n} d_{ij}(f_j(v_j^*(x)) - f_j(s_j^*(x))) \\
&+ \sum_{j=1}^{n} e_{ij}(f_j(v_j^*(x)) - f_j(s_j^*(x))),
\end{aligned} \tag{7.24}
$$

where $i = 1, 2, \ldots, n$.

Multiplying both sides of (7.24) by $v_i^*(x) - s_i^*(x)$ and then integrating with respect to x over the domain Ω, we have

$$
\begin{aligned}
&\int_{\Omega} (v_i^*(x) - s_i^*(x)) \sum_{k=1}^{\sigma} \frac{\partial}{\partial x_k}\left(a_{ik}\frac{\partial(v_i^*(x) - s_i^*(x))}{\partial x_k}\right) dx \\
&+ \int_{\Omega} (v_i^*(x) - s_i^*(x)) \sum_{j=1}^{n} d_{ij}(f_j(v_j^*(x)) - f_j(s_j^*(x)))dx \\
&+ \int_{\Omega} (v_i^*(x) - s_i^*(x)) \sum_{j=1}^{n} e_{ij}(f_j(v_j^*(x)) - f_j(s_j^*(x)))dx \\
&- c_i \int_{\Omega} (v_i^*(x) - s_i^*(x))^2 dx = 0,
\end{aligned}
$$

where $i = 1, 2, \ldots, n$.

From Lemma 1.5, Green's formula and Dirichlet boundary condition, we can obtain

$$
\begin{aligned}
0 = &\sum_{i=1}^{n} 2\lambda_i \Bigg\{ - \sum_{k=1}^{\sigma} \int_{\Omega} a_{ik}\left(\frac{\partial(v_i^*(x) - s_i^*(x))}{\partial x_k}\right)^2 dx \\
&+ \int_{\Omega} (v_i^*(x) - s_i^*(x)) \sum_{j=1}^{n} d_{ij}(f_j(v_j^*(x)) - f_j(s_j^*(x)))dx \\
&+ \int_{\Omega} (v_i^*(x) - s_i^*(x)) \sum_{j=1}^{n} e_{ij}(f_j(v_j^*(x)) - f_j(s_j^*(x)))dx
\end{aligned}
$$

$$-c_i \int_{\Omega} (v_i^*(x) - s_i^*(x))^2 dx \Big\}$$

$$\leqslant \sum_{i=1}^{n} \Big\{ 2\lambda_i \Big(-\sum_{k=1}^{\sigma} \frac{a_{ik}}{l_k^2} - \underline{c}_i \Big) + \lambda_i \sum_{j=1}^{n} (d_{ij}^* + e_{ij}^*) F_j$$

$$+ \sum_{j=1}^{n} \lambda_j (d_{ji}^* + e_{ji}^*) F_i \Big\} \int_{\Omega} (v_i^*(x) - s_i^*(x))^2 dx.$$

Therefore, by (7.23), we can get $v^*(x) = s^*(x)$ and the uniqueness of the equilibrium solution is proven.

Remark 7.24 Obviously, the conditions of Corollary 7.23 can be seen as a special case of Theorem 7.21.

7.5 Numerical Examples

In this section, an illustrative example is provided to verify the effectiveness of the proposed theoretical results.

Consider the following RDNN with time-varying delays and Dirichlet boundary conditions:

$$\begin{cases} \frac{\partial w_i(x_1,t)}{\partial t} = a_{i1} \frac{\partial^2 w_i(x_1,t)}{\partial x_1^2} - c_i w_i(x_1,t) + u_i(x_1,t) \\ \qquad + \sum_{j=1}^{2} d_{ij} f_j(w_j(x_1,t)) \\ \qquad + \sum_{j=1}^{2} e_{ij} f_j(w_j(x_1, t - \tau_j(t))), \\ y_i(x_1,t) = h_i w_i(x_1,t) + g_i u_i(x_1,t), \\ w_i(x_1,t) = i \times \sin(\pi x_1), \ (x_1,t) \in \Omega \times [-1, 0], \\ w_i(x_1,t) = 0, \ (x_1,t) \in \partial\Omega \times [-1, +\infty), \end{cases} \tag{7.25}$$

where $i = 1, 2$, $\Omega = \{x_1 \mid |x_1| < 1\}$, $a_{i1} = 0.5 + 0.5i$, $c_i = 6 + i$, $d_{ij} = \frac{1}{i+j} + 0.2$, $e_{ij} = \frac{1}{i+j} + 0.3$, $h_i = i$, $g_i = 1.5i$, $f_1(\xi) = f_2(\xi) = \frac{|\xi+1| - |\xi-1|}{10}$, $F_1 = F_2 = 0.2$, $\tau_1(t) = 1 - \frac{1}{2}e^{-t}$, $\tau_2(t) = 1 - \frac{1}{3}e^{-t}$, $\tau = 1$, $\delta = \frac{1}{2}$. By simple calculation with $\lambda_1 = \lambda_2 = 1$ and $\gamma = 3$, we have

$$\begin{pmatrix} \hat{M}_1 & \lambda_1 - h_1 \\ \lambda_1 - h_1 & \gamma - 2g_1 \end{pmatrix} = \begin{pmatrix} -\frac{2197}{150} & 0 \\ 0 & 0 \end{pmatrix} \leqslant 0,$$

$$\begin{pmatrix} \hat{M}_2 & \lambda_2 - h_2 \\ \lambda_2 - h_2 & \gamma - 2g_2 \end{pmatrix} = \begin{pmatrix} -\frac{5369}{300} & -1 \\ -1 & -3 \end{pmatrix} < 0,$$

that is, (7.4) holds. Hence, it follows from Theorem 7.6 that system (7.25) is input strictly passive in the sense of Definition 7.2.

It is easy to verify that condition (7.7) is satisfied if $\lambda_1 = \lambda_2 = 1$ and $\gamma = 3$. From Theorem 7.8, system (7.25) is also passive in the sense of Definition 7.4.

Moreover, if we take $\lambda_1 = \lambda_2 = \epsilon = 1$, then the condition (7.9) is satisfied. According to Theorem 7.10, RDNN (7.25) is globally exponentially stable.

Set $u_i(x_1, t) = \sin(\pi x_1 t)$ and the simulation results are shown in Fig. 7.1, in which $e_1(t) = \sum_{i=1}^{2} \int_0^t \int_\Omega y_i(x_1, s)u_i(x_1, s)dx_1ds - \frac{3}{2}\sum_{i=1}^{2}\int_0^t \int_\Omega u_i^2(x_1, s)dx_1ds$, $e_2(t) = 2\sum_{i=1}^{2} \int_0^t \int_\Omega y_i(x_1, s)u_i(x_1, s)dx_1ds + 3\sum_{i=1}^{2}\int_0^t \int_\Omega u_i^2(x_1, s)dx_1ds$.

In the following, we shall discuss the robust passivity and robust stability of RDNN (7.25) with the following parameter ranges:

$$\begin{cases} A_I := \{A = (a_{ik})_{2\times 1} : 1 \leqslant a_{11} \leqslant 1.5, 1.5 \leqslant a_{21} \leqslant 2\} \\ C_I := \{C = \mathrm{diag}(c_i) : 7 \leqslant c_1 \leqslant 8, 8 \leqslant c_2 \leqslant 9\} \\ D_I := \{D = (d_{ij})_{2\times 2} : \frac{1}{i+j} + 0.15 \leqslant d_{ij} \leqslant \frac{1}{i+j} + 0.2 \\ \qquad i, j = 1, 2\}, \\ E_I := \{E = (e_{ij})_{2\times 2} : \frac{1}{i+j} + 0.25 \leqslant e_{ij} \leqslant \frac{1}{i+j} + 0.3 \\ \qquad i, j = 1, 2\}, \\ H_I := \{H = \mathrm{diag}(h_i) : 1 \leqslant h_1 \leqslant 2, 2 \leqslant h_2 \leqslant 3\}, \\ G_I := \{G = \mathrm{diag}(g_i) : 1.5 \leqslant g_1 \leqslant 2, 3 \leqslant g_2 \leqslant 3.5\}. \end{cases}$$

We take $\lambda_1 = \lambda_2 = \gamma = 1$. Then $\gamma - 2\underline{g}_1 = -2 < 0, \gamma - 2\underline{g}_2 = -5 < 0, -\gamma - 2\underline{g}_1 = -4 < 0, -\gamma - 2\underline{g}_2 = -7 < 0$,

$$-2\lambda_1\frac{\underline{a}_{11}}{l_1^2} - 2\underline{c}_1\lambda_1 + \lambda_1\sum_{j=1}^{2}(d_{1j}^* + e_{1j}^*)F_j + \sum_{j=1}^{2}\lambda_j(d_{j1}^* + \frac{e_{j1}^*}{1-\delta})F_1 - \frac{\hat{m}_1^2}{\gamma - 2\underline{g}_1} = -\frac{2122}{150} < 0,$$

$$-2\lambda_2\frac{\underline{a}_{21}}{l_1^2} - 2\underline{c}_2\lambda_2 + \lambda_2\sum_{j=1}^{2}(d_{2j}^* + e_{2j}^*)F_j + \sum_{j=1}^{2}\lambda_j(d_{j2}^* + \frac{e_{j2}^*}{1-\delta})F_2 - \frac{\hat{m}_2^2}{\gamma - 2\underline{g}_2} = -\frac{5129}{300} < 0,$$

$$-2\lambda_1\frac{\underline{a}_{11}}{l_1^2} - 2\underline{c}_1\lambda_1 + \lambda_1\sum_{j=1}^{2}(d_{1j}^* + e_{1j}^*)F_j + \sum_{j=1}^{2}\lambda_j(d_{j1}^* + \frac{e_{j1}^*}{1-\delta})F_1 - \frac{\hat{m}_1^2}{-\gamma - 2\underline{g}_1} = -\frac{4319}{300} < 0,$$

$$-2\lambda_2\frac{\underline{a}_{21}}{l_1^2} - 2\underline{c}_2\lambda_2 + \lambda_2\sum_{j=1}^{2}(d_{2j}^* + e_{2j}^*)F_j + \sum_{j=1}^{2}\lambda_j(d_{j2}^* + \frac{e_{j2}^*}{1-\delta})F_2 - \frac{\hat{m}_2^2}{-\gamma - 2\underline{g}_2} = -\frac{36383}{2100} < 0.$$

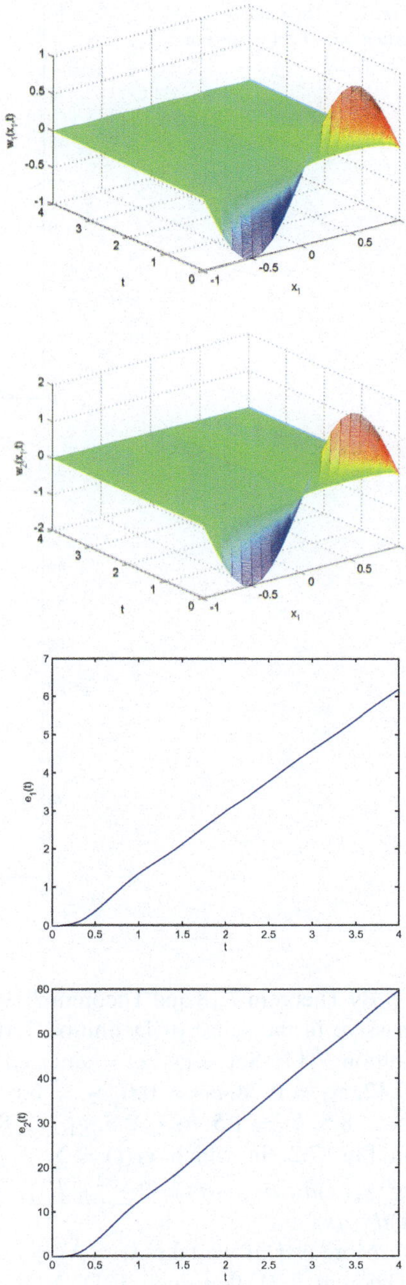

Fig. 7.1 The change processes of state variables $w_i(x_1, t)$ and $e_i(t)$, where $i = 1, 2$.

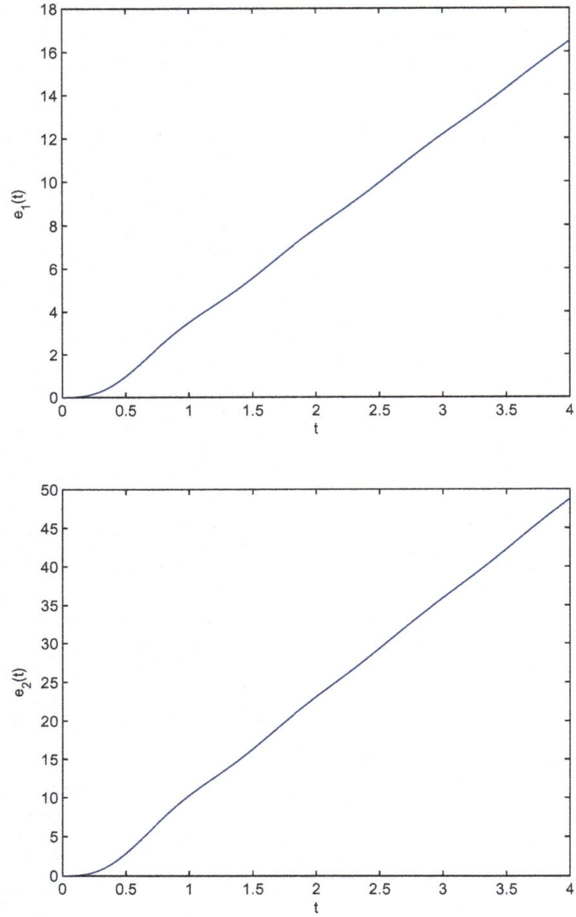

Fig. 7.2 The change processes of $e_i(t)$, where $i = 1, 2$.

By Theorem 7.18 and Theorem 7.19, we know that such IRDNN is input robustly passive in the sense of Definition 7.16 and robustly passive in the sense of Definition 7.17. Set $u_i(x_1, t) = \sin(\pi x_1 t)$, $d_{11} = 0.68$, $d_{12} = 0.51$, $d_{21} = 0.51$, $d_{22} = 0.42$, $e_{11} = 0.78$, $e_{12} = 0.6$, $e_{21} = 0.6$, $e_{22} = 0.53$, $a_{11} = 1.4$, $a_{21} = 1.9$, $c_1 = 7.5$, $c_2 = 8.5$, $h_1 = 1.5$, $h_2 = 2.5$, $g_1 = 1.8$, $g_2 = 3.3$. The simulation results are shown in Fig. 7.2, in which $e_1(t) = \sum_{i=1}^{2} \int_0^t \int_\Omega y_i(x_1, s) u_i(x_1, s) dx_1 ds - \frac{1}{2} \sum_{i=1}^{2} \int_0^t \int_\Omega u_i^2(x_1, s) dx_1 ds$, $e_2(t) = 2 \sum_{i=1}^{2} \int_0^t \int_\Omega y_i(x_1, s) u_i(x_1, s) dx_1 ds + \sum_{i=1}^{2} \int_0^t \int_\Omega u_i^2(x_1, s) dx_1 ds$.

Moreover, if we take $\lambda_1 = \lambda_2 = \epsilon = 1$, then (7.20) is satisfied. It follows from Theorem 7.21 that such IRDNN is robustly globally exponentially stable. Set $u_i(x_1, t) = 0$, $d_{11} = 0.68$, $d_{12} = 0.51$, $d_{21} = 0.51$, $d_{22} = 0.42$, $e_{11} = 0.78$, $e_{12} = 0.6$, $e_{21} = 0.6$, $e_{22} = 0.53$, $a_{11} = 1.3$, $a_{21} = 1.8$, $c_1 = 7.6$, $c_2 = 8.6$, $h_1 = 1.6$, $h_2 = 2.6$, $g_1 = 1.7$, $g_2 = 3.2$. The simulation results are shown in Fig. 7.3.

Fig. 7.3 Converged state variables $(w_i(x_1, t), i = 1, 2)$

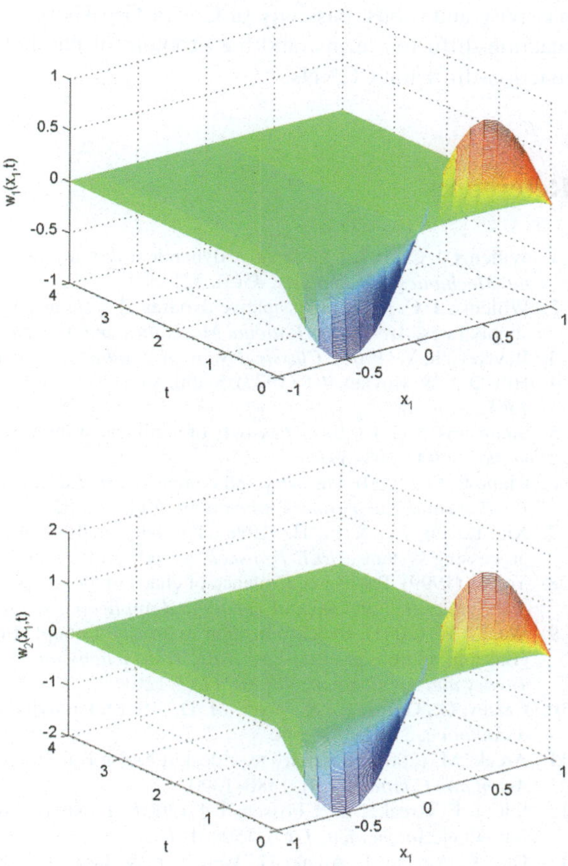

7.6 Conclusion

We have investigated the passivity and global exponential stability of RDNNs with Dirichlet boundary conditions. Several sufficient conditions have been obtained by utilizing the Lyapunov functional method combined with the inequality techniques. Furthermore, when the parameter uncertainties appear in RDNNs, robust passivity and robust global exponential stability conditions have also been derived. Illustrative simulations have been provided to verify the correctness and effectiveness of the obtained results.

As a well-known example of complex networks, food webs attract increasing attention of researchers from different fields in recent years. It is also common to consider the diffusion effects in food webs. We would like to point out that it is possible to apply our main results to food webs. In addition, we shall study the passivity and robust passivity of Cohen-Grossberg neural networks (CGNNs) with reaction-diffusion terms, and the problem of passivity-based controller design for reaction-diffusion CGNNs.

References

1. Willems, J. C. (1972). Dissipative dynamical systems part I: General theory. *Archive for Rational Mechanics and Analysis*, *45*(5), 321–351.
2. Willems, J. C. (1972). Dissipative dynamical systems part II: Linear systems with quadratic supply rates. *Archive for Rational Mechanics and Analysis*, *45*(5), 352–393.
3. Bevelevich, V. (1968). *Classical Network Synthesis*. New York, NY: Van Nostrand.
4. Hill, D. J., & Moylan, P. J. (1977). Stability results for nonlinear feedback systems. *Automatica*, *13*(4), 377–382.
5. Santosuosso, G. J. (1997). Passivity of nonlinear systems with input-output feedthrough. *Automatica*, *33*(4), 693–697.
6. Chua, L. O. (1999). Passivity and complexity. *IEEE Transactions on Circuits and Systems I: Fundamental Theory and Applications*, *46*(1), 71–82.
7. Xie, L., Fu, M., & Li, H. (1998). Passivity analysis and passification for uncertain signal processing systems. *IEEE Transactions on Signal Processing*, *46*(9), 2394–2403.
8. Yu, W. (1999). Passive equivalence of chaos in lorenz system. *IEEE Transactions on Circuits and Systems I: Fundamental Theory and Applications*, *46*(7), 876–878.
9. Wu, C. W. (2001). Synchronization in arrays of coupled nonlinear systems: Passivity, circle criterion, and observer design. *IEEE Transactions on Circuits and Systems I: Fundamental Theory and Applications*, *48*(10), 1257–1261.
10. Calcev, G., Gorez, R., & Neyer, M. D. (1998). Passivity approach to fuzzy control systems. *Automatica*, *34*(3), 339–344.
11. Arcak, M. (2007). Passivity as a design tool for group coordination. *IEEE Transactions on Automatic Control*, *52*(8), 1380–1390.
12. Ihle, I. F., Arcak, M., & Fossen, T. I. (2007). Passivity-based designs for synchronized path-following. *Automatica*, *43*(9), 1508–1518.
13. Fan, X., Alpcan, T., Arcak, M., Wen, T. J., & Basar, T. (2006). A passivity approach to game-theoretic CDMA power control. *Automatica*, *42*(11), 1837–1847.
14. Wen, J. T., & Arcak, M. (2004). A unifying passivity framework for network flow control. *IEEE Transactions on Automatic Control*, *49*(2), 162–174.
15. Dörfler, F., Johnsen, J. K., & Allgöwer, F. (2009). An introduction to interconnection and damping assignment passivity-based control in process engineering. *Journal of Process Control*, *19*(9), 1413–1426.
16. Mangold, M., Bück, A., & Hanke-Rauschenbach, R. (2010). Passivity based control of a distributed PEM fuel cell model. *Journal of Process Control*, *20*(3), 292–313.
17. Ayad, M. Y., Becherif, M., Henni, A., Aboubou, A., Wack, M., & Laghrouche, S. (2010). Passivity-based control applied to DC hybrid power source using fuel cell and supercapacitors. *Energy Conversion and Management*, *51*(7), 1468–1475.
18. Lestas, I., & Vinnicombe, G. (2010). Heterogeneity and scalability in group agreement protocols: Beyond small gain and passivity approaches. *Automatica*, *46*(7), 1141–1151.
19. Liu, P., Zhang, Q., Yang, X., Yang, L. (2008). Passivity and optimal control of descriptor biological complex systems. *IEEE Transactions on Automatic Control 53*, 122–125.
20. Uang, H. J. (2005). On the dissipativity of nonlinear systems: Fuzzy control approach. *Fuzzy Sets and Systems*, *156*(2), 185–207.
21. Chang, W. J., Ku, C. C., Huang, P. H., & Chang, W. (2009). Fuzzy controller design for passive continuous-time affine T-S fuzzy models with relaxed stability conditions. *ISA Transactions*, *48*(3), 295–303.

22. Li, C., & Liao, X. (2005). Passivity analysis of neural networks with time delay. *IEEE Transactions on Circuits and Systems II: Express Briefs*, *52*(8), 471–475.
23. Park, J. (2007). Further results on passivity analysis of delayed cellular neural networks. *Chaos, Solitons and Fractals*, *34*(5), 1546–1551.
24. Song, Q., & Wang, Z. (2010). New results on passivity analysis of uncertain neural networks with time-varying delays. *International Journal of Computer Mathematics*, *87*(3), 668–678.
25. Song, Q., Liang, J., & Wang, Z. (2009). Passivity analysis of discrete-time stochastic neural networks with time-varying delays. *Neurocomputing*, *72*(7–9), 1782–1788.
26. Zhu, J., Zhang, Q., & Yuan, Z. (2010). Delay-dependent passivity criterion for discrete-time delayed standard neural network model. *Neurocomputing*, *73*(7–9), 1384–1393.
27. Zhang, Z., Mou, S., Lam, J., & Gao, H. (2009). New passivity criteria for neural networks with time-varying delay. *Neural Networks*, *22*(7), 864–868.
28. Xu, S., Zheng, W. X., & Zou, Y. (2009). Passivity analysis of neural networks with time-varying delays. *IEEE Transactions on Circuits and Systems II: Express Briefs*, *56*(4), 325–329.
29. Chen, B., Li, H., Lin, C., & Zhou, Q. (2009). Passivity analysis for uncertain neural networks with discrete and distributed time-varying delays. *Physics Letters A*, *373*(14), 1242–1248.
30. Li, H., Wang, C., Shi, P., & Gao, H. (2010). New passivity results for uncertain discrete-time stochastic neural networks with mixed time delays. *Neurocomputing*, *73*(16–18), 3291–3299.
31. Balasubramaniam, P., & Nagamani, G. (2010). Passivity analysis of neural networks with Markovian jumping parameters and interval time-varying delays. *Nonlinear Analysis: Hybrid Systems*, *4*(4), 853–864.
32. Fu, J., Zhang, H., Ma, T., & Zhang, Q. (2010). On passivity analysis for stochastic neural networks with interval time-varying delay. *Neurocomputing*, *73*(4–6), 795–801.
33. Balasubramaniam, P., Nagamani, G., & Rakkiyappan, R. (2010). Global passivity analysis of interval neural networks with discrete and distributed delays of neutral type. *Neural Processing Letters*, *32*(2), 109–130.
34. Balasubramaniam, P., & Vidhya, C. (2010). Global asymptotic stability of stochastic BAM neural networks with distributed delays and reaction-diffusion terms. *Journal of Computational and Applied Mathematics*, *234*(12), 3458–3466.
35. Song, Q., & Wang, Z. (2009). Dynamical behaviors of fuzzy reaction-diffusion periodic cellular neural networks with variable coefficients and delays. *Applied Mathematical Modelling*, *33*(9), 3533–3545.
36. Hu, C., Jiang, H., & Teng, Z. (2010). Impulsive control and synchronization for delayed neural networks with reaction-diffusion terms. *IEEE Transactions on Neural Networks*, *21*(1), 67–81.
37. Chen, J. L., Lee, L. (2001). Passivity approach to feedback connection stability for discrete-time descriptor systems. In *Proceedings of the 40th IEEE Conference on Decision and Control (Cat. No.01CH37228)* (vol. 3, pp. 2865–2866).
38. Wang, L., & Gao, Y. (2006). Global exponential robust stability of reaction-diffusion interval neural networks with time-varying delays. *Physics Letters A*, *350*(5–6), 342–348.
39. Lu, J. G. (2007). Robust global exponential stability for interval reaction-diffusion hopfield neural networks with distributed delays. *IEEE Transactions on Circuits and Systems II: Express Briefs*, *54*(12), 1115–1119.
40. Balasubramaniam, P., & Ali, M. S. (2010). Robust stability of uncertain fuzzy cellular neural networks with time-varying delays and reaction diffusion terms. *Neurocomputing*, *74*(1–3), 439–446.
41. Lu, J. G. (2008). Global exponential stability and periodicity of reaction-diffusion delayed recurrent neural networks with Dirichlet boundary conditions. *Chaos, Solitons and Fractals*, *35*(1), 116–125.
42. Yao, J., Wang, H. O., Guan, Z. H., & Xu, W. (2009). Passive stability and synchronization of complex spatio-temporal switching networks with time delays. *Automatica*, *45*(7), 1721–1728.
43. Yao, J., Guan, Z. H., & Hill, D. J. (2009). Passivity-based control and synchronization of general complex dynamical networks. *Automatica*, *45*(9), 2107–2113.
44. Niculescu, S., & Lozano, R. (2001). On the passivity of linear delay systems. *IEEE Transactions on Automatic Control*, *46*(3), 460–464.

45. Wang, J. L., Yang, Z. C., & Wu, H. N. (2012). Passivity analysis of complex dynamical networks with multiple time-varying delays. *Journal of Engineering Mathematics*, *74*(1), 175–188.
46. Lu, J. G., & Lu, L. J. (2009). Global exponential stability and periodicity of reaction-diffusion recurrent neural networks with distributed delays and Dirichlet boundary conditions. *Chaos, Solitons and Fractals*, *39*(4), 1538–1549.

Chapter 8
Passivity of DRDNs with Application to a Food Web Model

8.1 Introduction

Recently, stability and control of delayed CDNs have received much attention. The main reasons for this are twofold. First, in reality, there usually are some time-delays in spreading and response due to the finite speeds of transmission and/or the traffic congestions, and most of delays are notable, thus time-delays should be modeled in order to simulate more realistic networks. Second, the existence of a delay in a network model may induce oscillations, instability or bad performances for the closed-loop schemes [1]. On the other hand, stability problems are often linked to the theory of dissipative systems which postulates the energy dissipated inside a dynamic system is less than the energy supplied from external source. Dissipativeness was initially introduced by Willems [2, 3], and further extended by Hill, Moylan, Byrnes et al. [4–8]. Dissipativity is characterized by supply rate and storage function, which respectively represent the energy supplied from outside the system and energy stored inside the system. Lyapunov function can be viewed as a generalisation of storage function for dynamical systems, thus in many cases stability and stabilization problems can be solved once the dissipativity property is assured [6, 8].

Passivity, one of the most useful forms of dissipativity, has been studied separately. The passivity theory was firstly proposed in the circuit analysis [9] and since then has found successful applications in diverse areas such as stability [10], complexity [11], signal processing [12], chaos control and synchronization [13, 14], and fuzzy control [15]. Moreover, in recent years, as a powerful tool, passivity has played an important role in energy management, process control, group coordination, networked control and so on [16–20]. These are the main reasons why passivity theory has become a very hot topic, and attracted increasing attention of researchers from different fields. In particular, special attention has been focused on the passivity problems of NNs [21–31] and complex networks with diffusive coupling [32–35]. Song, Liang and

The original version of this book was revised: The belated corrections have been incorporated. The corrections to this book are available at https://doi.org/10.1007/978-981-33-4287-3_10

J.-L. Wang et al., *Passivity of Complex Dynamical Networks*,
https://doi.org/10.1007/978-981-33-4287-3_8

Wang [22] investigated the passivity and robust passivity of discrete-time stochastic neural networks with time-varying delays. By constructing proper Lyapunov-Krasovskii functionals and employing a combination of the free-weighting matrix method and stochastic analysis technique, two delay-dependent passivity conditions were established. In [25], the problem of robust passivity analysis was studied for interval neural networks with discrete time-varying delays. Applying the delay decomposition approach, a robust passivity condition was derived in terms of linear matrix inequalities. Yao et al. [33] discussed a class of complex spatio-temporal switching networks with time-delays, and several criteria for globally asymptotically passive stability and synchronization were obtained. Wang, Yang and Wu [35] investigated input passivity and output passivity for a generalized complex network with non-linear, time-varying, non-symmetric and delayed coupling, and several sufficient conditions ensuring input passivity and output passivity were derived. It is well known that the passivity theory plays an important role in designing asymptotically stabilizing controller for nonlinear systems in recent years. To control the undesirable chaotic oscillations in power system, Wei and Luo [36] proposed a design method of passivity-based adaptive controllers. However, few authors have considered the problem of passivity-based controller design for NNs and complex networks with diffusive coupling. For instance, a passivity-based controller design method for Hopfield NNs was given in [37].

Practically, passivity concepts are not only important for the traditional network models without reaction diffusion, but also for DRDNs. Unfortunately, there are few works on the passivity of DRDNs [38], in which Wang, Wu and Guo investigated the passivity and stability of a class of RDNNs with time-varying delays and Dirichlet boundary conditions. By utilizing the Lyapunov functional method combined with the inequality techniques, some sufficient conditions ensuring the passivity and global exponential stability were derived. Furthermore, when the parameter uncertainties appear in RDNNs, several criteria for robust passivity and robust global exponential stability were also presented. In most existing works (see also the above mentioned references), there is a simplified assumption that the input and output variables are only dependent on the time. Practically, the input and output variables are not only dependent on the time, but also intensively dependent on space variable in many realistic networks, particularly, reaction-diffusion networks. It is, therefore, the purpose of this chapter is to pave a way for investigating the passivity problem of DRDNs, in which the input and output variables are varied with the time and space variables.

This chapter studies the passivity of a class of DRDNs by using the Lyapunov functional method, and also reveals the relationship between passivity and stability of the DRDNs. The food web is one of most well-known examples of complex networks and holds a central place in ecology. More recently, the reaction-diffusion models have been used to describe the dynamic changes of spatial density of species in a bounded spatial habitat by investigators [39–48]. Pang and Wang [42] investigated a strongly coupled system of partial differential equations (PDEs) which models the dynamics of a two-predator-one-prey ecosystem. In the real-life world, there are many species whose growth rate is not only related to the current number of population, but also to the past one of population [47]. Thus time-delay should be modeled in order to simulate more realistic food webs. In [47], Wang et al. proposed a class of 3-species Lotka-Volterra mutualism models with diffusion and delay effects. A condition was given to ensure the global asymptotic stability of the positive steady-state solution. Xu [48] investigated a reaction-diffusion predator-prey model with stage structure and nonlocal delay. As a natural extension of the existing models, we introduce a new CDN model with time-delay which describes a cooperative interaction of N-species that benefits each other in a bounded habitat Ω. Then we apply the obtained results to investigate the asymptotic stability and control design of the proposed network model.

8.2 Preliminaries

Definition 8.1 A system with supply rate ϑ is said to be dissipative if there exists a nonnegative function $S : \mathbb{R}^+ \to \mathbb{R}^+$, called the storage function, such that

$$\int_{t_0}^{t_p} \vartheta(u, y)dt \geqslant S(t_p) - S(t_0)$$

for any $t_p, t_0 \in \mathbb{R}^+$ and $t_p \geqslant t_0$, where $u(x, t) \in \mathbb{R}^n$ and $y(x, t) \in \mathbb{R}^n$ are the input and output of the system at time t and in space x, respectively, $(x, t) \in \Omega \times \mathbb{R}^+$.

Remark 8.2 In [3], Willems proposed a dissipativity notion for nonlinear ordinary differential equation (ODE) systems using storage function and supply rate. The Definition 8.1 is an extension of the definition of dissipativity in [3], and will play an important role in the stability analysis of PDE systems.

Definition 8.3 A system is said to be passive if it is dissipative with respect to

$$\vartheta(u, y) = \int_{\Omega} y^T(x, t)u(x, t)dx,$$

where $u(x, t) \in \mathbb{R}^n$ and $y(x, t) \in \mathbb{R}^n$ are the input and output of the system at time t and in space x, respectively, $(x, t) \in \Omega \times \mathbb{R}^+$.

Definition 8.4 A system is said to be strictly passive if it is dissipative with respect to

$$\vartheta(u, y) = \int_{\Omega} y^T(x, t)u(x, t)dx - \gamma_1 \int_{\Omega} u^T(x, t)u(x, t)dx$$
$$-\gamma_2 \int_{\Omega} y^T(x, t)y(x, t)dx$$

for $\gamma_1 \geqslant 0, \gamma_2 \geqslant 0, \gamma_1 + \gamma_2 > 0$, where $u(x, t) \in \mathbb{R}^n$ and $y(x, t) \in \mathbb{R}^n$ are the input and output of the system at time t and in space x, respectively, $(x, t) \in \Omega \times \mathbb{R}^+$.

The system is said to be input strictly passive if $\gamma_1 > 0$ and output strictly passive if $\gamma_2 > 0$.

Remark 8.5 Many authors have studied the passivity of delay ordinary differential equation (DODE) systems, in which the input and output variables are only dependent on the time. Unfortunately, the input and output variables in many systems are varied with the time and space variables. But this case can not be dealt with by traditional passivity theory. Obviously, the Definitions 8.4 and 8.6 can be viewed as the extension of existing ones [1, 32–35].

Consider the following delay partial differential equation (DPDE) system:

$$\begin{cases} \frac{\partial z(x,t)}{\partial t} = D\Delta z(x,t) + g(z(x,t-\tau)) + f(z(x,t)), \\ z(x,t) = \phi(x,t), \ (x,t) \in \Omega \times [-\tau, 0], \\ z(x,t) = 0, \ (x,t) \in \partial\Omega \times [-\tau, +\infty), \end{cases} \tag{8.1}$$

where $x = (x_1, x_2, \ldots, x_\sigma)^T \in \Omega \subset \mathbb{R}^\sigma$; $\Omega = \{x = (x_1, x_2, \ldots, x_\sigma)^T \mid |x_k| < l_k, k = 1, 2, \ldots, \sigma\}$ is a bounded compact set with smooth boundary $\partial\Omega$ and $\text{mes}\Omega > 0$ in space $\mathbb{R}^\sigma$; $z(x,t) = (z_1(x,t), z_2(x,t), \ldots, z_n(x,t))^T \in \mathbb{R}^n$ is the state variable of the system at time t and in space x; $\Delta = \sum_{k=1}^{\sigma} \frac{\partial^2}{\partial x_k^2}$ is the Laplace operator on Ω; $\tau > 0$ represents the time-delay and $D = \text{diag}(d_1, d_2, \ldots, d_n) > 0$; $\phi(x,t) \in \mathbb{R}^n$ is bounded and continuous on $\Omega \times [-\tau, 0]$; the functions $f(\cdot) \in \mathbb{R}^n$ and $g(\cdot) \in \mathbb{R}^n$ satisfy $f(0) = 0, g(0) = 0$, and the following assumption:

(A1) The functions $f(\cdot)$ and $g(\cdot)$ are continuous, and there exist two positive constants ρ_1 and ρ_2 such that

$$||f(\xi_1) - f(\xi_2)|| \leqslant \rho_1||\xi_1 - \xi_2||, \quad ||g(\xi_1) - g(\xi_2)| \leqslant \rho_2||\xi_1 - \xi_2||$$

for any $\xi_1, \xi_2 \in \mathbb{R}^n$,

Definition 8.6 Let $z(x,t,\phi)$ be the state trajectory of system (8.1) with initial condition $\phi(x,t)$. The equilibrium solution $z^* = 0$ of system (8.1) is said to be asymptotically stable if any $\epsilon > 0$, there exists $\delta(\epsilon)$ such that $\|\phi(\cdot,0)\|_\tau < \delta(\epsilon)$ implies $\|z(\cdot,t,\phi)\|_2 < \epsilon$ for $t \geqslant 0$, and there is a $b_0 > 0$ such that $\|\phi(\cdot,0)\|_\tau < b_0$ implies $\|z(\cdot,t,\phi)\|_2 \to 0$ as $t \to +\infty$.

To our knowledge, there are few works on the asymptotic stability (in the sense of Definition 8.6) of DPDE systems. It is well known that Lyapunov functional is a very convenient tool to analyze the asymptotic stability of DODE systems [49–51]. In the following, we give a sufficient condition for the asymptotic stability of the equilibrium solution $z^* = 0$ of system (8.1).

Lemma 8.7 (see [52]) *Suppose that $v_1, v_2, v_3 : \mathbb{R}^+ \to \mathbb{R}^+$ are continuous and strictly monotonically nondecreasing functions, $v_1(s), v_2(s), v_3(s)$ are positive for $s > 0$ with $v_1(0) = v_2(0) = 0$. If there is a continuous functional $V : \mathbb{R}^+ \times C \to \mathbb{R}^+$ such that*

$$v_1(\|z(\cdot,t)\|_2) \leqslant V(t, z_t(x)) \leqslant v_2(\|z(\cdot,t)\|_\tau),$$
$$\dot{V}(t, z_t(x)) \leqslant -v_3(\|z(\cdot,t)\|_2),$$

where $z_t(x) \triangleq z(x, t+\theta)$, $-\tau \leqslant \theta \leqslant 0$, $\dot{V}$ *is the derivative of* V *along the solution of system* (8.1)*, then the equilibrium solution* $z^* = 0$ *of system* (8.1) *is asymptotically stable.*

8.3 On the Passivity of a Class of DRDNs

Given the importance of passivity theory for traditional network models, it seems quite natural to generalize the passivity theory to DRDNs. However, this has not received much attention until now with few results appearing on the topic. Moreover, in most existing works on the passivity of systems, they always assume that the input and output variables are only dependent on the time. Therefore, it is important and interesting to investigate the passivity of DRDNs, in which the input and output variables are varied with the time and space variables. The reaction-diffusion network to be studied in this section is described by

$$\begin{cases} \frac{\partial w(x,t)}{\partial t} = D\Delta w(x,t) + f(w(x,t)) + Ag(w(x,t)) + Eu(x,t) \\ \qquad\qquad +Bg(w(x,t-\tau)), \\ y(x,t) = Fw(x,t) + Hu(x,t), \\ w(x,t) = \Phi(x,t), (x,t) \in \Omega \times [-\tau, 0], \\ w(x,t) = 0, (x,t) \in \partial\Omega \times [-\tau, +\infty), \end{cases} \tag{8.2}$$

where $x = (x_1, x_2, \ldots, x_\sigma)^T \in \Omega \subset \mathbb{R}^\sigma$ and $\Omega = \{x = (x_1, x_2, \ldots, x_\sigma)^T \mid |x_k| < l_k, k = 1, 2, \ldots, \sigma\}$ is a bounded compact set with smooth boundary $\partial\Omega$ and mes$\Omega > 0$ in space $\mathbb{R}^\sigma$; $w(x,t) = (w_1(x,t), w_2(x,t), \ldots, w_n(x,t))^T \in \mathbb{R}^n$, $u(x,t) \in \mathbb{R}^n$ and $y(x,t) \in \mathbb{R}^n$ are the state, input and output of the network at time t and in space x, respectively; $\Delta = \sum_{k=1}^{\sigma} \frac{\partial^2}{\partial x_k^2}$ is the Laplace operator on Ω; $\tau > 0$ represents the time-delay, $D = \text{diag}(d_1, d_2, \ldots, d_n) > 0$, and $A, B, E, F, H \in \mathbb{R}^{n\times n}$ are known real matrices; $\Phi(x,t) = (\phi_1(x,t), \phi_2(x,t), \ldots, \phi_n(x,t))^T$, and $\phi_i(x,t) (i = 1, 2, \ldots, n)$ is bounded and continuous on $\Omega \times [-\tau, 0]$; the functions $f(w(x,t)) = (f_1(w_1(x,t)),$ $f_2(w_2(x,t)), \ldots, f_n(w_n(x,t)))^T \in \mathbb{R}^n$ and $g(w(x,t)) = (g_1(w_1(x,t)), g_2(w_2(x,t)), \ldots, g_n(w_n(x,t)))^T \in \mathbb{R}^n$ satisfy $f(0) = 0$, $g(0) = 0$, and the following assumption:

(A2) There exist positive constants ρ_j and $\upsilon_j (j = 1, 2, \ldots, n)$ such that

$$|f_j(\xi_1) - f_j(\xi_2)| \leqslant \rho_j |\xi_1 - \xi_2|,$$
$$|g_j(\xi_1) - g_j(\xi_2)| \leqslant \upsilon_j |\xi_1 - \xi_2|$$

for any $\xi_1, \xi_2 \in \mathbb{R}$.

Remark 8.8 As is well known, in the modeling of food webs, it is sometimes necessary to consider the diffusion effects because the population is usually not homogeneously distributed and the different population densities of predators and preys may cause different population movements. On the other hand, the diffusion effects can not be avoided in NNs once electrons transport in a nonuniform electromagnetic field [53]. In addition, the diffusion phenomena can not be ignored in chemical reactions. Therefore, the reaction-diffusion models have been used to describe these diffusion phenomena by investigators [38, 53–55]. Obviously, these reaction-diffusion systems can be represented by (8.2) through an appropriate choice of the parameters.

For convenient analysis, we let

$$\rho = \mathrm{diag}(\rho_1, \rho_2, \ldots, \rho_n), \upsilon = \mathrm{diag}(\upsilon_1, \upsilon_2, \ldots, \upsilon_n).$$

Theorem 8.9 *If there exist matrices* $\mathbb{R}^{n\times n} \ni P = \mathrm{diag}(p_1, p_2, \ldots, p_n) > 0$, $\mathbb{R}^{n\times n} \ni Q > 0$, $\mathbb{R}^{n\times n} \ni M_1 > 0$, $\mathbb{R}^{n\times n} \ni M_2 > 0$, $\mathbb{R}^{n\times n} \ni M_3 > 0$ *and a scalar* $\gamma \geqslant 0$ *such that*

$$\begin{pmatrix} \chi & PE - F^T & 0 \\ E^T P - F & -H - H^T - \gamma I_n & 0 \\ 0 & 0 & \Theta \end{pmatrix} \leqslant 0, \tag{8.3}$$

where $\chi = \sum_{k=1}^{\sigma} \frac{-2\lambda_m(PD)}{l_k^2} I_n + PM_1P + PAM_2A^TP + PBM_3B^TP + \lambda_M(M_1^{-1})\rho^2 + \lambda_M(M_2^{-1})\upsilon^2 + Q$, $\Theta = \lambda_M(M_3^{-1})\upsilon^2 - Q$, *then network* (8.2) *satisfies the following inequality:*

$$\int_{t_0}^{t_p}\int_{\Omega} y^T(x,t)u(x,t)dxdt \geqslant \frac{1}{2}(V(t_p) - V(t_0)) - \frac{\gamma}{2}\int_{t_0}^{t_p}\int_{\Omega} u^T(x,t)u(x,t)dxdt,$$

where $t_p, t_0 \in \mathbb{R}^+$, $t_p \geqslant t_0$, *and*

$$V(t) = \int_{\Omega} w^T(x,t)Pw(x,t)dx + \int_{t-\tau}^{t}\int_{\Omega} w^T(x,s)Qw(x,s)dxds. \tag{8.4}$$

Proof Calculating the derivative of $V(t)$ along the solution of network (8.2), we have

$$\begin{aligned}\dot{V}(t) &= \int_{\Omega} w^T(x,t)P\frac{\partial w(x,t)}{\partial t}dx + \int_{\Omega} w^T(x,t)Qw(x,t)dx \\ &\quad - \int_{\Omega} w^T(x,t-\tau)Qw(x,t-\tau)dx \\ &= 2\int_{\Omega} w^T(x,t)P[D\Delta w(x,t) + f(w(x,t)) + Ag(w(x,t)) \\ &\quad + Bg(w(x,t-\tau)) + Eu(x,t)]dx + \int_{\Omega} w^T(x,t)Qw(x,t)dx\end{aligned}$$

$$
\begin{aligned}
&- \int_\Omega w^T(x,t-\tau)Qw(x,t-\tau)dx \\
&\leqslant 2\int_\Omega w^T(x,t)PD\Delta w(x,t)dx + \int_\Omega w^T(x,t)PM_1Pw(x,t)dx \\
&\quad + \int_\Omega f^T(w(x,t))M_1^{-1}f(w(x,t))dx + 2\int_\Omega w^T(x,t)PEu(x,t)dx \\
&\quad + \int_\Omega w^T(x,t)PAM_2A^TPw(x,t)dx \\
&\quad + \int_\Omega g^T(w(x,t))M_2^{-1}g(w(x,t))dx \\
&\quad + \int_\Omega w^T(x,t)PBM_3B^TPw(x,t)dx \\
&\quad + \int_\Omega g^T(w(x,t-\tau))M_3^{-1}g(w(x,t-\tau))dx \\
&\quad + \int_\Omega w^T(x,t)Qw(x,t)dx - \int_\Omega w^T(x,t-\tau)Qw(x,t-\tau)dx.
\end{aligned}
$$

From Green's formula and the Dirichlet boundary condition, we then have

$$
\int_\Omega w_i(x,t)\Delta w_i(x,t)dx = -\sum_{k=1}^{\sigma}\int_\Omega \left(\frac{\partial w_i(x,t)}{\partial x_k}\right)^2 dx. \tag{8.5}
$$

It follows from (8.5) and Lemma 1.5 that

$$
\begin{aligned}
\int_\Omega w^T(x,t)PD\Delta w(x,t)dx &= -\sum_{k=1}^{\sigma}\int_\Omega \left(\frac{\partial w(x,t)}{\partial x_k}\right)^T PD\frac{\partial w(x,t)}{\partial x_k}dx \\
&\leqslant -\lambda_m(PD)\sum_{k=1}^{\sigma}\int_\Omega \left(\frac{\partial w(x,t)}{\partial x_k}\right)^T \frac{\partial w(x,t)}{\partial x_k}dx \\
&\leqslant \sum_{k=1}^{\sigma}\frac{-\lambda_m(PD)}{l_k^2}\int_\Omega w^T(x,t)w(x,t)dx.
\end{aligned}
$$

Therefore,

$$
\begin{aligned}
&\dot{V}(t) - 2\int_{\Omega} y^T(x,t)u(x,t)dx - \gamma\int_{\Omega} u^T(x,t)u(x,t)dx \\
&\leqslant \int_{\Omega} w^T(x,t)\Bigg(\sum_{k=1}^{\sigma}\frac{-2\lambda_m(PD)}{l_k^2}I_n + PAM_2A^TP + PM_1P + PBM_3B^TP \\
&+\lambda_M(M_1^{-1})\rho^2 + Q + \lambda_M(M_2^{-1})\upsilon^2\Bigg)w(x,t)dx \\
&+\int_{\Omega} w^T(x,t-\tau)(-Q + \lambda_M(M_3^{-1})\upsilon^2)w(x,t-\tau)dx \\
&+2\int_{\Omega} w^T(x,t)(PE - F^T)u(x,t)dx \\
&+\int_{\Omega} u^T(x,t)(-H - H^T - \gamma I_n)u(x,t)dx \\
&= \int_{\Omega} \xi^T(x,t)\begin{pmatrix} \chi & PE - F^T & 0 \\ E^TP - F & -H - H^T - \gamma I_n & 0 \\ 0 & 0 & \Theta \end{pmatrix}\xi(x,t)dx,
\end{aligned}
$$

where $\xi(x,t) = (w^T(x,t), u^T(x,t), w^T(x,t-\tau))^T$. From (8.3), we can get

$$
\dot{V}(t) - \gamma\int_{\Omega} u^T(x,t)u(x,t)dx \leqslant 2\int_{\Omega} y^T(x,t)u(x,t)dx. \tag{8.6}
$$

By integrating (8.6) with respect to t over the time period t_0 to t_p, we can obtain

$$
2\int_{t_0}^{t_p}\int_{\Omega} y^T(x,t)u(x,t)dxdt \geqslant V(t_p) - V(t_0) - \gamma\int_{t_0}^{t_p}\int_{\Omega} u^T(x,t)u(x,t)dxdt.
$$

Namely,

$$
\int_{t_0}^{t_p}\int_{\Omega} y^T(x,t)u(x,t)dxdt \geqslant \frac{1}{2}(V(t_p) - V(t_0)) - \frac{\gamma}{2}\int_{t_0}^{t_p}\int_{\Omega} u^T(x,t)u(x,t)dxdt
$$

for any $t_p, t_0 \in \mathbb{R}^+$ and $t_p \geqslant t_0$. The proof is completed. Obviously, if $\gamma = 0$, then network (8.2) is passive in the sense of Definition 8.3.

Remark 8.10 In the existing works, there is another widely accepted definition of passivity [24, 28]:

A system with input $u(t)$ and output $y(t)$ where $u(t), y(t) \in \mathbb{R}^n$ is called passive if there exists a scalar $\gamma \geqslant 0$ such that

$$2\int_0^{t_p} y^T(t)u(t)dt \geqslant -\gamma \int_0^{t_p} u^T(t)u(t)dt$$

for all $t_p \geqslant 0$ and for all solution $x(t, 0)$. Correspondingly, we introduce the following concept of passivity:

A system is said to be passive if it is dissipative with respect to

$$\vartheta(u, y) = 2\int_\Omega y^T(x, t)u(x, t)dx + \gamma \int_\Omega u^T(x, t)u(x, t)dx,$$

where $u(x, t) \in \mathbb{R}^n$ and $y(x, t) \in \mathbb{R}^n$ are the input and output of the system at time t and in space x, respectively, $(x, t) \in \Omega \times \mathbb{R}^+$. Obviously, network (8.2) is passive in the sense of the above definition if condition (8.3) holds.

By the similar proof of Theorem 8.9, we can obtain the following conclusions. Here we omit their proof to avoid the repetition.

Theorem 8.11 *The network* (8.2) *is input strictly passive if there exist matrices* $\mathbb{R}^{n\times n} \ni P = \mathrm{diag}(p_1, p_2, \ldots, p_n) > 0$, $\mathbb{R}^{n\times n} \ni Q > 0$, $\mathbb{R}^{n\times n} \ni M_1 > 0$, $\mathbb{R}^{n\times n} \ni M_2 > 0$, $\mathbb{R}^{n\times n} \ni M_3 > 0$ *and a scalar* $\gamma > 0$ *such that*

$$\begin{pmatrix} \chi & PE - F^T & 0 \\ E^T P - F & -H - H^T + \gamma I_n & 0 \\ 0 & 0 & \Theta \end{pmatrix} \leqslant 0, \tag{8.7}$$

where $\chi = \sum_{k=1}^{\sigma} \frac{-2\lambda_m(PD)}{l_k^2} I_n + PM_1P + PAM_2A^TP + PBM_3B^TP + \lambda_M(M_1^{-1})\rho^2 + \lambda_M(M_2^{-1})\upsilon^2 + Q$, $\Theta = \lambda_M(M_3^{-1})\upsilon^2 - Q$.

Theorem 8.12 *The network* (8.2) *is output strictly passive if there exist matrices* $\mathbb{R}^{n\times n} \ni P = \mathrm{diag}(p_1, p_2, \ldots, p_n) > 0$, $Q > 0$, $\mathbb{R}^{n\times n} \ni M_1 > 0$, $\mathbb{R}^{n\times n} \ni M_2 > 0$, $\mathbb{R}^{n\times n} \ni M_3 > 0$ *and a scalar* $\gamma > 0$ *such that*

$$\begin{pmatrix} \hat{\chi} & \Xi & 0 \\ \Xi^T & -H - H^T + \gamma H^T H & 0 \\ 0 & 0 & \Theta \end{pmatrix} \leqslant 0, \tag{8.8}$$

where $\hat{\chi} = \sum_{k=1}^{\sigma} \frac{-2\lambda_m(PD)}{l_k^2} I_n + PM_1P + PAM_2A^TP + PBM_3B^TP + \lambda_M(M_1^{-1})\rho^2 + \lambda_M(M_2^{-1})\upsilon^2 + Q + \gamma F^T F$, $\Xi = PE - F^T + \gamma F^T H$, $\Theta = \lambda_M(M_3^{-1})\upsilon^2 - Q$.

Remark 8.13 The main difficulty for passivity analysis of reaction-diffusion network (8.2) comes from the reaction-diffusion terms, which can not be dealt with by those techniques used in traditional network models without reaction diffusion. By employing Green's formula and Lemma 1.5, several passivity criteria are proposed in this chapter, which are dependent on the reaction-diffusion terms.

In the above, some sufficient conditions ensuring the passivity, input strict passivity and output strict passivity are obtained. It is well known that passivity is a significant theory to handle the problem of stability analysis of nonlinear ODE systems. In [4], Hill and Moylan discussed the stability of a class of nonlinear passive systems. Under the assumption that system is zero-state detectable, they proved that passive and input strictly passive systems are stable, while output strictly passive systems are asymptotically stable. Practically, these results can also be extended to PDE systems.

Theorem 8.14 *Suppose that functional $V : \mathbb{R}^+ \times C \to \mathbb{R}^+$ is continuously differentiable and satisfies the following condition*

$$v_1(\|w(\cdot,t)\|_2) \leqslant V(t, w_t(x)) \leqslant v_2(\|w(\cdot,t)\|_\tau), \tag{8.9}$$

where $w_t(x) \triangleq w(x, t+\theta), -\tau \leqslant \theta \leqslant 0,\ v_1, v_2 : \mathbb{R}^+ \to \mathbb{R}^+$ are continuous and strictly monotonically nondecreasing functions, $v_1(s)$ and $v_2(s)$ are positive for $s > 0$ with $v_1(0) = v_2(0) = 0$. Then

(i) network (8.2) is asymptotically stable if it is output strictly passive with respect to storage function V and matrix F is nonsingular;
(ii) network (8.2) is stable if it is passive or input strictly passive with respect to storage function V.

Proof If network (8.2) is output strictly passive with respect to storage function V, then there obviously exists a positive constant γ such that

$$\begin{aligned} V(t+\varepsilon) - V(t) \leqslant & \int_t^{t+\varepsilon}\int_\Omega y^T(x,s)u(x,s)dxds \\ & -\gamma \int_t^{t+\varepsilon}\int_\Omega y^T(x,s)y(x,s)dxds \end{aligned} \tag{8.10}$$

for any $t \in \mathbb{R}^+$ and $\varepsilon > 0$. In what follows, we can derive from (8.10) that

$$\begin{aligned} \frac{V(t+\varepsilon) - V(t)}{\varepsilon} \leqslant & \frac{\int_t^{t+\varepsilon}\int_\Omega y^T(x,s)u(x,s)dxds}{\varepsilon} \\ & -\gamma \frac{\int_t^{t+\varepsilon}\int_\Omega y^T(x,s)y(x,s)dxds}{\varepsilon}. \end{aligned} \tag{8.11}$$

By taking limit $\epsilon \to 0$ in (8.11), we have

$$\dot{V}(t) \leqslant \int_{\Omega} y^T(x,t)u(x,t)dx - \gamma \int_{\Omega} y^T(x,t)y(x,t)dx.$$

Letting $u(x,t) = 0$, we can get

$$\dot{V}(t) \leqslant -\gamma\lambda_m(F^T F)\|w(\cdot,t)\|_2^2.$$

Therefore, we can obtain from Lemma 8.7 that network (8.2) is asymptotically stable.

By the similar proof of conclusion (i), we can get the conclusion (ii). Here we omit its proof to avoid the repetition. The proof is completed.

Remark 8.15 The passivity theory plays an important role in designing stabilizing controllers for asymptotic stability of nonlinear ODE systems, e.g., power system model [36], Hopfield neural networks [37], stochastic systems [56], and so on. In the next section, it will be seen that passivity theory can also be used to design asymptotically stabilizing controller for PDE systems.

8.4 Modeling, Analysis and Control of Food Webs

In this section, we propose a new complex network model to describe the dynamic changes of spatial density of species in a bounded habitat. Then, a sufficient condition for asymptotic stability is derived using Lyapunov method and inequality techniques. Furthermore, we address a design problem for passivity-based controllers of the proposed network model.

8.4.1 Model and Preliminaries

Food webs are among the most well-known examples of complex networks and hold a central place in ecology to study the dynamics of animal populations. A food web can be characterized by a model of complex network, in which a node represents a species. Species are usually inhomogeneously distributed and the different population densities of predators and preys may cause different population movements, thus it is important and interesting to investigate their spatial density in order to better protect and control their population. In such a case, the node state will represent the spatial density of species. We introduce a new network model which is totally different from some existing complex network models. Our model is described by the following PDEs:

$$
\begin{aligned}
\frac{\partial w_i(x,t)}{\partial t} &= d_i \Delta w_i(x,t) + a_i w_i(x,t) + u_i(x,t) + \sum_{j=1}^{N} G_{ij} f_j(w_j(x,t)) \\
&\quad + \sum_{j=1}^{N} \hat{G}_{ij} g_j(w_j(x,t-\tau)), \qquad (8.12)
\end{aligned}
$$

where $i = 1, 2, \ldots, N$, N is the number of species in the food web; $x = (x_1, x_2, \ldots, x_\sigma)^T \in \Omega \subset \mathbb{R}^\sigma$ and $\Omega = \{x = (x_1, x_2, \ldots, x_\sigma)^T \mid |x_k| < l_k, k = 1, 2, \ldots, \sigma\}$ is a bounded domain with smooth boundary $\partial\Omega$ and mes$\Omega > 0$ in space $\mathbb{R}^\sigma$; $\tau > 0$ is an arbitrary but bounded constant representing the time-delay; $w_i(x,t) \in \mathbb{R}$ represents the spatial density of the ith species, $(x,t) \in \Omega \times (0, +\infty)$; $u_i(x,t)$ is the control input; the constant d_i is the diffusion coefficient of the ith species and is hence assumed to be positive; $\Delta = \sum_{k=1}^{\sigma} \frac{\partial^2}{\partial x_k^2}$ is the Laplace diffusion operator on Ω; a_i is the natural growth rate of the ith species; the functions $f_j(\cdot)$ and $g_j(\cdot) \in \mathbb{R}$, describing the interaction relations between species i and j, are continuous; $G = (G_{ij})_{N\times N}$ and $\hat{G} = (\hat{G}_{ij})_{N\times N}$ represent the interaction strength between species for non-delayed configuration and delayed one, respectively, where G_{ij} and $\hat{G}_{ij}$ are defined as follows: $G_{ij}, \hat{G}_{ij} (i \neq j)$ are nonnegative with $G_{ij} + \hat{G}_{ij} > 0$; G_{ii} is a negative constant and $\hat{G}_{ii} = 0$.

Let $w(x,t) = (w_1(x,t), w_2(x,t), \ldots, w_N(x,t))^T$. The initial value and boundary value conditions associated with the network (8.12) are given in form

$$
\begin{aligned}
& w(x,t) = \Phi(x,t), (x,t) \in \Omega \times [-\tau, 0], \qquad (8.13) \\
& \Phi(x,t) = (\phi_1(x,t), \phi_2(x,t), \ldots, \phi_N(x,t))^T, \\
& w(x,t) = 0, (x,t) \in \partial\Omega \times [-\tau, +\infty), \qquad (8.14)
\end{aligned}
$$

where $\phi_i(x,t) \geqslant 0$ is Holder continuous on $\overline{\Omega} \times [-\tau, 0]$, and satisfies the condition $\phi_i(x,t) = 0$ for $t \in [-\tau, 0]$ and $x \in \partial\Omega$. Further, $\phi_i(x, 0) \not\equiv 0$.

Remark 8.16 G_{ii} and $\hat{G}_{ii}$ represent intraspecific competition within species i, G_{ij} and $\hat{G}_{ij} (i \neq j)$ denote the interspecific cooperations among species. In the ecological sense, network (8.12) describes a cooperative interaction of N-species that benefits each other in a bounded habitat Ω.

Remark 8.17 It should be noted that network (8.12) represents a more general CDN than that considered in the literature in the following sense:

(1) The coupling configurations are not restricted to the symmetric and irreducible connections. Moreover, the case that the coupling configurations are related to the current states and the delayed states is also taken into account.
(2) In this network model, the state of node is dependent on the time and space, which can better reflect the dynamical behavior of real systems in some circumstances.
(3) The CDN consisting of nonidentical nodes may represent a more general and practical network than the models, which have been discussed in the literature.

(4) The coupling matrices in this model are not assumed to be diffusive. To our knowledge, the coupling matrices are often assumed to be diffusive in the existing literature.

In this section, we assume that the functions $f_j(\cdot)$ and $g_j(\cdot)$ satisfy $f_j(0) = g_j(0) = 0$, and the following property:

(A3) There exist positive constants ϱ_j and ν_j such that

$$0 \leqslant \frac{f_j(\xi_1) - f_j(\xi_2)}{\xi_1 - \xi_2} \leqslant \varrho_j,$$
$$0 \leqslant \frac{g_j(\xi_1) - g_j(\xi_2)}{\xi_1 - \xi_2} \leqslant \nu_j$$

for any $\xi_1, \xi_2 \in \mathbb{R}$, $\xi_1 \neq \xi_2$.

Let $u_i(x,t) = 0$, $i = 1, 2, \ldots, N$, we can get from (8.12) that

$$\frac{\partial w_i(x,t)}{\partial t} - d_i \Delta w_i(x,t) = \mathfrak{F}_i(w(x,t), w(x,t-\tau)), \tag{8.15}$$

where $\mathfrak{F}_i(w(x,t), w(x,t-\tau)) = a_i w_i(x,t) + \sum_{j=1}^{N} G_{ij} f_j(w_j(x,t)) + \sum_{j=1}^{N} \hat{G}_{ij} g_j(w_j(x,t-\tau))$, $i = 1, 2, \ldots, N$.

In what follows, we introduce some useful definitions and lemmas.

Definition 8.18 A vector function $\mathfrak{F}(w, v) = (\mathfrak{F}_1(w, v), \mathfrak{F}_2(w, v), \ldots, \mathfrak{F}_N(w, v))^T$ defined for w, v, in a subset Λ of $\mathbb{R}^N$ is said to be mixed quasimonotone if for each $i = 1, 2, \ldots, N$, there exist nonnegative integers c_i, e_i, k_i and m_i with $c_i + e_i = N - 1$ and $k_i + m_i = N$ such that for every $w = (w_i, [w]_{c_i}, [w]_{e_i})^T$ and $v = ([v]_{k_i}, [v]_{m_i})^T$ in Λ, $\mathfrak{F}_i(w, v)$ is monotone nondecreasing in $[w]_{c_i}$ and $[v]_{k_i}$ and monotone nonincreasing in $[w]_{e_i}$ and $[v]_{m_i}$. The function $\mathfrak{F}(w, v)$ is said to be quasimonotone nondecreasing in Λ if $e_i = m_i = 0$ for all i.

Definition 8.19 A pair of smooth functions $\tilde{w}(x,t) = (\tilde{w}_1(x,t), \tilde{w}_2(x,t), \ldots, \tilde{w}_N(x,t))^T$, $\hat{w}(x,t) = (\hat{w}_1(x,t), \hat{w}_2(x,t), \ldots, \hat{w}_N(x,t))^T$ are called coupled upper and lower solutions of (8.13)–(8.15) if $\tilde{w}(x,t) \geqslant \hat{w}(x,t)$ on $\overline{\Omega} \times [-\tau, +\infty)$ and if for each $i = 1, 2, \ldots, N$, $\tilde{w}_i(x,t)$ and $\hat{w}_i(x,t)$ satisfy the differential inequalities

$$\frac{\partial \tilde{w}_i(x,t)}{\partial t} - d_i \Delta \tilde{w}_i(x,t) \geqslant \mathfrak{F}_i(\tilde{w}_i(x,t), [\tilde{w}(x,t)]_{N-1}, [\tilde{w}(x,t-\tau)]_N),$$
$$\frac{\partial \hat{w}_i(x,t)}{\partial t} - d_i \Delta \hat{w}_i(x,t) \leqslant \mathfrak{F}_i(\hat{w}_i(x,t), [\hat{w}(x,t)]_{N-1}, [\hat{w}(x,t-\tau)]_N),$$
$$x \in \Omega, t > 0$$
$$\hat{w}_i(x,t) \leqslant \phi_i(x,t) \leqslant \tilde{w}_i(x,t), \text{ in } \Omega \times [-\tau, 0],$$
$$\hat{w}_i(x,t) \leqslant 0 \leqslant \tilde{w}_i(x,t), \text{ on } \partial\Omega \times [-\tau, +\infty).$$

Lemma 8.20 *Assume that* **(A3)** *holds and* $\tilde{w}(x,t) = (\tilde{w}_1(x,t), \tilde{w}_2(x,t), \ldots, \tilde{w}_N(x,t))^T$, $\hat{w}(x,t) = (\hat{w}_1(x,t), \hat{w}_2(x,t), \ldots, \hat{w}_N(x,t))^T$ *are a pair of coupled upper and lower solutions of system* (8.13)–(8.15), *then there exists a unique solution* $w^*(x,t) = (w_1^*(x,t), w_2^*(x,t), \ldots, w_N^*(x,t))^T$ *in* $\langle \hat{w}(x,t), \tilde{w}(x,t)\rangle = \{\varpi(x,t) \in \mathbb{R}^N, (x,t) \in \overline{\Omega} \times [-\tau, +\infty) \mid \hat{w}(x,t) \leqslant \varpi(x,t) \leqslant \tilde{w}(x,t)\}$ *for system* (8.13)–(8.15).

Proof Denote $\mathfrak{F}(w(x,t), w(x,t-\tau)) = (\mathfrak{F}_1(w(x,t), w(x,t-\tau)), \mathfrak{F}_2(w(x,t), w(x,t-\tau)), \ldots, \mathfrak{F}_N(w(x,t), w(x,t-\tau)))^T$. According to **(A3)**, we can easily derive that $\mathfrak{F}(w(x,t), w(x,t-\tau))$ is quasimonotone nondecreasing in $\langle \hat{w}(x,t), \tilde{w}(x,t)\rangle$. By the similar proof of Theorem 8.9 in [57], we can obtain the existence and uniqueness of the solution for system (8.13)–(8.15). The proof of Lemma 8.20 is completed.

Obviously, 0 is a lower solution of system (8.13)–(8.15). According to Lemma 8.20 , we can get the following conclusion.

Lemma 8.21 *Let* **(A3)** *hold. If there exists a nonnegative smooth function* $\tilde{w}(x,t) = (\tilde{w}_1(x,t), \tilde{w}_2(x,t), \ldots, \tilde{w}_N(x,t))^T$ *such that*

$$\begin{aligned}
\frac{\partial \tilde{w}_i(x,t)}{\partial t} - d_i \Delta \tilde{w}_i(x,t) &\geqslant \mathfrak{F}_i(\tilde{w}_i(x,t), [\tilde{w}(x,t)]_{N-1}, [\tilde{w}(x,t-\tau)]_N), \\
&\qquad x \in \Omega, t > 0 \\
\phi_i(x,t) &\leqslant \tilde{w}_i(x,t), \text{ in } \Omega \times [-\tau, 0], \\
0 &\leqslant \tilde{w}_i(x,t), \text{ on } \partial\Omega \times [-\tau, +\infty),
\end{aligned}$$

then there exists a unique solution $w^*(x,t) = (w_1^*(x,t), w_2^*(x,t), \ldots, w_N^*(x,t))^T$ *in* $\langle 0, \tilde{w}(x,t)\rangle$ *for system* (8.13)–(8.15).

Remark 8.22 It is obvious that bounded activation functions $f_j(\cdot)$ and $g_j(\cdot)$ always guarantee the existence of $\tilde{w}(x,t)$. That is, system (8.15) has a unique nonnegative solution for given initial value and boundary value conditions if **(A3)** holds and $f_j(\cdot), g_j(\cdot)$ are bounded functions. Practically, this assumption has been widely used in analyzing the stability of DRDNs [58, 59].

8.4.2 Asymptotic Stability Analysis

In this subsection, we shall discuss the asymptotic stability of system (8.15). By constructing suitable Lyapunov functional and utilizing some inequality techniques, a sufficient condition ensuring the asymptotic stability of equilibrium solution for system (8.15) is obtained.

Suppose that $w^*(x) = (w_1^*(x), w_2^*(x), \ldots, w_N^*(x))^T$ $(w_i^*(x) \not\equiv 0, i = 1, 2, \ldots, N)$ is a nonnegative equilibrium solution of system (8.15), then it satisfies

$$\begin{cases} 0 \quad = d_i \Delta w_i^*(x) + \sum_{j=1}^{N} G_{ij} f_j(w_j^*(x)) + a_i w_i^*(x) \\ \qquad + \sum_{j=1}^{N} \hat{G}_{ij} g_j(w_j^*(x)), \\ w_i^*(x) = 0, \quad x \in \partial\Omega, \end{cases} \tag{8.16}$$

where $i = 1, 2, \ldots, N$.

Suppose $w(x, t, \Phi) = (w_1(x, t, \Phi), w_2(x, t, \Phi), \ldots, w_N(x, t, \Phi))^T$ is an arbitrary solution of system (8.15) with initial condition $\Phi(x, t)$, and define $z(x, t) = w(x, t, \Phi) - w^*(x)$, $\Phi_z(x, t) = \Phi(x, t) - w^*(x)$, then the dynamics of difference vector $z(x, t) = (z_1(x, t), z_2(x, t), \ldots, z_N(x, t))^T$ is governed by the following equations:

$$\begin{aligned} \frac{\partial z_i(x, t)}{\partial t} &= d_i \Delta z_i(x, t) + \sum_{j=1}^{N} G_{ij} \hat{f}_j(z_j(x, t)) + a_i z_i(x, t) \\ &\quad + \sum_{j=1}^{N} \hat{G}_{ij} \hat{g}_j(z_j(x, t - \tau)), \end{aligned} \tag{8.17}$$

where $i = 1, 2, \ldots, N$, $\hat{f}_j(z_j(x, t)) = f_j(w_j(x, t, \Phi)) - f_j(w_j^*(x))$, $\hat{g}_j(z_j(x, t - \tau)) = g_j(w_j(x, t - \tau, \Phi)) - g_j(w_j^*(x))$.

For convenient analysis, we let

$$\begin{aligned} \varrho &= \mathrm{diag}(\varrho_1, \varrho_2, \ldots, \varrho_N), \quad \nu = \mathrm{diag}(\nu_1, \nu_2, \ldots, \nu_N) \\ D &= \mathrm{diag}(d_1, d_2, \ldots, d_N), \quad A = \mathrm{diag}(a_1, a_2, \ldots, a_N). \end{aligned}$$

Theorem 8.23 *The system* (8.15) *is asymptotically stable if there exist matrices* $\mathbb{R}^{N\times N} \ni P = \mathrm{diag}(p_1, p_2, \ldots, p_N) > 0$, $\mathbb{R}^{N\times N} \ni Q > 0$, $\mathbb{R}^{N\times N} \ni W_1 > 0$ *and* $\mathbb{R}^{N\times N} \ni W_2 > 0$ *such that*

$$\lambda_M(W_2^{-1})\nu^2 - Q \leqslant 0, \tag{8.18}$$

$$\begin{aligned} &\sum_{k=1}^{\sigma} \frac{-2\lambda_m(PD)}{l_k^2} I_N + PA + A^T P + PGW_1G^T P + P\hat{G}W_2\hat{G}^T P \\ &+ \lambda_M(W_1^{-1})\varrho^2 + Q < 0. \end{aligned} \tag{8.19}$$

Proof Firstly, we can rewrite system (8.17) in a compact form as follows:

$$\begin{cases} \frac{\partial z(x,t)}{\partial t} = D\Delta z(x, t) + G\hat{f}(z(x, t)) + Az(x, t) + \hat{G}\hat{g}(z(x, t - \tau)), \\ z(x, t) = 0, \quad (x, t) \in \partial\Omega \times [-\tau, +\infty), \end{cases} \tag{8.20}$$

where $\hat{f}(z(x, t)) = (\hat{f}_1(z_1(x, t)), \hat{f}_2(z_2(x, t)), \ldots, \hat{f}_N(z_N(x, t)))^T$, $\hat{g}(z(x, t - \tau)) = (\hat{g}_1(z_1(x, t - \tau)), \hat{g}_2(z_2(x, t - \tau)), \ldots, \hat{g}_N(z_N(x, t - \tau)))^T$.

Next, construct the following Lyapunov functional for system (8.20):

$$V(t) = \int_{\Omega} z^T(x,t)Pz(x,t)dx + \int_{t-\tau}^{t}\int_{\Omega} z^T(x,s)Qz(x,s)dxds.$$

Calculating the derivative of $V(t)$ along the solution $z(x,t)$ of system (8.20), we have

$$\begin{aligned}
\dot{V}(t) &= 2\int_{\Omega} z^T(x,t)P\frac{\partial z(x,t)}{\partial t}dx + \int_{\Omega} z^T(x,t)Qz(x,t)dx \\
&\quad - \int_{\Omega} z^T(x,t-\tau)Qz(x,t-\tau)dx \\
&= 2\int_{\Omega} z^T(x,t)P[D\Delta z(x,t) + Az(x,t) + G\hat{f}(z(x,t)) \\
&\quad + \hat{G}\hat{g}(z(x,t-\tau))]dx + \int_{\Omega} z^T(x,t)Qz(x,t)dx \\
&\quad - \int_{\Omega} z^T(x,t-\tau)Qz(x,t-\tau)dx \\
&\leqslant \sum_{k=1}^{\sigma}\frac{-2\lambda_m(PD)}{l_k^2}\int_{\Omega} z^T(x,t)z(x,t)dx + \int_{\Omega} z^T(x,t)(PA \\
&\quad + A^TP)z(x,t)dx + \int_{\Omega} \hat{f}^T(z(x,t))W_1^{-1}\hat{f}(z(x,t))dx \\
&\quad + \int_{\Omega} z^T(x,t)PGW_1G^TPz(x,t)dx \\
&\quad + \int_{\Omega} \hat{g}^T(z(x,t-\tau))W_2^{-1}\hat{g}(z(x,t-\tau))dx \\
&\quad + \int_{\Omega} z^T(x,t)P\hat{G}W_2\hat{G}^TPz(x,t)dx + \int_{\Omega} z^T(x,t)Qz(x,t)dx \\
&\quad - \int_{\Omega} z^T(x,t-\tau)Qz(x,t-\tau)dx \\
&\leqslant \int_{\Omega} z^T(x,t)\bigg(\sum_{k=1}^{\sigma}\frac{-2\lambda_m(PD)}{l_k^2}I_N + PA + A^TP + PGW_1G^TP \\
&\quad + P\hat{G}W_2\hat{G}^TP + \lambda_M(W_1^{-1})\varrho^2 + Q\bigg)z(x,t)dx \\
&\quad + \int_{\Omega} z^T(x,t-\tau)(\lambda_M(W_2^{-1})\nu^2 - Q)z(x,t-\tau)dx.
\end{aligned}$$

It follows from (8.18) that

$$\dot{V}(t) \leqslant \int_{\Omega} z^T(x,t)Sz(x,t)dx \leqslant \lambda_M(S)\|z(\cdot,t)\|_2^2,$$

where $S = \sum_{k=1}^{\sigma} \frac{-2\lambda_m(PD)}{l_k^2} I_N + PA + A^T P + PGW_1G^T P + P\hat{G}W_2\hat{G}^T P + \lambda_M (W_1^{-1})\varrho^2 + Q$. Therefore, we can derive from (8.19) and Lemma 8.7 that system (8.15) is asymptotically stable. The proof is completed.

Remark 8.24 In this subsection, we assume that $w^*(x)$ is a nonnegative equilibrium solution of system (8.15). Practically, the existence of nontrivial nonnegative equilibrium solution is an important and interesting problem and will become our future investigative direction.

8.4.3 Passivity-based Control of Food Web Model

In this subsection, we shall consider the problem of passivity-based controller design for food web model (8.12). By making use of passivity and Lyapunov theory, a criterion for existence of the controller is given.

Suppose $w(x,t,\Phi)$ is an arbitrary solution of system (8.12) with initial condition $\Phi(x,t)$. Let $u(x,t) = (u_1(x,t), u_2(x,t), \ldots, u_N(x,t))^T$, $\hat{z}(x,t) = w(x,t,\Phi) - w^*(x)$ and $\Phi_{\hat{z}}(x,t) = \Phi(x,t) - w^*(x)$, then the dynamics of the difference vector $\hat{z}(x,t) = (\hat{z}_1(x,t), \hat{z}_2(x,t), \ldots, \hat{z}_N(x,t))^T$ is governed by the following equation:

$$\begin{cases} \frac{\partial \hat{z}(x,t)}{\partial t} = D\Delta\hat{z}(x,t) + G\hat{f}(\hat{z}(x,t)) + u(x,t) + A\hat{z}(x,t) \\ \qquad\qquad +\hat{G}\hat{g}(\hat{z}(x,t-\tau)), \\ \hat{z}(x,t) = \Phi_{\hat{z}}(x,t), \quad (x,t) \in \Omega \times [-\tau, 0], \\ \hat{z}(x,t) = 0, \quad (x,t) \in \partial\Omega \times [-\tau, +\infty), \end{cases} \tag{8.21}$$

where $w^*(x)$, D, A denote the same meaning as that in system (8.20), and

$$\begin{aligned} \hat{f}_j(\hat{z}_j(x,t)) &= f_j(w_j(x,t,\Phi)) - f_j(w_j^*(x)), \\ \hat{g}_j(\hat{z}_j(x,t-\tau)) &= g_j(w_j(x,t-\tau,\Phi)) - g_j(w_j^*(x)), \\ \hat{f}(\hat{z}(x,t)) &= (\hat{f}_1(\hat{z}_1(x,t)), \hat{f}_2(\hat{z}_2(x,t)), \ldots, \hat{f}_N(\hat{z}_N(x,t)))^T, \\ \hat{g}(\hat{z}(x,t-\tau)) &= (\hat{g}_1(\hat{z}_1(x,t-\tau)), \hat{g}_2(\hat{z}_2(x,t-\tau)), \ldots, \hat{g}_N(\hat{z}_N(x,t-\tau)))^T. \end{aligned}$$

Theorem 8.25 *If there exist matrices* $\mathbb{R}^{N\times N} \ni P = \mathrm{diag}(p_1, p_2, \ldots, p_N) > 0$, $\mathbb{R}^{N\times N} \ni Q > 0$, $\mathbb{R}^{N\times N} \ni M_1 > 0$, $\mathbb{R}^{N\times N} \ni M_2 > 0$, $\mathbb{R}^{N\times N} \ni K = \mathrm{diag}(k_1, k_2, \ldots, k_N) \leqslant 0$ *and a scalar* $\gamma > 0$ *such that*

$$\lambda_M(M_2^{-1})\nu^2 - Q \leqslant 0, \tag{8.22}$$

$$\begin{pmatrix} \chi_1 + \lambda_M(M_1^{-1})\varrho^2 & P \\ P & -\frac{1}{\gamma}I_N \end{pmatrix} \leqslant 0, \tag{8.23}$$

where $\chi_1 = \sum_{k=1}^{\sigma} \frac{-2\lambda_m(PD)}{l_k^2} I_N + P(A+K) + (A+K)^T P + Q + PGM_1G^TP + P\hat{G}M_2\hat{G}^TP$, $\varrho = \mathrm{diag}(\varrho_1, \varrho_2, \ldots, \varrho_N)$, $\nu = \mathrm{diag}(\nu_1, \nu_2, \ldots, \nu_N)$, *then the system* (8.21) *is output strictly passive and internally asymptotically stable through the feedback controller*

$$u(x,t) = K\hat{z}(x,t) + v(x,t), \tag{8.24}$$

where $v(x,t) \in \mathbb{R}^N$ *is the external input.*

Proof First, by applying controller (8.24) into system (8.21), the resulting closed-loop dynamic system is of the form:

$$\frac{\partial \hat{z}(x,t)}{\partial t} = D\Delta\hat{z}(x,t) + v(x,t) + (A+K)\hat{z}(x,t) + G\hat{f}(\hat{z}(x,t)) + \hat{G}\hat{g}(\hat{z}(x,t-\tau)). \tag{8.25}$$

In what follows, we establish the passivity of system (8.25). An output of system (8.25)is defined as $y(x,t) = P\hat{z}(x,t)$ for the passivity scheme.

Construct a Lyapunov functional for system (8.25) as follows:

$$V(t) = \int_\Omega \hat{z}^T(x,t)P\hat{z}(x,t)dx + \int_{t-\tau}^{t}\int_\Omega \hat{z}^T(x,s)Q\hat{z}(x,s)dxds.$$

Then the derivative $\dot{V}(t)$ of $V(t)$ is given by

$$\begin{aligned}
\dot{V}(t) &= 2\int_\Omega \hat{z}^T(x,t)P\frac{\partial\hat{z}(x,t)}{\partial t}dx + \int_\Omega \hat{z}^T(x,t)Q\hat{z}(x,t)dx \\
&\quad - \int_\Omega \hat{z}^T(x,t-\tau)Q\hat{z}(x,t-\tau)dx \\
&= 2\int_\Omega \hat{z}^T(x,t)P[D\Delta\hat{z}(x,t) + (A+K)\hat{z}(x,t) + G\hat{f}(\hat{z}(x,t)) \\
&\quad + \hat{G}\hat{g}(\hat{z}(x,t-\tau)) + v(x,t)]dx + \int_\Omega \hat{z}^T(x,t)Q\hat{z}(x,t)dx \\
&\quad - \int_\Omega \hat{z}^T(x,t-\tau)Q\hat{z}(x,t-\tau)dx
\end{aligned}$$

$$
\begin{aligned}
&\leqslant \int_{\Omega} \hat{z}^T(x,t)\Big[\sum_{k=1}^{\sigma} \frac{-2\lambda_m(PD)}{l_k^2} I_N + P(A+K) + (A+K)^T P \\
&+Q\Big]\hat{z}(x,t)dx + \int_{\Omega} \hat{f}^T(\hat{z}(x,t))M_1^{-1}\hat{f}(\hat{z}(x,t))dx \\
&+ \int_{\Omega} \hat{z}^T(x,t)PGM_1G^T P\hat{z}(x,t)dx \\
&+ \int_{\Omega} \hat{g}^T(\hat{z}(x,t-\tau))M_2^{-1}\hat{g}(\hat{z}(x,t-\tau))dx \\
&+ \int_{\Omega} \hat{z}^T(x,t)P\hat{G}M_2\hat{G}^T P\hat{z}(x,t)dx \\
&- \int_{\Omega} \hat{z}^T(x,t-\tau)Q\hat{z}(x,t-\tau)dx + 2\int_{\Omega} \hat{z}^T(x,t)Pv(x,t)dx \\
&\leqslant \int_{\Omega} \hat{z}^T(x,t)\Big(\sum_{k=1}^{\sigma} \frac{-2\lambda_m(PD)}{l_k^2} I_N + P(A+K) + (A+K)^T P + Q \\
&+PGM_1G^T P + P\hat{G}M_2\hat{G}^T P + \lambda_M(M_1^{-1})\varrho^2\Big)\hat{z}(x,t)dx \\
&+2\int_{\Omega} \hat{z}^T(x,t)Pv(x,t)dx \\
&+ \int_{\Omega} \hat{z}^T(x,t-\tau)(\lambda_M(M_2^{-1})\nu^2 - Q)\hat{z}(x,t-\tau)dx.
\end{aligned}
$$

On the other hand, it follows from (8.23) that

$$\chi_1 + \lambda_M(M_1^{-1})\varrho^2 \leqslant -\gamma P^2. \tag{8.26}$$

Then, we can derive from (8.22) and (8.26) that

$$\dot{V}(t) \leqslant -\gamma \int_{\Omega} \hat{z}^T(x,t)P^T P\hat{z}(x,t)dx + 2\int_{\Omega} \hat{z}^T(x,t)Pv(x,t)dx. \tag{8.27}$$

When external input $v(x,t) = 0$, the system (8.25) is asymptotically stable because

$$\dot{V}(t) \leqslant -\gamma \int_{\Omega} \hat{z}^T(x,t)P^T P\hat{z}(x,t)dx \leqslant -\gamma\lambda_m(P^2)\|\hat{z}(\cdot,t)\|_2^2. \tag{8.28}$$

In the case that $v(x,t) \not\equiv 0$, we integrate the inequality (8.27) with respect to t over the time period t_0 to t_p,

$$\int_{t_0}^{t_p} \dot{V}(t)dt \leqslant -\gamma \int_{t_0}^{t_p}\int_{\Omega} \hat{z}^T(x,t)P^T P\hat{z}(x,t)dxdt$$
$$+2\int_{t_0}^{t_p}\int_{\Omega} \hat{z}^T(x,t)Pv(x,t)dxdt.$$

Namely,

$$\int_{t_0}^{t_p}\int_{\Omega} y^T(x,t)v(x,t)dxdt - \frac{\gamma}{2}\int_{t_0}^{t_p}\int_{\Omega} y^T(x,t)y(x,t)dxdt \geqslant \frac{V(t_p)}{2} - \frac{V(t_0)}{2}.$$

Therefore, system (8.21) is output strictly passive from the external input $v(x,t)$ to the output $y(x,t)$ through the state feedback controller (8.24). This completes the proof.

8.5 Numerical Example

In this section, a numerical example is provided to verify the effectiveness of the proposed theoretical results.

Consider the following network model:

$$\begin{cases} \frac{\partial w_i(x,t)}{\partial t} = d_i \Delta w_i(x,t) + a_i w_i(x,t) + u_i(x,t) + \sum_{j=1}^{4} G_{ij} f_j(w_j(x,t)) \\ \qquad + \sum_{j=1}^{4} \hat{G}_{ij} g_j(w_j(x,t-0.1)), \\ w_i(x,t) = 0.1i\cos(\pi x),\ (x,t) \in \Omega \times [-0.1, 0], \\ w_i(x,t) = 0,\ (x,t) \in \partial\Omega \times [-0.1, +\infty), \end{cases} \tag{8.29}$$

where $i = 1, 2, 3, 4$, $\Omega = \{x \mid -0.5 < x < 0.5\}$, $d_1 = d_2 = d_3 = 0.05$, $d_4 = 0.075$, $a_1 = 0.05\pi^2 - 0.55$, $a_2 = 0.05\pi^2 - 0.525$, $a_3 = 0.05\pi^2 - 0.575$, $a_4 = 0.075\pi^2 - 0.775$, $G_{11} = -0.0125$, $G_{12} = 0.05$, $G_{13} = 0.025$, $G_{14} = 0.075$, $G_{21} = 0.025$, $G_{22} = -0.0125$, $G_{23} = 0.05$, $G_{24} = 0.025$, $G_{31} = 0.025$, $G_{32} = 0.075$, $G_{33} = -0.0125$, $G_{34} = 0.025$, $G_{41} = 0.225$, $G_{42} = 0.225$, $G_{43} = 0.2$, $G_{44} = -0.025$, $\hat{G}_{11} = 0$, $\hat{G}_{12} = 0.075$, $\hat{G}_{13} = 0.075$, $\hat{G}_{14} = 0.025$, $\hat{G}_{21} = 0.075$, $\hat{G}_{22} = 0$, $\hat{G}_{23} = 0.0625$, $\hat{G}_{24} = 0.05$, $\hat{G}_{31} = 0.075$, $\hat{G}_{32} = 0.075$, $\hat{G}_{33} = 0$, $\hat{G}_{34} = 0.05$, $\hat{G}_{41} = 0.075$, $\hat{G}_{42} = 0.075$, $\hat{G}_{43} = 0.125$, $\hat{G}_{44} = 0$, $f_j(\xi) = \frac{|\xi+1|-|\xi-1|}{2}$, $g_j(\xi) = |\xi+1| - |\xi-1|$.

Obviously, $f_j(\cdot)$ and $g_j(\cdot)$ satisfy $f_j(0) = g_j(0) = 0$ and the assumption **(A3)** with $\varrho_j = 1$ and $\nu_j = 2$. Since $f_j(\cdot)$ and $g_j(\cdot)$ are bounded functions, we can easily find some positive constants $\sigma_i \geqslant 1$, $\lambda_i, i = 1, 2, 3, 4$, such that

$$\begin{cases} \frac{\mathrm{d}(\sigma_i e^{\lambda_i(t+0.1)})}{\mathrm{d}t} \geqslant a_i\sigma_i e^{\lambda_i(t+0.1)} + \sum_{j=1}^4 \hat{G}_{ij} g_j(\sigma_j e^{\lambda_j t}) \\ \qquad + \sum_{j=1}^4 G_{ij} f_j(\sigma_j e^{\lambda_j(t+0.1)}), \\ 0.1i\cos(\pi x) \leqslant \sigma_i e^{\lambda_i(t+0.1)}, \text{ in } \Omega \times [-0.1, 0], \\ 0 \leqslant \sigma_i e^{\lambda_i(t+0.1)}, \text{ on } \partial\Omega \times [-0.1, +\infty). \end{cases}$$

Therefore, we can derive from Lemma 8.21 that system (8.29) has a unique nonnegative solution under the condition that $u_i(x, t) = 0, i = 1, 2, 3, 4$.

On the other hand, it is easy to verify that (8.16) is satisfied if $w^*(x) = (0.2\cos(\pi x), 0.2\cos(\pi x), 0.2\cos(\pi x), 0.3\cos(\pi x))^T$. This shows that $(0.2\cos(\pi x), 0.2\cos(\pi x), 0.2\cos(\pi x), 0.3\cos(\pi x))^T$ is a nonnegative equilibrium solution of system (8.29). In what follows, we shall design the passivity-based controller for network model (8.29).

We design the following feedback controller for system (8.29):

$$\begin{pmatrix} u_1 \\ u_2 \\ u_3 \\ u_4 \end{pmatrix} = \begin{pmatrix} -0.65 & 0 & 0 & 0 \\ 0 & -0.55 & 0 & 0 \\ 0 & 0 & -0.85 & 0 \\ 0 & 0 & 0 & -0.95 \end{pmatrix} \times \begin{pmatrix} w_1(x,t) - 0.2\cos(\pi x) \\ w_2(x,t) - 0.2\cos(\pi x) \\ w_3(x,t) - 0.2\cos(\pi x) \\ w_4(x,t) - 0.3\cos(\pi x) \end{pmatrix} + \begin{pmatrix} v_1 \\ v_2 \\ v_3 \\ v_4 \end{pmatrix}, \tag{8.30}$$

where $v_i(x, t) \in \mathbb{R}, i = 1, 2, 3, 4$, is the external input.

By simple calculation with $M_1 = M_2 = 8I_4$, $P = I_4$, $Q = 0.5I_4$ and $\gamma = 0.05$, we can obtain that conditions (8.22) and (8.23) are satisfied. Therefore, it follows from Theorem 8.25 that system (8.29) is output strictly passive and internally asymptotically stable through the feedback controller (8.30).

The results of numerical simulation are shown in Figs. 8.1 8.2 and 8.3. From Fig. 8.1, we clearly see that $(w_1(x, t), w_2(x, t), w_3(x, t), w_4(x, t))^T$ is very close to $(0.2\cos(\pi x), 0.2\cos(\pi x), 0.2\cos(\pi x), 0.3\cos(\pi x))^T$ when the time t increases gradually to 2 under the designed controller, and that state is maintained along with the increasing of the time. Figure 8.2 visualizes the change processes of the output variables $y_1(x, t)$, $y_2(x, t)$, $y_3(x, t)$ and $y_4(x, t)$ in time interval [0, 8]. Figure 8.3 shows the change process of $e(t)$, which reflects the output strict passivity of closed-loop dynamic system.

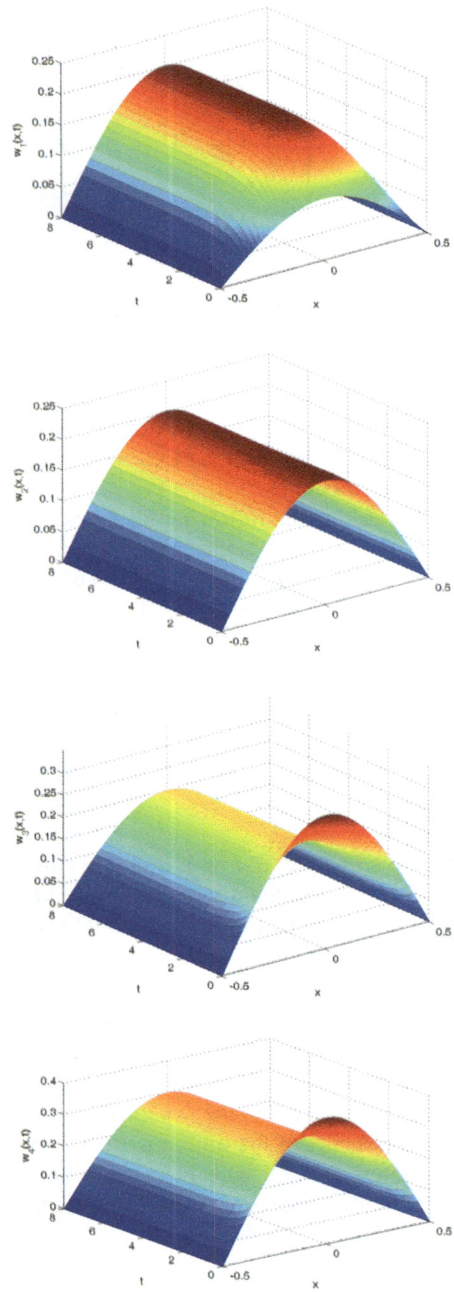

Fig. 8.1 The change processes of state variables of system (8.29) under the feedback controller (8.30) in time interval [0, 8], where $v_1(x,t) = v_2(x,t) = v_3(x,t) = v_4(x,t) = 0$.

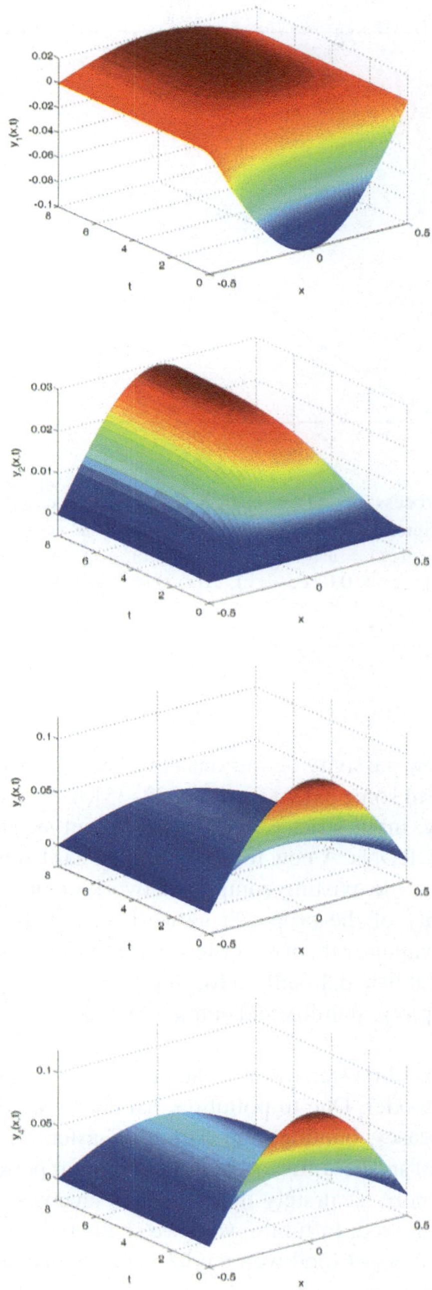

Fig. 8.2 The change processes of output variables $y_1(x, t) = w_1(x, t) - 0.2\cos(\pi x)$, $y_2(x, t) = w_2(x, t) - 0.2\cos(\pi x)$, $y_3(x, t) = w_3(x, t) - 0.2\cos(\pi x)$, $y_4(x, t) = w_4(x, t) - 0.3\cos(\pi x)$ of system (8.29) under the feedback controller (8.30) in time interval [0, 8], where $v_1(x, t) = 0.01\cos(\pi x)$, $v_2(x, t) = 0.02\cos(\pi x)$, $v_3(x, t) = 0.02\cos(\pi x)$, $v_4(x, t) = 0.03\cos(\pi x)$.

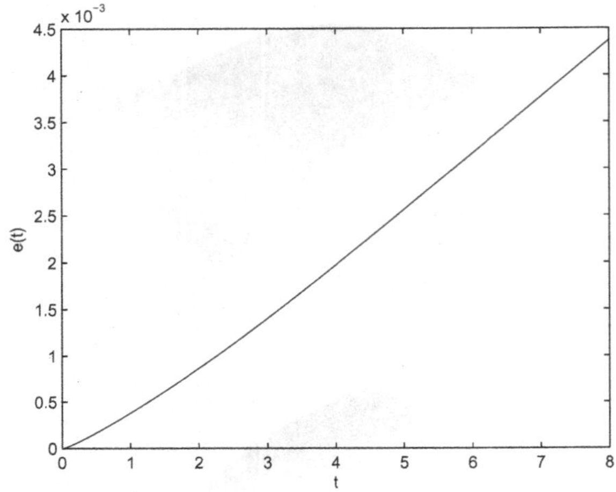

Fig. 8.3 The change process of $e(t) = \int_0^t \int_\Omega \sum_{i=1}^4 y_i(x,s)v_i(x,s)dxds - 0.025 \int_0^t \int_\Omega \sum_{i=1}^4 y_i^2(x,s)dxds$ in time interval [0, 8], where $y_1(x,t) = w_1(x,t) - 0.2\cos(\pi x)$, $y_2(x,t) = w_2(x,t) - 0.2\cos(\pi x)$, $y_3(x,t) = w_3(x,t) - 0.2\cos(\pi x)$, $y_4(x,t) = w_4(x,t) - 0.3\cos(\pi x)$, $v_1(x,t) = 0.01\cos(\pi x)$, $v_2(x,t) = 0.02\cos(\pi x)$, $v_3(x,t) = 0.02\cos(\pi x)$, $v_4(x,t) = 0.03\cos(\pi x)$.

8.6 Conclusion

In this chapter, two new passivity definitions have been proposed, which generalize the previous concepts to some extent (e.g., [1, 32–35],). Some sufficient conditions ensuring the passivity, input strict passivity and output strict passivity have been derived for a class of DRDNs. A new food web model has been introduced, which is totally different from some existing complex network models. We have considered the asymptotic stability of the proposed network model. In addition, by making use of passivity and Lyapunov theory, a criterion for existence of the passivity-based controller has been established. Finally, a four-node network model has been given to demonstrate the asymptotic stability and output strict passivity by using the proposed control method.

This chapter is only a first step towards the problem of passivity-based controller design for food web model. Due to pollution, habitat loss, competition with other species, predator, diseases, hunting, and natural disasters or other reasons, sharp change of the density of species often occurs in a very short period of time. Therefore, in order to describe more accurately the dynamics changes of species, impulsive effects should be considered. In future work we shall study the design problem for passivity-based controllers of food web model with impulsive effects and reaction-diffusion terms.

References

1. Niculescu, S., & Lozano, R. (2001). On the passivity of linear delay systems. *IEEE Transactions on Automatic Control*, *46*(3), 460–464.
2. Willems, J. C. (1972). Dissipative dynamical systems part II: Linear systems with quadratic supply rates. *Archive for Rational Mechanics and Analysis*, *45*(5), 352–393.
3. Willems, J. C. (1972). Dissipative dynamical systems part I: General theory. *Archive for Rational Mechanics and Analysis*, *45*(5), 321–351.
4. Hill, D., & Moylan, P. (1976). The stability of nonlinear dissipative systems. *IEEE Transactions on Automatic Control*, *21*(5), 708–711.
5. Hill, D. J., & Moylan, P. J. (1977). Stability results for nonlinear feedback systems. *Automatica*, *13*(4), 377–382.
6. Byrnes, C. I., Isidori, A., & Willems, J. C. (1991). Passivity, feedback equivalence, and the global stabilization of minimum phase nonlinear systems. *IEEE Transactions on Automatic Control*, *36*(11), 1228–1240.
7. Pota, H. R., & Moylan, P. J. (1993). Stability of locally dissipative interconnected systems. *IEEE Transactions on Automatic Control*, *38*(2), 308–312.
8. Zhao, J., & Hill, D. J. (2008). Dissipativity theory for switched systems. *IEEE Transactions on Automatic Control*, *53*(4), 941–953.
9. Bevelevich, V. (1968). *Classical Network Synthesis*. New York, NY: Van Nostrand.
10. Santosuosso, G. J. (1997). Passivity of nonlinear systems with input-output feedthrough. *Automatica*, *33*(4), 693–697.
11. Chua, L. O. (1999). Passivity and complexity. *IEEE Transactions on Circuits and Systems I: Fundamental Theory and Applications*, *46*(1), 71–82.
12. Xie, L., Fu, M., & Li, H. (1998). Passivity analysis and passification for uncertain signal processing systems. *IEEE Transactions on Signal Processing*, *46*(9), 2394–2403.
13. Yu, W. (1999). Passive equivalence of chaos in lorenz system. *IEEE Transactions on Circuits and Systems I: Fundamental Theory and Applications*, *46*(7), 876–878.
14. Wu, C. W. (2001). Synchronization in arrays of coupled nonlinear systems: Passivity, circle criterion, and observer design. *IEEE Transactions on Circuits and Systems I: Fundamental Theory and Applications*, *48*(10), 1257–1261.
15. Calcev, G., Gorez, R., & Neyer, M. D. (1998). Passivity approach to fuzzy control systems. *Automatica*, *34*(3), 339–344.
16. Arcak, M. (2007). Passivity as a design tool for group coordination. *IEEE Transactions on Automatic Control*, *52*(8), 1380–1390.
17. Wen, J. T., & Arcak, M. (2004). A unifying passivity framework for network flow control. *IEEE Transactions on Automatic Control*, *49*(2), 162–174.
18. Ayad, M. Y., Becherif, M., Henni, A., Aboubou, A., Wack, M., & Laghrouche, S. (2010). Passivity-based control applied to DC hybrid power source using fuel cell and supercapacitors. *Energy Conversion and Management*, *51*(7), 1468–1475.
19. Lestas, I., & Vinnicombe, G. (2010). Heterogeneity and scalability in group agreement protocols: Beyond small gain and passivity approaches. *Automatica*, *46*(7), 1141–1151.
20. Liu, P., Zhang, Q., Yang, X., Yang, L. (2008). Passivity and optimal control of descriptor biological complex systems. *IEEE Transactions on Automatic Control 53*, 122–125.
21. Song, Q., & Wang, Z. (2010). New results on passivity analysis of uncertain neural networks with time-varying delays. *International Journal of Computer Mathematics*, *87*(3), 668–678.
22. Song, Q., Liang, J., & Wang, Z. (2009). Passivity analysis of discrete-time stochastic neural networks with time-varying delays. *Neurocomputing*, *72*(7–9), 1782–1788.
23. Balasubramaniam, P., & Nagamani, G. (2010). Passivity analysis of neural networks with Markovian jumping parameters and interval time-varying delays. *Nonlinear Analysis: Hybrid Systems*, *4*(4), 853–864.
24. Balasubramaniam, P., Nagamani, G., & Rakkiyappan, R. (2010). Global passivity analysis of interval neural networks with discrete and distributed delays of neutral type. *Neural Processing Letters*, *32*(2), 109–130.

25. Balasubramaniam, P., & Nagamani, G. (2011). A delay decomposition approach to delay-dependent passivity analysis for interval neural networks with time-varying delay. *Neurocomputing*, *74*(10), 1646–1653.
26. Fu, J., Zhang, H., Ma, T., & Zhang, Q. (2010). On passivity analysis for stochastic neural networks with interval time-varying delay. *Neurocomputing*, *73*(4–6), 795–801.
27. Zhu, J., Zhang, Q., & Yuan, Z. (2010). Delay-dependent passivity criterion for discrete-time delayed standard neural network model. *Neurocomputing*, *73*(7–9), 1384–1393.
28. Zhang, Z., Mou, S., Lam, J., & Gao, H. (2009). New passivity criteria for neural networks with time-varying delay. *Neural Networks*, *22*(7), 864–868.
29. Li, H., Wang, C., Shi, P., & Gao, H. (2010). New passivity results for uncertain discrete-time stochastic neural networks with mixed time delays. *Neurocomputing*, *73*(16–18), 3291–3299.
30. Wu, Z., Shi, P., Su, H., & Chu, J. (2011). Passivity analysis for discrete-time stochastic Markovian jump neural networks with mixed time delays. *IEEE Transactions on Neural Networks*, *22*(10), 1566–1575.
31. Xu, S., Zheng, W. X., & Zou, Y. (2009). Passivity analysis of neural networks with time-varying delays. *IEEE Transactions on Circuits and Systems II: Express Briefs*, *56*(4), 325–329.
32. Yao, J., Guan, Z. H., & Hill, D. J. (2009). Passivity-based control and synchronization of general complex dynamical networks. *Automatica*, *45*(9), 2107–2113.
33. Yao, J., Wang, H. O., Guan, Z. H., & Xu, W. (2009). Passive stability and synchronization of complex spatio-temporal switching networks with time delays. *Automatica*, *45*(7), 1721–1728.
34. Wang, J. L., Wu, H. N., & Yang, Z. C. (2011). Passivity analysis of impulsive complex networks. *International Journal of Automation and Computing*, *8*(4), 484–489.
35. Wang, J. L., Yang, Z. C., & Wu, H. N. (2012). Passivity analysis of complex dynamical networks with multiple time-varying delays. *Journal of Engineering Mathematics*, *74*(1), 175–188.
36. Wei, D. Q., Luo, X. S. (2007). Passivity-based adaptive control of chaotic oscillations in power system. *Chaos, Solitons and Fractals*, *31*(3), 665–671.
37. Ji, D. H., Koo, J. H., Won, S. C., Lee, S. M., & Park, J. H. (2011). Passivity-based control for Hopfield neural networks using convex representation. *Applied Mathematics and Computation*, *217*(13), 6168–6175.
38. Wang, J., Wu, H., & Guo, L. (2011). Passivity and stability analysis of reaction-diffusion neural networks with Dirichlet boundary conditions. *IEEE Transactions on Neural Networks*, *22*(12), 2105–2116.
39. Ryu, K., & Ahn, I. (2005). Positive solutions for ratio-dependent predator-prey interaction systems. *Journal of Differential Equations*, *218*(1), 117–135.
40. Peng, R., & Shi, J. (2009). Non-existence of non-constant positive steady states of two Holling type-II predator-prey systems: Strong interaction case. *Journal of Differential Equations*, *247*(3), 866–886.
41. Wang, J., Shi, J., & Wei, J. (2011). Dynamics and pattern formation in a diffusive predator-prey system with strong Allee effect in prey. *Journal of Differential Equations*, *251*(4), 1276–1304.
42. Pang, P. Y. H., & Wang, M. (2004). Strategy and stationary pattern in a three-species predator-prey model. *Journal of Differential Equations*, *200*(2), 245–273.
43. Aly, S., Kim, I., & Sheen, D. (2011). Turing instability for a ratio-dependent predator-prey model with diffusion. *Applied Mathematics and Computation*, *217*(17), 7265–7281.
44. Zhu, L. M., Wang, A. L., Liu, Y. J., & Wang, B. (2010). Stationary patterns of a predator-prey model with spatial effect. *Applied Mathematics and Computation*, *216*(12), 3620–3626.
45. Duque, C., Kiss, K., & Lizana, M. (2009). On the dynamics of an n-dimensional ratio-dependent predator-prey system with diffusion. *Applied Mathematics and Computation*, *208*(1), 98–105.
46. Wang, Y. M. (2009). Numerical solutions of a Michaelis-Menten-type ratio-dependent predator-prey system with diffusion. *Applied Numerical Mathematics*, *59*(5), 1075–1093.
47. Wang, C. Y., Wang, S., Yang, F. P., & Li, L. R. (2010). Global asymptotic stability of positive equilibrium of three-species Lotka-Volterra mutualism models with diffusion and delay effects. *Applied Mathematical Modelling*, *34*(12), 4278–4288.
48. Xu, R. (2006). A reaction-diffusion predator-prey model with stage structure and nonlocal delay. *Applied Mathematics and Computation*, *175*(2), 984–1006.

49. Wu, Z. G., Shi, P., Su, H., & Chu, J. (2013). Stochastic synchronization of Markovian jump neural networks with time-varying delay using sampled data. *IEEE Transactions on Cybernetics*, *43*(6), 1796–1806.
50. Zhao, H., Chen, Q., & Xu, S. (2009). H_∞ guaranteed cost control for uncertain Markovian jump systems with mode-dependent distributed delays and input delays. *Journal of the Franklin Institute*, *346*(10), 945–957.
51. Zhang, D., Lu, L., Wang, Q. G., & Ong, C. J. (2012). Estimator design for discrete-time switched neural networks with asynchronous switching and time-varying delay. *IEEE Transactions on Neural Networks and Learning Systems*, *23*(5), 827–834.
52. Wang, J. L., Wu, H. N., & Guo, L. (2012). Pinning control of spatially and temporally complex dynamical networks with time-varying delays. *Nonlinear Dynamics*, *70*(2), 1657–1674.
53. Hu, C., Jiang, H., & Teng, Z. (2010). Impulsive control and synchronization for delayed neural networks with reaction-diffusion terms. *IEEE Transactions on Neural Networks*, *21*(1), 67–81.
54. Lu, J. G., & Lu, L. J. (2009). Global exponential stability and periodicity of reaction-diffusion recurrent neural networks with distributed delays and Dirichlet boundary conditions. *Chaos, Solitons and Fractals*, *39*(4), 1538–1549.
55. Wang, J. L., & Wu, H. N. (2011). Stability analysis of impulsive parabolic complex networks. *Chaos, Solitons and Fractals*, *44*(11), 1020–1034.
56. Shen, H., Xu, S., Lu, J., & Zhou, J. (2012). Passivity-based control for uncertain stochastic jumping systems with mode-dependent round-trip time delays. *Journal of the Franklin Institute*, *349*(5), 1665–1680.
57. Pao, C. V. (1995). Coupled nonlinear parabolic systems with time delays. *Journal of Mathematical Analysis and Applications*, *196*(1), 237–265.
58. Lu, J. G. (2007). Robust global exponential stability for interval reaction-diffusion hopfield neural networks with distributed delays. *IEEE Transactions on Circuits and Systems II: Express Briefs*, *54*(12), 1115–1119.
59. Pan, J., & Zhan, Y. (2011). On periodic solutions to a class of non-autonomously delayed reaction-diffusion neural networks. *Communications in Nonlinear Science and Numerical Simulation*, *16*(1), 414–422.

Chapter 9
Passivity and Synchronization of CURDNNs with Multiple Time-Delays

9.1 Introduction

In recent decades, NNs have been a subject of extensive research owing to their wide applications in signal processing, optimization, image processing, pattern classification, etc. In a complex network, if each node represents one NN and each edge indicates that one NN has an impact on the other one, we refer to this network as a CNNs. CNNs have gained considerable attention in recent years because of their useful applications in harmonic oscillation generation, secure communication, and chaos generators design. As is well known, NNs can be realized through electric circuits, and the reaction-diffusion phenomenon [1–3] can not be avoided in electric circuits. Therefore, the investigating CRDNNs can be interesting and challenging.

As is widely known, the applications of CNNs are closely associated with their dynamical behaviors. Synchronization [4–8] is one of the most significant dynamical behaviors that has become a hot research topic in CNNs [9–16]. In [9], the authors proposed a delayed CNNs with impulsive effects and studied the globally exponential synchronization problem for such network by utilizing impulsive delay differential inequality. The local synchronization for a Markovian nonlinear CNNs with uncertainties was discussed in [15]. Nevertheless, in the above-mentioned studies [9–16], the authors did not consider the reaction-diffusion phenomenon. In recent years, some investigations into the synchronization of CRDNNs have been reported [17–25]. Wang et al. [18] presented a hybrid coupled CRDNNs. Subsequently, by taking advantage of the Lyapunov stability theory, the adaptive synchronization and $\mathcal{H}_\infty$ synchronization problems of such network were studied. Two CRDNNs were introduced by Xu et al. [24], and the synchronization problem for the proposed networks was investigated by exploiting the adaptive and pinning control methods.

In addition to synchronization, passivity [26] is a vital dynamical behavior in CNNs that has gained considerable attention. In [27], the authors presented a switched coupled uncertain NNs, and derived several delay-independent and delay-dependent

---

The original version of this book was revised: The belated corrections have been incorporated. The corrections to this book are available at https://doi.org/10.1007/978-981-33-4287-3_10

J.-L. Wang et al., *Passivity of Complex Dynamical Networks*,
https://doi.org/10.1007/978-981-33-4287-3_9

passivity criteria for such a network model. Moreover, by considering the output and input vectors that possess diverse dimensions in many networks, Ren et al. [28] examined the passivity of CNNs with diverse output and input dimensions. More recently, several studies on the passivity in CRDNNs have been reported [29–33]. In [33], Huang et al. proposed a nonlinear CRDNNs, and discussed its passivity and pinning passivity problems. Nevertheless, in the papers [29–33], the output and input dimensions were always assumed to be the same. However, the passivity of CRDNNs with output and input vectors of diverse dimensions has seldom been considered [34, 35]. In [34], Wang et al. studied the passivity of CRDNNs with fixed coupling and diverse dimensions of output and input vectors, and several sufficient conditions were developed to ensure that the proposed network is passive.

Moreover, many real-world networks such as public traffic roads networks, communication networks, social networks and so on, are supposed to be described by network models that have multiple coupling time-delays. Recently, there has been growing concern about the dynamical behaviors of complex networks with multiple time-delays [36–38]. Wang [37] et al. presented a CNNs with multiple time-delays, and studied the FTP problem of the network by designing appropriate controller. Qin et al. [38] introduced a class of multiple delayed complex networks with uncertain inner coupling matrices, and designed an appropriate adaptive controller that guarantees the proposed network can achieve synchronization and $\mathcal{H}_\infty$ synchronization. To our knowledge, the synchronization and passivity problems in CRDNNs with multiple time-delays have never been considered.

In this chapter, we propose a new CRDNNs model with multiple time-delays and uncertain parameters. The main contributions of this chapter are as follows. First, by using of some inequality techniques and the Lyapunov functional, we analyze the passivity of the proposed network model. Second, in order to guarantee that the network model is passive, an adaptive controller is also developed. Third, by using the similar methods, we respectively study the synchronization and adaptive synchronization of CRDNNs with multiple time-delays.

9.2 Network Model and Preliminaries

9.2.1 *Definitions*

Definition 9.1 (*see* [39]) If there exists a nonnegative function $P : [0, +\infty) \to [0, +\infty)$ satisfying

$$\int_{t_0}^{t_q} \nu(u, y)\, dt \geqslant P(t_q) - P(t_0)$$

for any $t_q, t_0 \in [0, +\infty)$ and $t_q \geqslant t_0$, the system with supply rate $\nu(u, y)$ is dissipative, where $u(x, t) \in \mathbb{R}^q$ is the input of the system and $y(x, t) \in \mathbb{R}^m$ is output of the system, and $(x, t) \in \Omega \times [0, +\infty)$.

Definition 9.2 If the system is dissipative with

$$\nu(u, y) = \int_{\Omega} y^T(x,t)Qu(x,t)dx,$$

the system is passive, where $Q \in \mathbb{R}^{m\times q}$ is a constant matrix, $u(x,t) \in \mathbb{R}^q$ is input of the system and $y(x,t) \in \mathbb{R}^m$ is output of the system.

Definition 9.3 If the system is dissipative with

$$\begin{aligned}\nu(u, y) = &\int_{\Omega} y^T(x,t)Qu(x,t)dx - \int_{\Omega} u^T(x,t)F_1u(x,t)dx \\ &- \int_{\Omega} y^T(x,t)F_2y(x,t)dx,\end{aligned}$$

in which $Q \in \mathbb{R}^{m\times q}$, $F_1 \in \mathbb{R}^{q\times q}$ and $F_2 \in \mathbb{R}^{m\times m}$ not only satisfy $F_1 \geqslant 0$ and $F_2 \geqslant 0$, but also require $\lambda_m(F_1) + \lambda_m(F_2) > 0$, then we consider the system to be strictly passive, where $u(x,t) \in \mathbb{R}^q$ is input of the system and $y(x,t) \in \mathbb{R}^m$ is output of the system.

If $F_1 > 0$, the system is input strictly passive, and if $F_2 > 0$, the system is output strictly passive.

9.2.2 Network Model

The CRDNNs with multiple time-delays is considered as follows:

$$\begin{aligned}\frac{\partial w_i(x,t)}{\partial t} = &C\Delta w_i(x,t) - Dw_i(x,t) + Hf(w_i(x,t)) + J + Bu_i(x,t) \\ &+a\sum_{j=1}^{N} A_{ij}\Gamma w_j(x,t-\tau_1) + a\sum_{j=1}^{N} A_{ij}\Gamma w_j(x,t-\tau_2) + \cdots \\ &+a\sum_{j=1}^{N} A_{ij}\Gamma w_j(x,t-\tau_\eta),\end{aligned} \tag{9.1}$$

where $i = 1, 2, \ldots, N$, $x = (x_1, x_2, \ldots, x_\sigma)^T \in \Omega \subset \mathbb{R}^\sigma$; $\Delta = \sum_{k=1}^{\sigma}(\partial^2/\partial x_k^2)$ is Laplace diffusion operator; $w_i(x,t) = (w_{i1}(x,t), w_{i2}(x,t), \ldots, w_{in}(x,t))^T \in \mathbb{R}^n$ is the state vector of node i; $\mathbb{R}^{n\times n} \ni C = \text{diag}(c_1, c_2, \ldots, c_n) > 0$; $\mathbb{R}^{n\times n} \ni D = \text{diag}(d_1, d_2, \ldots, d_n) > 0$; $H = (h_{ij})_{n\times n} \in \mathbb{R}^{n\times n}$; $J = (J_1, J_2, \ldots, J_n)^T \in \mathbb{R}^n$; $u_i(x,t) \in \mathbb{R}^q$ is the input vector of node i; $B \in \mathbb{R}^{n\times q}$; $0 < a \in \mathbb{R}$ represents the coupling strength; $0 < \Gamma \in \mathbb{R}^{n\times n}$ is the inner coupling matrix; $f(w_i(x,t)) =$

$(f_1(w_{i1}(x,t)),\ f_2(w_{i2}(x,t)),\ldots,\ f_n(w_{in}(x,t)))^T \in \mathbb{R}^n$; $\tau_s, s = 1, 2, \ldots, \eta$ represent the time-delays; $A = (A_{ij})_{N\times N} \in \mathbb{R}^{N\times N}$ is the outer coupling matrix satisfying the following conditions: if a link exists from node i to node j, $\mathbb{R} \ni A_{ij} > 0 (i \neq j)$; or else, $\mathbb{R} \ni A_{ij} = 0 (i \neq j)$; and

$$A_{ii} = -\sum_{\substack{j=1\\ j\neq i}}^{N} A_{ij}, \quad i = 1, 2, \ldots, N.$$

In this work, it is necessary that the network is connected.

Remark 9.4 Some authors have recently studied the passivity and synchronization problems for CRDNNs [17–25, 29–35]. However, in some existing studies [18, 19, 29, 32, 33], authors always assume that the outer coupling matrix is symmetric. In this paper, we remove this constraint and present a CRDNNs with multiple time-delays, which is completely different from the network models in existence [17–25, 29–35].

As is known, parameter fluctuation is ubiquitous in NNs. Moreover, these deviations are bounded. Therefore, in this study, we assume that c_i, d_i and h_{ij} are internalized as follows:

$$\begin{cases} C_I := \{C = \mathrm{diag}(c_i) : \underline{C} \leqslant C \leqslant \overline{C}, i.e., \underline{c}_i \leqslant c_i \leqslant \overline{c}_i, \\ \qquad i = 1, 2, \ldots, n, \forall C \in C_I\}, \\ D_I := \{D = \mathrm{diag}(d_i) : \underline{D} \leqslant D \leqslant \overline{D}, i.e., \underline{d}_i \leqslant d_i \leqslant \overline{d}_i, \\ \qquad i = 1, 2, \ldots, n, \forall D \in D_I\}, \\ H_I := \{H = (h_{ij})_{n\times n} : \underline{H} \leqslant H \leqslant \overline{H}, i.e., \\ \qquad \underline{h}_{ij} \leqslant h_{ij} \leqslant \overline{h}_{ij}, i, j = 1, 2, \ldots, n, \forall H \in H_I\}. \end{cases}$$

The boundary value and initial value conditions for network (9.1) are given as follows:

$$\begin{aligned} w_i(x,t) &= \Upsilon_i(x,t), \quad (x,t) \in \Omega \times [-\tau, 0], \\ w_i(x,t) &= 0, \quad (x,t) \in \partial\Omega \times [-\tau, +\infty), \end{aligned}$$

in which $\tau = \max_{s\in\{1,2,\ldots,\eta\}} \{\tau_s\}$, $\Upsilon_i(x,t)$ is continuous and bounded on $\Omega \times [-\tau, 0]$.

In this chapter, the function $f_j(\cdot)(j = 1, 2, \ldots, n)$ is always assumed to satisfy the Lipschitz condition. Namely, there is a positive constant θ_j satisfying

$$|f_j(\varphi_1) - f_j(\varphi_2)| \leqslant \theta_j |\varphi_1 - \varphi_2|$$

for any $\varphi_1, \varphi_2 \in \mathbb{R}$.

Define

$$\hat{h}_{ij} = \max\left\{|\underline{h}_{ij}|, |\overline{h}_{ij}|\right\},$$
$$\varsigma = \sum_{i=1}^{n}\sum_{j=1}^{n}\hat{h}_{ij}^2,$$
$$\Theta = \mathrm{diag}(\theta_1^2, \theta_2^2, \ldots, \theta_n^2).$$

9.3 Passivity of CRDNNs with Multiple Time-Delays

9.3.1 Passivity Analysis

On the basis of Lemma 1.3, there exists a positive vector $\kappa = (\kappa_1, \kappa_2, \ldots, \kappa_N)^T \in \mathbb{R}^N$ such that $\kappa^T A = 0$ and

$$\sum_{j=1}^{N}\tilde{A}_{ij} = \sum_{j=1}^{N}\tilde{A}_{ji} = 0$$

for all $i = 1, 2, \ldots, N$, where $\Psi = \mathrm{diag}(\kappa_1, \kappa_2, \ldots, \kappa_N)$ and $\tilde{A} = (\tilde{A}_{ij})_{N\times N} = \Psi A + A^T\Psi$.

Letting $\delta_i = \kappa_i/(\kappa_1 + \kappa_2 + \cdots + \kappa_N)$, $i = 1, 2, \ldots, N$, we obtain

$$A_{ii} = -\frac{\sum_{\substack{j=1\\ j\neq i}}^{N}\left(\delta_i A_{ij} + \delta_j A_{ji}\right)}{2\delta_i},$$
$$\sum_{j=1}^{N}\delta_j A_{ji} = 0, \quad i = 1, 2, \ldots, N.$$

Letting $\tilde{w}(x,t) = \sum_{i=1}^{N}\delta_i w_i(x,t)$, we obtain

$$\begin{aligned}
\frac{\partial\tilde{w}(x,t)}{\partial t} &= \sum_{i=1}^{N}\delta_i\frac{\partial w_i(x,t)}{\partial t}\\
&= \sum_{i=1}^{N}\delta_i C\Delta w_i(x,t) - \sum_{i=1}^{N}\delta_i D w_i(x,t) + \sum_{i=1}^{N}\delta_i J + \sum_{i=1}^{N}\delta_i H f(w_i(x,t))s\\
&\quad + \sum_{i=1}^{N}\delta_i B u_i(x,t) + a\sum_{s=1}^{\eta}\sum_{j=1}^{N}\left(\sum_{i=1}^{N}\delta_i A_{ij}\right)\Gamma w_j(x, t-\tau_s)\\
&= C\Delta\tilde{w}(x,t) - D\tilde{w}(x,t) + H\sum_{i=1}^{N}\delta_i f(w_i(x,t)) + J + B\sum_{i=1}^{N}\delta_i u_i(x,t).
\end{aligned}$$

Taking $\vartheta_i(x,t) = (\vartheta_{i1}(x,t), \vartheta_{i2}(x,t), \ldots, \vartheta_{in}(x,t))^T = w_i(x,t) - \tilde{w}(x,t)$, $i = 1, 2, \ldots, N$, we have

$$\begin{aligned}\frac{\partial \vartheta_i(x,t)}{\partial t} &= C\Delta\vartheta_i(x,t) - D\vartheta_i(x,t) + Hf(w_i(x,t)) - H\sum_{l=1}^{N}\delta_l f(w_l(x,t)) \\ &\quad + Bu_i(x,t) + a\sum_{s=1}^{\eta}\sum_{j=1}^{N} A_{ij}\Gamma\vartheta_j(x,t-\tau_s) \\ &\quad - B\sum_{l=1}^{N}\delta_l u_l(x,t), \quad i = 1, 2, \ldots, N. \end{aligned} \tag{9.2}$$

For network (9.2), we define the output vector $y_i(x,t) \in \mathbb{R}^m$ as follows:

$$y_i(x,t) = E\vartheta_i(x,t) + Fu_i(x,t), \tag{9.3}$$

where $E \in \mathbb{R}^{m\times n}$ and $F \in \mathbb{R}^{m\times q}$ are the known matrices.

Denote

$$\begin{aligned}\vartheta(x,t) &= \left(\vartheta_1^T(x,t), \vartheta_2^T(x,t), \ldots, \vartheta_N^T(x,t)\right)^T, \\ y(x,t) &= \left(y_1^T(x,t), y_2^T(x,t), \ldots, y_N^T(x,t)\right)^T, \\ u(x,t) &= \left(u_1^T(x,t), u_2^T(x,t), \ldots, u_N^T(x,t)\right)^T, \\ \delta &= \mathrm{diag}(\delta_1, \delta_2, \ldots, \delta_N).\end{aligned}$$

Theorem 9.5 *If there are matrices $P \in \mathbb{R}^{mN\times qN}$ and $0 < Q_s \in \mathbb{R}^{nN\times nN}$ $(s = 1, 2, \ldots, \eta)$ such that*

$$\begin{pmatrix} W_1 & \Xi_1 \\ \Xi_1^T & -P^T(I_N\otimes F) - (I_N\otimes F^T)P \end{pmatrix} \leqslant 0, \tag{9.4}$$

the network (9.2) is passive, where $W_1 = \delta\otimes\left(-\sum_{k=1}^{\sigma}\frac{2}{l_k^2}\underline{C} - 2\underline{D} + \varsigma I_n + \Theta\right) + a\sum_{s=1}^{\eta} Q_s + a\sum_{s=1}^{\eta}\left[(\delta A)\otimes\Gamma\right]Q_s^{-1}\left[(A^T\delta)\otimes\Gamma\right]$, $\Xi_1 = \delta\otimes B - (I_N\otimes E^T)P$.

Proof The Lyapunov functional for network (9.2) is chosen as follows:

$$\begin{aligned}V_1(t) &= a\sum_{s=1}^{\eta}\int_{t-\tau_s}^{t}\int_{\Omega}\vartheta^T(x,h)Q_s\vartheta(x,h)dxdh \\ &\quad + \sum_{i=1}^{N}\delta_i\int_{\Omega}\vartheta_i^T(x,t)\vartheta_i(x,t)dx.\end{aligned} \tag{9.5}$$

Then, one gets

$$
\begin{aligned}
\dot{V}_1(t) &= 2\sum_{i=1}^{N}\delta_i\int_{\Omega}\vartheta_i^T(x,t)\frac{\partial\vartheta_i(x,t)}{\partial t}dx + a\sum_{s=1}^{\eta}\int_{\Omega}\vartheta^T(x,t)Q_s\vartheta(x,t)dx \\
&\quad -a\sum_{s=1}^{\eta}\int_{\Omega}\vartheta^T(x,t-\tau_s)Q_s\vartheta(x,t-\tau_s)dx \\
&= 2\sum_{i=1}^{N}\delta_i\int_{\Omega}\vartheta_i^T(x,t)\Big[C\Delta\vartheta_i(x,t) - D\vartheta_i(x,t) + Hf(w_i(x,t)) \\
&\quad -Hf(\tilde{w}(x,t)) + Hf(\tilde{w}(x,t)) - H\sum_{l=1}^{N}\delta_l f(w_l(x,t)) \\
&\quad -B\sum_{l=1}^{N}\delta_l u_l(x,t) + Bu_i(x,t) + a\sum_{s=1}^{\eta}\sum_{j=1}^{N}A_{ij}\Gamma\vartheta_j(x,t-\tau_s)\Big]dx \\
&\quad -a\sum_{s=1}^{\eta}\int_{\Omega}\vartheta^T(x,t-\tau_s)Q_s\vartheta(x,t-\tau_s)dx \\
&\quad +a\sum_{s=1}^{\eta}\int_{\Omega}\vartheta^T(x,t)Q_s\vartheta(x,t)dx.
\end{aligned}
$$

Obviously,

$$
\begin{aligned}
\sum_{i=1}^{N}\delta_i\vartheta_i(x,t) &= \sum_{i=1}^{N}\delta_i\left(w_i(x,t) - \sum_{j=1}^{N}\delta_j w_j(x,t)\right) \\
&= \sum_{i=1}^{N}\delta_i w_i(x,t) - \sum_{j=1}^{N}\delta_j\left(\sum_{i=1}^{N}\delta_i\right)w_j(x,t) \\
&= \sum_{i=1}^{N}\delta_i w_i(x,t) - \sum_{j=1}^{N}\delta_j w_j(x,t) \\
&= 0.
\end{aligned}
$$

Then, we can get

$$
\begin{aligned}
\sum_{i=1}^{N}\delta_i\vartheta_i^T(x,t)\Big(Hf(\tilde{w}(x,t)) - H\sum_{l=1}^{N}\delta_l f(w_l(x,t))\Big) &= 0, \\
\sum_{i=1}^{N}\delta_i\vartheta_i^T(x,t)\Big(B\sum_{l=1}^{N}\delta_l u_l(x,t)\Big) &= 0. \qquad (9.6)
\end{aligned}
$$

Moreover,

$$
\begin{aligned}
&2\sum_{i=1}^{N}\delta_i\vartheta_i^T(x,t)H\left(f(w_i(x,t))-f(\tilde{w}(x,t))\right)\\
&\leqslant\sum_{i=1}^{N}\delta_i\vartheta_i^T(x,t)\left(HH^T+\Theta\right)\vartheta_i(x,t).
\end{aligned}
\tag{9.7}
$$

On the other hand, in view of the boundary condition and Green's formula, we have

$$
\int_{\Omega}\vartheta_{il}(x,t)\Delta\vartheta_{il}(x,t)dx=-\sum_{k=1}^{\sigma}\int_{\Omega}\left(\frac{\partial\vartheta_{il}(x,t)}{\partial x_k}\right)^2dx,
$$

where $i=1,2,\ldots,N,\ l=1,2,\ldots,n.$

On the basis of Lemma (1.5), we obtain

$$
\begin{aligned}
&\int_{\Omega}\vartheta_i^T(x,t)C\Delta\vartheta_i(x,t)dx\\
&=\sum_{l=1}^{n}c_l\int_{\Omega}\vartheta_{il}(x,t)\Delta\vartheta_{il}(x,t)dx\\
&=-\sum_{l=1}^{n}c_l\sum_{k=1}^{\sigma}\int_{\Omega}\left(\frac{\partial\vartheta_{il}(x,t)}{\partial x_k}\right)^2dx\\
&\leqslant-\sum_{k=1}^{\sigma}\frac{1}{l_k^2}\sum_{l=1}^{n}c_l\int_{\Omega}\vartheta_{il}^2(x,t)dx\\
&=-\sum_{k=1}^{\sigma}\frac{1}{l_k^2}\int_{\Omega}\vartheta_i^T(x,t)C\vartheta_i(x,t)dx.
\end{aligned}
\tag{9.8}
$$

From (9.6)–(9.8), we obtain

$$
\begin{aligned}
\dot{V}_1(t)\leqslant{}&2a\sum_{s=1}^{\eta}\sum_{i=1}^{N}\sum_{j=1}^{N}\delta_iA_{ij}\int_{\Omega}\vartheta_i^T(x,t)\Gamma\vartheta_j(x,t-\tau_s)dx\\
&+\sum_{i=1}^{N}\delta_i\int_{\Omega}\vartheta_i^T(x,t)\Bigg(-\sum_{k=1}^{\sigma}\frac{2}{l_k^2}C-2D+HH^T\\
&+\Theta\Bigg)\vartheta_i(x,t)dx+2\sum_{i=1}^{N}\delta_i\int_{\Omega}\vartheta_i^T(x,t)Bu_i(x,t)dx
\end{aligned}
$$

$$
\begin{aligned}
&-a\sum_{s=1}^{\eta}\int_{\Omega}\vartheta^T(x,t-\tau_s)Q_s\vartheta(x,t-\tau_s)dx \\
&+a\sum_{s=1}^{\eta}\int_{\Omega}\vartheta^T(x,t)Q_s\vartheta(x,t)dx \\
\leqslant &\int_{\Omega}\vartheta^T(x,t)\Big[\delta\otimes\Big(-\sum_{k=1}^{\sigma}\frac{2}{l_k^2}\underline{C}-2\underline{D}+\varsigma I_n+\Theta\Big)\Big]\vartheta(x,t)dx \\
&+2\int_{\Omega}\vartheta^T(x,t)(\delta\otimes B)u(x,t)dx \\
&+2a\sum_{s=1}^{\eta}\int_{\Omega}\vartheta^T(x,t)\big[(\delta A)\otimes\Gamma\big]\vartheta(x,t-\tau_s)dx \\
&-a\sum_{s=1}^{\eta}\int_{\Omega}\vartheta^T(x,t-\tau_s)Q_s\vartheta(x,t-\tau_s)dx \\
&+a\sum_{s=1}^{\eta}\int_{\Omega}\vartheta^T(x,t)Q_s\vartheta(x,t)dx \\
\leqslant &\int_{\Omega}\vartheta^T(x,t)\Big[\delta\otimes\Big(-\sum_{k=1}^{\sigma}\frac{2}{l_k^2}\underline{C}-2\underline{D}+\varsigma I_n+\Theta\Big)\Big]\vartheta(x,t)dx \\
&+a\sum_{s=1}^{\eta}\int_{\Omega}\vartheta^T(x,t)\big[(\delta A)\otimes\Gamma\big]Q_s^{-1}\big[(A^T\delta)\otimes\Gamma\big]\vartheta(x,t)dx \\
&+2\int_{\Omega}\vartheta^T(x,t)(\delta\otimes B)u(x,t)dx+a\sum_{s=1}^{\eta}\int_{\Omega}\vartheta^T(x,t)Q_s\vartheta(x,t)dx. \quad (9.9)
\end{aligned}
$$

By (9.9), we obtain

$$
\begin{aligned}
&\dot{V}_1(t)-2\int_{\Omega}y^T(x,t)Pu(x,t)dx \\
\leqslant &\int_{\Omega}\vartheta^T(x,t)\Big[\delta\otimes\Big(-\sum_{k=1}^{\sigma}\frac{2}{l_k^2}\underline{C}-2\underline{D}+\varsigma I_n+\Theta\Big)\Big]\vartheta(x,t)dx \\
&+a\sum_{s=1}^{\eta}\int_{\Omega}\vartheta^T(x,t)Q_s\vartheta(x,t)dx+2\int_{\Omega}\vartheta^T(x,t)(\delta\otimes B)u(x,t)dx
\end{aligned}
$$

$$+a\sum_{s=1}^{\eta}\int_{\Omega}\vartheta^T(x,t)[(\delta A)\otimes\Gamma]Q_s^{-1}[(A^T\delta)\otimes\Gamma]\vartheta(x,t)dx$$
$$-2\int_{\Omega}\left[\vartheta^T(x,t)(I_N\otimes E^T)+u^T(x,t)(I_N\otimes F^T)\right]Pu(x,t)dx$$
$$=\int_{\Omega}\xi^T(x,t)\begin{pmatrix}W_1 & \Xi_1\\ \Xi_1^T & W_2\end{pmatrix}\xi(x,t)dx, \tag{9.10}$$

where $\xi(x,t)=\left(\vartheta^T(x,t),u^T(x,t)\right)^T$, $W_2=-P^T(I_N\otimes F)-(I_N\otimes F^T)P$.

Then, from (9.4) and (9.10), we can infer that

$$2\int_{\Omega}y^T(x,t)Pu(x,t)dx\geqslant\dot{V}_1(t). \tag{9.11}$$

By (9.11), we obtain

$$\int_{t_0}^{t_q}\int_{\Omega}y^T(x,t)Pu(x,t)dxdt\geqslant\frac{V_1(t_q)}{2}-\frac{V_1(t_0)}{2}$$

for any $t_q,t_0\in[0,+\infty)$ and $t_q\geqslant t_0$.

Using the similar proof method in Theorem 9.5, the following conclusions can be made.

Theorem 9.6 *If there are matrices $P\in\mathbb{R}^{mN\times qN}, 0<Q_s\in\mathbb{R}^{nN\times nN}$ $(s=1,2,\ldots,\eta)$ and $0<F_1\in\mathbb{R}^{qN\times qN}$ such that*

$$\begin{pmatrix}W_1 & \Xi_1\\ \Xi_1^T & F_1-P^T(I_N\otimes F)-(I_N\otimes F^T)P\end{pmatrix}\leqslant 0,$$

the network (9.2) is input strictly passive, where $W_1=\delta\otimes\big(-\sum_{k=1}^{\sigma}\frac{2}{l_k^2}\underline{C}-2\underline{D}+\varsigma I_n+\Theta\big)+a\sum_{s=1}^{\eta}Q_s+a\sum_{s=1}^{\eta}\left[(\delta A)\otimes\Gamma\right]Q_s^{-1}\left[(A^T\delta)\otimes\Gamma\right]$, $\Xi_1=\delta\otimes B-(I_N\otimes E^T)P$.

Theorem 9.7 *If there are matrices $P\in\mathbb{R}^{mN\times qN}, 0<Q_s\in\mathbb{R}^{nN\times nN}$ $(s=1,2,\ldots,\eta)$ and $0<F_2\in\mathbb{R}^{mN\times mN}$ such that*

$$\begin{pmatrix}W_3 & \Xi_2\\ \Xi_2^T & W_4\end{pmatrix}\leqslant 0,$$

the network (9.2) is output strictly passive, where $W_3=\delta\otimes\big(-\sum_{k=1}^{\sigma}\frac{2}{l_k^2}\underline{C}-2\underline{D}+\varsigma I_n+\Theta\big)+a\sum_{s=1}^{\eta}Q_s+a\sum_{s=1}^{\eta}\left[(\delta A)\otimes\Gamma\right]Q_s^{-1}\left[(A^T\delta)\otimes\Gamma\right]+(I_N\otimes E^T)F_2(I_N$

$\otimes E)$, $W_4 = (I_N \otimes F^T)F_2(I_N \otimes F) - P^T(I_N \otimes F) - (I_N \otimes F^T)P$, $\Xi_2 = \delta \otimes B - (I_N \otimes E^T)P + (I_N \otimes E^T)F_2(I_N \otimes F)$.

Remark 9.8 In [40], the authors presented a criterion for synchronization of CRDNNs by employing the output strict passivity. Similarly, it is easy to prove that an output-strictly passive network (9.2) is also synchronized when $E^T E > 0$.

9.3.2 *Adaptive Control for Passivity*

To guarantee the passivity of network (9.1), the following adaptive state feedback controller is developed:

$$\begin{aligned} v_i(x,t) &= a\sum_{j=1}^{N} S_{ij}(t)\Gamma w_j(x,t), \\ \dot{S}_{ij}(t) &= \zeta_{ij}\delta_j \int_{\Omega} \left(w_i(x,t) - w_j(x,t)\right)^T \Gamma\left(w_i(x,t) - w_j(x,t)\right)dx, i \neq j, \end{aligned} \tag{9.12}$$

in which $S(t) = \left(S_{ij}(t)\right)_{N\times N} \in \mathbb{R}^{N\times N}$, $S_{ii}(t) = -\sum_{\substack{j=1\\ j\neq i}}^{N} S_{ij}(t)$, $S_{ij}(0) = \frac{A_{ij}}{\rho}$, ρ is a positive real number that is sufficiently large, δ_j represents the same meaning as in Sect. 9.3.1, $\mathbb{R} \ni \zeta_{ij} = \zeta_{ji} \geqslant 0$, and $\zeta_{ij} = 0$ if and only if $A_{ij} \times A_{ji} = 0$.

Then, we obtain

$$\begin{aligned} \frac{\partial w_i(x,t)}{\partial t} &= C\Delta w_i(x,t) - Dw_i(x,t) + Hf(w_i(x,t)) + J + Bu_i(x,t) \\ &\quad + a\sum_{j=1}^{N} A_{ij}\Gamma w_j(x,t-\tau_1) + a\sum_{j=1}^{N} A_{ij}\Gamma w_j(x,t-\tau_2) + \cdots \\ &\quad + a\sum_{j=1}^{N} A_{ij}\Gamma w_j(x,t-\tau_\eta) + a\sum_{j=1}^{N} S_{ij}(t)\Gamma w_j(x,t), \end{aligned} \tag{9.13}$$

where $i = 1, 2, \ldots, N$.

Obviously, $(\delta_1, \delta_2, \ldots, \delta_N)S(0) = 0$ and

$$S_{ii}(0) = -\frac{\sum_{\substack{j=1\\ j\neq i}}^{N} \left(\delta_i S_{ij}(0) + \delta_j S_{ji}(0)\right)}{2\delta_i}, \quad i = 1, 2, \ldots, N. \tag{9.14}$$

Using the adaptive law (9.12), we can acquire

$$\sum_{j=1}^{N} \delta_j \dot{S}_{ji}(t) = \sum_{j=1}^{N} \delta_i \dot{S}_{ij}(t) = 0.$$

Then,

$$\dot{S}_{ii}(t) = -\frac{\sum_{\substack{j=1\\ j\neq i}}^{N} \left(\delta_i \dot{S}_{ij}(t) + \delta_j \dot{S}_{ji}(t)\right)}{2\delta_i}, \tag{9.15}$$

$$\sum_{j=1}^{N} \delta_j S_{ji}(t) = 0, \tag{9.16}$$

where $i = 1, 2, \ldots, N$.

Furthermore, from (9.14) and (9.15), we have

$$S_{ii}(t) = -\frac{\sum_{\substack{j=1\\ j\neq i}}^{N} \left(\delta_i S_{ij}(t) + \delta_j S_{ji}(t)\right)}{2\delta_i}, \quad i = 1, 2, \ldots, N.$$

Letting $\tilde{w}(x,t) = \sum_{i=1}^{N} \delta_i w_i(x,t)$, we obtain

$$\begin{aligned}
\frac{\partial \tilde{w}(x,t)}{\partial t} &= \sum_{i=1}^{N} \delta_i \frac{\partial w_i(x,t)}{\partial t} \\
&= C\sum_{i=1}^{N} \delta_i \Delta w_i(x,t) - D\sum_{i=1}^{N} \delta_i w_i(x,t) + H\sum_{i=1}^{N} \delta_i f(w_i(x,t)) \\
&\quad + B\sum_{i=1}^{N} \delta_i u_i(x,t) + a\sum_{s=1}^{\eta}\sum_{j=1}^{N}\left(\sum_{i=1}^{N} \delta_i A_{ij}\right) \Gamma w_j(x, t-\tau_s) \\
&\quad + \sum_{i=1}^{N} \delta_i J + \sum_{i=1}^{N} \delta_i v_i(x,t) \\
&= C\Delta\tilde{w}(x,t) - D\tilde{w}(x,t) + H\sum_{i=1}^{N} \delta_i f(w_i(x,t)) + J \\
&\quad + B\sum_{i=1}^{N} \delta_i u_i(x,t) + \sum_{i=1}^{N} \delta_i v_i(x,t).
\end{aligned}$$

Define $\vartheta_i(x,t) = (\vartheta_{i1}(x,t), \vartheta_{i2}(x,t), \ldots, \vartheta_{in}(x,t))^T = w_i(x,t) - \tilde{w}(x,t), i = 1, 2, \ldots, N$. Then, we obtain

$$\begin{cases} \frac{\partial \vartheta_i(x,t)}{\partial t} = C\Delta\vartheta_i(x,t) - D\vartheta_i(x,t) + Hf(w_i(x,t)) \\ \qquad - H\sum_{l=1}^{N}\delta_l f(w_l(x,t)) + Bu_i(x,t) \\ \qquad - B\sum_{l=1}^{N}\delta_l u_l(x,t) - \sum_{l=1}^{N}\delta_l v_l(x,t) \\ \qquad + a\sum_{s=1}^{\eta}\sum_{j=1}^{N}A_{ij}\Gamma\vartheta_j(x,t-\tau_s) \\ \qquad + a\sum_{j=1}^{N}S_{ij}(t)\Gamma\vartheta_j(x,t), \\ \dot{S}_{ij}(t) = \zeta_{ij}\delta_j\int_{\Omega}\left(\vartheta_i(x,t) - \vartheta_j(x,t)\right)^T\Gamma\big(\vartheta_i(x,t) \\ \qquad - \vartheta_j(x,t)\big)dx, i \neq j, \ \ i = 1, 2, \ldots, N. \end{cases} \tag{9.17}$$

Choose the output vector $y_i(t) \in \mathbb{R}^m$ for network (9.17) as follows:

$$y_i(x,t) = E\vartheta_i(x,t) + Fu_i(x,t), \tag{9.18}$$

where $E \in \mathbb{R}^{m\times n}$ and $F \in \mathbb{R}^{m\times q}$ are the known matrices.

Denote

$$\begin{aligned} \vartheta(x,t) &= \left(\vartheta_1^T(x,t), \vartheta_2^T(x,t), \ldots, \vartheta_N^T(x,t)\right)^T, \\ y(x,t) &= \left(y_1^T(x,t), y_2^T(x,t), \ldots, y_N^T(x,t)\right)^T, \\ u(x,t) &= \left(u_1^T(x,t), u_2^T(x,t), \ldots, u_N^T(x,t)\right)^T, \\ \delta &= \operatorname{diag}(\delta_1, \delta_2, \ldots, \delta_N), \\ \tilde{S}_{ij}(t) &= \begin{cases} \delta_i S_{ij}(t) + \delta_j S_{ji}(t), & \text{if } (i,j) \text{ and } (j,i) \in \mathcal{B}; \\ -\sum_{\substack{l=1\\ l\neq i}}^{N}\tilde{S}_{il}(t), & \text{if } i = j; \\ 0, & \text{otherwise}. \end{cases} \\ G_{ij} &= \begin{cases} \delta_i S_{ij}(0), & \text{if } (i,j) \in \mathcal{B} \text{ and } (j,i) \notin \mathcal{B}; \\ \delta_j S_{ji}(0), & \text{if } (j,i) \in \mathcal{B} \text{ and } (i,j) \notin \mathcal{B}; \\ -\sum_{\substack{l=1\\ l\neq i}}^{N} G_{il}, & \text{if } i = j; \\ 0, & \text{otherwise}. \end{cases} \end{aligned}$$

Theorem 9.9 *If there are matrices $P \in \mathbb{R}^{mN\times qN}$ and $\hat{S} = \left(\hat{S}_{ij}\right)_{N\times N} \in \mathbb{R}^{N\times N}$ satisfying*

$$\begin{pmatrix} W_5 & \Xi_1 \\ \Xi_1^T & -P^T(I_N\otimes F) - (I_N\otimes F^T)P \end{pmatrix} \leqslant 0, \tag{9.19}$$

where $W_5 = \delta\otimes\left(-\sum_{k=1}^{\sigma}\frac{2}{l_k^2}\underline{C} - 2\underline{D} + \varsigma I_n + \Theta\right) + a\eta\left[(\delta A)\otimes\Gamma\right]\left[(A^T\delta)\otimes\Gamma\right] + a\eta I_{nN} + a\left(\hat{S} + G\right)\otimes\Gamma$, $\Xi_1 = \delta\otimes B - (I_N\otimes E^T)P$, $G = (G_{ij})_{N\times N}$, $\hat{S}_{ii} = -\sum_{\substack{j=1\\ j\neq i}}^{N}$

$\hat{S}_{ij}$, $i = 1, 2, \ldots, N$, $\hat{S}_{ij} = \hat{S}_{ji}(i \neq j)$ are nonnegative real numbers, and $\hat{S}_{ij} = 0(i \neq j)$ is equivalent to $A_{ij} \times A_{ji} = 0$, the network (9.17) is passive.

Proof The Lyapunov functional for the network (9.17) is chosen as follows:

$$
\begin{aligned}
V_2(t) = & \sum_{i=1}^{N} \sum_{\substack{(i,j)\in\mathcal{B} \\ (j,i)\in\mathcal{B}}} \frac{a}{8\zeta_{ij}\delta_i\delta_j} \left(\delta_i S_{ij}(t) + \delta_j S_{ji}(t) - \hat{S}_{ij}\right)^2 \\
& + a \sum_{s=1}^{\eta} \int_{t-\tau_s}^{t} \int_{\Omega} \vartheta^T(x, h)\vartheta(x, h) dx dh \\
& + \sum_{i=1}^{N} \delta_i \int_{\Omega} \vartheta_i^T(x, t)\vartheta_i(x, t) dx.
\end{aligned}
$$

Then, we have

$$
\begin{aligned}
\dot{V}_2(t) = & \, 2\sum_{i=1}^{N} \delta_i \int_{\Omega} \vartheta_i^T(x, t) \frac{\partial \vartheta_i(x, t)}{\partial t} dx - a \sum_{s=1}^{\eta} \int_{\Omega} \vartheta^T(x, t-\tau_s)\vartheta(x, t-\tau_s) dx \\
& + \sum_{i=1}^{N} \sum_{\substack{(i,j)\in\mathcal{B} \\ (j,i)\in\mathcal{B}}} \frac{a}{4\zeta_{ij}\delta_i\delta_j} \left(\delta_i S_{ij}(t) + \delta_j S_{ji}(t) - \hat{S}_{ij}\right)\left(\delta_i \dot{S}_{ij}(t) + \delta_j \dot{S}_{ji}(t)\right) \\
& + a \sum_{s=1}^{\eta} \int_{\Omega} \vartheta^T(x, t)\vartheta(x, t) dx \\
= & \, 2\sum_{i=1}^{N} \delta_i \int_{\Omega} \vartheta_i^T(x, t) \Big[C\Delta\vartheta_i(x, t) - D\vartheta_i(x, t) + Hf(w_i(x, t)) \\
& - Hf(\tilde{w}(x, t)) + Hf(\tilde{w}(x, t)) - H\sum_{l=1}^{N} \delta_l f(w_l(x, t)) - B\sum_{l=1}^{N} \delta_l u_l(x, t) \\
& - \sum_{l=1}^{N} \delta_l v_l(x, t) + a\sum_{j=1}^{N} S_{ij}(t)\Gamma\vartheta_j(x, t) + Bu_i(x, t) \\
& + a\sum_{s=1}^{\eta}\sum_{j=1}^{N} A_{ij}\Gamma\vartheta_j(x, t-\tau_s) \Big] dx + a\eta \int_{\Omega} \vartheta^T(x, t)\vartheta(x, t) dx \\
& + \frac{a}{2} \sum_{i=1}^{N} \sum_{\substack{(i,j)\in\mathcal{B} \\ (j,i)\in\mathcal{B}}} \left(\delta_i S_{ij}(t) + \delta_j S_{ji}(t) - \hat{S}_{ij}\right) \int_{\Omega} (\vartheta_i(x, t) \\
& - \vartheta_j(x, t))^T \Gamma(\vartheta_i(x, t) - \vartheta_j(x, t)) dx
\end{aligned}
$$

$$
\begin{aligned}
&-a\sum_{s=1}^{\eta}\int_{\Omega}\vartheta^T(x,t-\tau_s)\vartheta(x,t-\tau_s)dx \\
\leqslant\ & 2a\sum_{s=1}^{\eta}\sum_{i=1}^{N}\sum_{j=1}^{N}\delta_i A_{ij}\int_{\Omega}\vartheta_i^T(x,t)\Gamma\vartheta_j(x,t-\tau_s)dx \\
&+\sum_{i=1}^{N}\delta_i\int_{\Omega}\vartheta_i^T(x,t)\Big(-\sum_{k=1}^{\sigma}\frac{2}{l_k^2}C-2D+HH^T+\Theta\Big)\vartheta_i(x,t)dx \\
&+2\sum_{i=1}^{N}\delta_i\int_{\Omega}\vartheta_i^T(x,t)Bu_i(x,t)dx+\frac{a}{2}\sum_{i=1}^{N}\sum_{\substack{(i,j)\in\mathcal{B}\\(j,i)\in\mathcal{B}}}\Big(\delta_i S_{ij}(t)+\delta_j S_{ji}(t) \\
&-\hat{S}_{ij}\Big)\int_{\Omega}\big(\vartheta_i(x,t)-\vartheta_j(x,t)\big)^T\Gamma\big(\vartheta_i(x,t)-\vartheta_j(x,t)\big)dx \\
&+2a\sum_{i=1}^{N}\sum_{j=1}^{N}\delta_i S_{ij}(t)\int_{\Omega}\vartheta_i^T(x,t)\Gamma\vartheta_j(x,t)dx \\
&-a\sum_{s=1}^{\eta}\int_{\Omega}\vartheta^T(x,t-\tau_s)\vartheta(x,t-\tau_s)dx+a\eta\int_{\Omega}\vartheta^T(x,t)\vartheta(x,t)dx \\
\leqslant\ & \int_{\Omega}\vartheta^T(x,t)\Big[\delta\otimes\Big(-\sum_{k=1}^{\sigma}\frac{2}{l_k^2}\underline{C}-2\underline{D}+\varsigma I_n+\Theta\Big)\Big]\vartheta(x,t)dx \\
&+a\eta\int_{\Omega}\vartheta^T(x,t)[(\delta A)\otimes\Gamma][(A^T\delta)\otimes\Gamma]\vartheta(x,t)dx \\
&+\frac{a}{2}\sum_{i=1}^{N}\sum_{\substack{(i,j)\in\mathcal{B}\\(j,i)\in\mathcal{B}}}\Big(\delta_i S_{ij}(t)+\delta_j S_{ji}(t)-\hat{S}_{ij}\Big)\int_{\Omega}\big(\vartheta_i(x,t) \\
&-\vartheta_j(x,t)\big)^T\Gamma\big(\vartheta_i(x,t)-\vartheta_j(x,t)\big)dx+2\int_{\Omega}\vartheta^T(x,t)(\delta\otimes B)u(x,t)dx \\
&+2a\sum_{i=1}^{N}\sum_{j=1}^{N}\delta_i S_{ij}(t)\int_{\Omega}\vartheta_i^T(x,t)\Gamma\vartheta_j(x,t)dx+a\eta\int_{\Omega}\vartheta^T(x,t)\vartheta(x,t)dx.
\end{aligned}
$$

Moreover,

$$
\begin{aligned}
&\sum_{i=1}^{N}\sum_{\substack{(i,j)\in\mathcal{B}\\(j,i)\in\mathcal{B}}}\Big(\delta_i S_{ij}(t)+\delta_j S_{ji}(t)-\hat{S}_{ij}\Big)\big(\vartheta_i(x,t) \\
&-\vartheta_j(x,t)\big)^T\Gamma\big(\vartheta_i(x,t)-\vartheta_j(x,t)\big)
\end{aligned}
$$

$$= -2\sum_{i=1}^{N}\sum_{j=1}^{N}\left(\tilde{S}_{ij}(t) - \hat{S}_{ij}\right)\vartheta_i^T(x,t)\Gamma\vartheta_j(x,t).$$

Therefore, we have

$$\begin{aligned}
\dot{V}_2(t) \leqslant &\int_{\Omega}\vartheta^T(x,t)\left[\delta\otimes\left(-\sum_{k=1}^{\sigma}\frac{2}{l_k^2}\underline{C} - 2\underline{D} + \varsigma I_n + \Theta\right)\right]\vartheta(x,t)dx \\
&+2\int_{\Omega}\vartheta^T(x,t)(\delta\otimes B)u(x,t)dx + a\eta\int_{\Omega}\vartheta^T(x,t)\vartheta(x,t)dx \\
&-a\sum_{i=1}^{N}\sum_{j=1}^{N}\left(\tilde{S}_{ij}(t) - \hat{S}_{ij}\right)\int_{\Omega}\vartheta_i^T(x,t)\Gamma\vartheta_j(x,t)dx \\
&+a\eta\int_{\Omega}\vartheta^T(x,t)[(\delta A)\otimes\Gamma][(A^T\delta)\otimes\Gamma]\vartheta(x,t)dx \\
&+a\sum_{i=1}^{N}\sum_{j=1}^{N}\left(\tilde{S}_{ij}(t) + G_{ij}\right)\int_{\Omega}\vartheta_i^T(x,t)\Gamma\vartheta_j(x,t)dx \\
= &\int_{\Omega}\vartheta^T(x,t)\left[\delta\otimes\left(-\sum_{k=1}^{\sigma}\frac{2}{l_k^2}\underline{C} - 2\underline{D} + \varsigma I_n + \Theta\right)\right]\vartheta(x,t)dx \\
&+a\eta\int_{\Omega}\vartheta^T(x,t)\vartheta(x,t)dx + 2\int_{\Omega}\vartheta^T(x,t)(\delta\otimes B)u(x,t)dx \\
&+a\eta\int_{\Omega}\vartheta^T(x,t)[(\delta A)\otimes\Gamma][(A^T\delta)\otimes\Gamma]\vartheta(x,t)dx \\
&+a\int_{\Omega}\vartheta^T(x,t)\left[(\hat{S} + G)\otimes\Gamma\right]\vartheta(x,t)dx. \qquad (9.20)
\end{aligned}$$

By (9.20), we obtain

$$\begin{aligned}
&\dot{V}_2(t) - 2\int_{\Omega}y^T(x,t)Pu(x,t)dx \\
\leqslant &\int_{\Omega}\vartheta^T(x,t)\left\{\delta\otimes\left(-\sum_{k=1}^{\sigma}\frac{2}{l_k^2}\underline{C} - 2\underline{D} + \varsigma I_n + \Theta\right)\right. \\
&\left.+a\eta[(\delta A)\otimes\Gamma][(A^T\delta)\otimes\Gamma] + a\eta I_{nN} + a(\hat{S} + G)\otimes\Gamma\right\}\vartheta(x,t)dx
\end{aligned}$$

$$-2\int_{\Omega}\vartheta^T(x,t)(I_N\otimes E^T)Pu(x,t)dx$$
$$-\int_{\Omega}u^T(x,t)\big[P^T(I_N\otimes F)+(I_N\otimes F^T)P\big]u(x,t)dx$$
$$+2\int_{\Omega}\vartheta^T(x,t)(\delta\otimes B)u(x,t)$$
$$=\int_{\Omega}\xi^T(x,t)\begin{pmatrix}W_5 & \Xi_1\\ \Xi_1^T & W_6\end{pmatrix}\xi(x,t)dx, \tag{9.21}$$

where $\xi(x,t)=\left(\vartheta^T(x,t),u^T(x,t)\right)^T$, $W_6=-P^T(I_N\otimes F)-(I_N\otimes F^T)P$.
By (9.19) and (9.21), we have

$$\dot V_2(t)\leqslant 2\int_{\Omega}y^T(x,t)Pu(x,t)dx. \tag{9.22}$$

From (9.22), we obtain

$$\int_{t_0}^{t_q}\int_{\Omega}y^T(x,t)Pu(x,t)dxdt\geqslant\frac{V_2(t_q)}{2}-\frac{V_2(t_0)}{2}$$

for any $t_q,t_0\in[0,+\infty)$ and $t_q\geqslant t_0$.
Using the similar proof method in Theorem 9.9, the following conclusions can be made.

Theorem 9.10 *If there are matrices $P\in\mathbb{R}^{mN\times qN}$, $\hat S=\left(\hat S_{ij}\right)_{N\times N}\in\mathbb{R}^{N\times N}$ and $0<F_1\in\mathbb{R}^{qN\times qN}$ satisfying*

$$\begin{pmatrix}W_5 & \Xi_1\\ \Xi_1^T & F_1-P^T(I_N\otimes F)-(I_N\otimes F^T)P\end{pmatrix}\leqslant 0,$$

where $W_5=\delta\otimes\left(-\sum_{k=1}^{\sigma}\frac{2}{l_k^2}\underline{C}-2\underline{D}+\varsigma I_n+\Theta\right)+a\eta\big[(\delta A)\otimes\Gamma\big]\big[(A^T\delta)\otimes\Gamma\big]$ $+a\eta I_{nN}+a\left(\hat S+G\right)\otimes\Gamma$, $\Xi_1=\delta\otimes B-(I_N\otimes E^T)P$, $G=(G_{ij})_{N\times N}$, $\hat S_{ii}=-\sum_{\substack{j=1\\ j\neq i}}^{N}\hat S_{ij}$, $i=1,2,\ldots,N$, $\hat S_{ij}=\hat S_{ji}(i\neq j)$ *are nonnegative real numbers, and* $\hat S_{ij}=0(i\neq j)$ *is equivalent to* $A_{ij}\times A_{ji}=0$, *the network (9.17) is input strictly passive.*

Theorem 9.11 *If there are matrices $P\in\mathbb{R}^{mN\times qN}$, $\hat S=\left(\hat S_{ij}\right)_{N\times N}\in\mathbb{R}^{N\times N}$ and $0<F_2\in\mathbb{R}^{mN\times mN}$ satisfying*

$$\begin{pmatrix} W_6 & \Xi_2 \\ \Xi_2^T & W_7 \end{pmatrix} \leqslant 0,$$

where $W_6 = \delta \otimes \big(-\sum_{k=1}^{\sigma} \frac{2}{l_k^2}\underline{C} - 2\underline{D} + \varsigma I_n + \Theta\big) + a\eta\big[(\delta A) \otimes \Gamma\big]\big[(A^T\delta) \otimes \Gamma\big] + a\eta I_{nN} + a\big(\hat{S} + G\big) \otimes \Gamma + (I_N \otimes E^T)F_2(I_N \otimes E)$, $W_7 = (I_N \otimes F^T)F_2(I_N \otimes F) - P^T(I_N \otimes F) - (I_N \otimes F^T)P$, $\Xi_2 = \delta \otimes B - (I_N \otimes E^T)P + (I_N \otimes E^T) F_2(I_N \otimes F)$, $G = (G_{ij})_{N\times N}$, $\hat{S}_{ii} = -\sum_{\substack{j=1\\ j\neq i}}^{N} \hat{S}_{ij}$, $i = 1, 2, \ldots, N$, $\hat{S}_{ij} = \hat{S}_{ji} (i \neq j)$ *are nonnegative real numbers, and* $\hat{S}_{ij} = 0 (i \neq j)$ *is equivalent to* $A_{ij} \times A_{ji} = 0$, *the network (9.17) is output strictly passive.*

Remark 9.12 In Theorem 9.11, an output strict passivity criterion for network (9.1) is given by utilizing the designed adaptive controller (9.12). Similarly, it is easy to prove that network (9.1) is also synchronized when the condition of Theorem 9.11 is satisfied and $E^T E > 0$.

9.4 Synchronization of CRDNNs with Multiple Time-Delays

9.4.1 Synchronization Analysis

Letting $u_i(x, t) = 0$ in network model (9.1), one has

$$\begin{aligned} \frac{\partial w_i(x,t)}{\partial t} &= C\Delta w_i(x,t) - Dw_i(x,t) + Hf(w_i(x,t)) + J \\ &\quad + a\sum_{j=1}^{N} A_{ij}\Gamma w_j(x, t-\tau_1) + a\sum_{j=1}^{N} A_{ij}\Gamma w_j(x, t-\tau_2) + \cdots \\ &\quad + a\sum_{j=1}^{N} A_{ij}\Gamma w_j(x, t-\tau_\eta), \end{aligned} \tag{9.23}$$

where $i = 1, 2, \ldots, N$.

On the basis of Lemma 1.3, there exists a positive vector $\kappa = (\kappa_1, \kappa_2, \ldots, \kappa_N)^T \in \mathbb{R}^N$ such that $\kappa^T A = 0$ and

$$\sum_{j=1}^{N} \tilde{A}_{ij} = \sum_{j=1}^{N} \tilde{A}_{ji} = 0$$

for all $i = 1, 2, \ldots, N$, where $\Psi = \mathrm{diag}(\kappa_1, \kappa_2, \ldots, \kappa_N)$ and $\tilde{A} = (\tilde{A}_{ij})_{N\times N} = \Psi A + A^T \Psi$.

Setting $\delta_i = \kappa_i/(\kappa_1 + \kappa_2 + \cdots + \kappa_N)$, $i = 1, 2, \ldots, N$, we obtain

$$A_{ii} = -\frac{\sum\limits_{\substack{j=1\\j\neq i}}^{N} \left(\delta_i A_{ij} + \delta_j A_{ji}\right)}{2\delta_i},$$

$$\sum_{j=1}^{N} \delta_j A_{ji} = 0, \quad i = 1, 2, \ldots, N.$$

Taking $\tilde{w}(x,t) = \sum\limits_{i=1}^{N} \delta_i w_i(x,t)$ and $\vartheta_i(x,t) = (\vartheta_{i1}(x,t), \vartheta_{i2}(x,t), \ldots, \vartheta_{in}(x, t))^T = w_i(x,t) - \tilde{w}(x,t)$, we have

$$\begin{aligned}\frac{\partial \vartheta_i(x,t)}{\partial t} &= C\Delta\vartheta_i(x,t) - D\vartheta_i(x,t) + Hf(w_i(x,t)) \\ &+a\sum_{s=1}^{\eta}\sum_{j=1}^{N} A_{ij}\Gamma\vartheta_j(x,t-\tau_s) - H\sum_{l=1}^{N}\delta_l f(w_l(x,t)), \end{aligned} \tag{9.24}$$

where $i = 1, 2, \ldots, N$.

Define

$$\begin{aligned}\delta &= \mathrm{diag}(\delta_1, \delta_2, \ldots, \delta_N),\\ \vartheta(x,t) &= \left(\vartheta_1^T(x,t), \vartheta_2^T(x,t), \ldots, \vartheta_N^T(x,t)\right)^T.\end{aligned}$$

Definition 9.13 (*see* [41]) The network model (9.23) is synchronized if

$$\lim_{t\to+\infty} \left\| w_i(\cdot,t) - \sum_{l=1}^{N} \delta_l w_l(\cdot,t) \right\|_2 = 0, \quad i = 1, 2, \ldots, N.$$

Theorem 9.14 *If there are matrices* $0 < Q_s \in \mathbb{R}^{nN\times nN}$, $s = 1, 2, \ldots, \eta$, *such that*

$$U + a\sum_{s=1}^{\eta} Q_s < 0, \tag{9.25}$$

the network (9.23) is synchronized, where $U = \delta \otimes \left(-\sum\limits_{k=1}^{\sigma} \frac{2}{l_k^2}\underline{C} - 2\underline{D} + \varsigma I_n + \Theta\right) + a\sum\limits_{s=1}^{\eta}\left[(\delta A)\otimes\Gamma\right]Q_s^{-1}\left[(A^T\delta)\otimes\Gamma\right]$.

Proof The Lyapunov functional for network (9.24) is chosen as follows:

$$V_1(t) = a\sum_{s=1}^{\eta}\int_{t-\tau_s}^{t}\int_{\Omega} \vartheta^T(x,h)Q_s\vartheta(x,h)dxdh$$

$$+\sum_{i=1}^{N}\delta_i\int_{\Omega}\vartheta_i^T(x,t)\vartheta_i(x,t)dx. \tag{9.26}$$

Then, we have

$$\begin{aligned}
\dot{V}_1(t) &= 2\sum_{i=1}^{N}\delta_i\int_{\Omega}\vartheta_i^T(x,t)\frac{\partial\vartheta_i(x,t)}{\partial t}dx - a\sum_{s=1}^{\eta}\int_{\Omega}\vartheta^T(x,t-\tau_s)Q_s\vartheta(x,t-\tau_s)dx \\
&\quad + a\sum_{s=1}^{\eta}\int_{\Omega}\vartheta^T(x,t)Q_s\vartheta(x,t)dx \\
&= 2\sum_{i=1}^{N}\delta_i\int_{\Omega}\vartheta_i^T(x,t)\Big[C\Delta\vartheta_i(x,t) - D\vartheta_i(x,t) + Hf(w_i(x,t)) \\
&\quad - Hf(\tilde{w}(x,t)) + Hf(\tilde{w}(x,t)) + a\sum_{s=1}^{\eta}\sum_{j=1}^{N}A_{ij}\Gamma\vartheta_j(x,t-\tau_s) \\
&\quad - H\sum_{l=1}^{N}\delta_l f(w_l(x,t))\Big]dx - a\sum_{s=1}^{\eta}\int_{\Omega}\vartheta^T(x,t-\tau_s)Q_s\vartheta(x,t-\tau_s)dx \\
&\quad + a\sum_{s=1}^{\eta}\int_{\Omega}\vartheta^T(x,t)Q_s\vartheta(x,t)dx \\
&\leqslant 2a\sum_{s=1}^{\eta}\sum_{i=1}^{N}\sum_{j=1}^{N}\delta_i A_{ij}\int_{\Omega}\vartheta_i^T(x,t)\Gamma\vartheta_j(x,t-\tau_s)dx \\
&\quad + \sum_{i=1}^{N}\delta_i\int_{\Omega}\vartheta_i^T(x,t)\Big(-\sum_{k=1}^{\sigma}\frac{2}{l_k^2}C - 2D + HH^T + \Theta\Big)\vartheta_i(x,t)dx \\
&\quad + a\sum_{s=1}^{\eta}\int_{\Omega}\vartheta^T(x,t)Q_s\vartheta(x,t)dx - a\sum_{s=1}^{\eta}\int_{\Omega}\vartheta^T(x,t-\tau_s)Q_s\vartheta(x,t-\tau_s)dx \\
&\leqslant \int_{\Omega}\vartheta^T(x,t)\Big[\delta\otimes\Big(-\sum_{k=1}^{\sigma}\frac{2}{l_k^2}\underline{C} - 2\underline{D} + \varsigma I_n + \Theta\Big)\Big]\vartheta(x,t)dx \\
&\quad + a\sum_{s=1}^{\eta}\int_{\Omega}\vartheta^T(x,t)[(\delta A)\otimes\Gamma]Q_s^{-1}[(A^T\delta)\otimes\Gamma]\vartheta(x,t)dx \\
&\quad + a\sum_{s=1}^{\eta}\int_{\Omega}\vartheta^T(x,t)Q_s\vartheta(x,t)dx
\end{aligned}$$

$$
\begin{aligned}
&= \int_{\Omega} \vartheta^T(x,t)\Big\{\delta \otimes \Big(-\sum_{k=1}^{\sigma} \frac{2}{l_k^2}\underline{C} - 2\underline{D} + \varsigma I_n + \Theta\Big) \\
&\quad + a\sum_{s=1}^{\eta} [(\delta A)\otimes \Gamma] Q_s^{-1} [(A^T\delta)\otimes \Gamma] + a\sum_{s=1}^{\eta} Q_s \Big\} \vartheta(x,t)dx \\
&\leqslant \varrho \|\vartheta(\cdot,t)\|_2^2,
\end{aligned} \tag{9.27}
$$

where $\varrho = \lambda_M \left(U + a\sum_{s=1}^{\eta} Q_s\right)$.

From (9.27) and the definition of $V_1(t)$, we can infer that $V_1(t)$ is bounded and nonincreasing. Therefore, $\lim\limits_{t\to+\infty} V_1(t)$ exists and satisfies $\lim\limits_{t\to+\infty} V_1(t) \geqslant 0$. Moreover, according to (9.27), we obtain

$$
||\vartheta(\cdot,t)||_2^2 \leqslant \frac{\dot{V}_1(t)}{\varrho}. \tag{9.28}
$$

From (9.28), it is easy to derived that $\lim\limits_{t\to+\infty} \int_0^t ||\vartheta(\cdot,s)||_2^2 ds$ exists and is a nonnegative real number. Because $0 \leqslant \tau_s \leqslant \tau$, we have

$$
\begin{aligned}
0 &\leqslant \lim_{t\to+\infty} \int_{t-\tau_s}^{t} \int_{\Omega} \vartheta^T(x,h) Q_s \vartheta(x,h) dx dh \\
&\leqslant \lambda_M(Q_s) \lim_{t\to+\infty} \int_{t-\tau}^{t} \int_{\Omega} \vartheta^T(x,h)\vartheta(x,h) dx dh \\
&= \lambda_M(Q_s) \lim_{t\to+\infty} \int_{t-\tau}^{t} \|\vartheta(\cdot,h)\|_2^2 dh \\
&= 0, \quad s = 1, 2, \ldots, \eta.
\end{aligned}
$$

Consequently, it can be concluded that $\lim\limits_{t\to+\infty} \sum\limits_{i=1}^{N} \delta_i \int_{\Omega} \vartheta_i^T(x,t)\vartheta_i(x,t)dx$ exists. Assume that

$$
\lim_{t\to+\infty} \sum_{i=1}^{N} \delta_i \int_{\Omega} \vartheta_i^T(x,t)\vartheta_i(x,t)dx = \iota > 0.
$$

Apparently, we can easily seek out a real number $\mu > 0$ satisfying

$$
\sum_{i=1}^{N} \delta_i \int_{\Omega} \vartheta_i^T(x,t)\vartheta_i(x,t)dx > \frac{\iota}{2} \text{ for } t \geqslant \mu.
$$

Then, we obtain

$$||\vartheta(\cdot,t)||_2^2 > \frac{\iota}{2\lambda_M(\delta)}, \quad t \geqslant \mu. \tag{9.29}$$

Using (9.27) and (9.30), wen can acquire

$$\dot{V}_1(t) < \frac{\iota\varrho}{2\lambda_M(\delta)}, \quad t \geqslant \mu. \tag{9.30}$$

By (9.30), we can obtain

$$\begin{aligned} -V_1(\mu) \leqslant V_1(+\infty) - V_1(\mu) &= \int_{\mu}^{+\infty} \dot{V}_1(t)dt \\ &< \int_{\mu}^{+\infty} \frac{\iota\varrho}{2\lambda_M(\delta)}dt \\ &= -\infty. \end{aligned}$$

This results in a contradiction, so

$$\lim_{t\to+\infty} \sum_{i=1}^{N} \delta_i \int_{\Omega} \vartheta_i^T(x,t)\vartheta_i(x,t)dx = 0.$$

Then, we obtain

$$\lim_{t\to+\infty} ||\vartheta(\cdot,t)||_2 = 0.$$

Therefore, the network (9.23) achieves synchronization.

9.4.2 Adaptive Control for Synchronization

To ensure the network (9.23) can achieve synchronization, the following adaptive state feedback controller is developed:

$$\begin{aligned} v_i(x,t) &= a\sum_{j=1}^{N} S_{ij}(t)\Gamma w_j(x,t), \\ \dot{S}_{ij}(t) &= \zeta_{ij}\delta_j \int_{\Omega} \big(w_i(x,t) - w_j(x,t)\big)^T \Gamma\big(w_i(x,t) \\ &\quad - w_j(x,t)\big)dx, \ i \neq j, \end{aligned} \tag{9.31}$$

in which $S(t) = \left(S_{ij}(t)\right)_{N\times N} \in \mathbb{R}^{N\times N}$, $S_{ii}(t) = -\sum_{\substack{j=1\\ j\neq i}}^{N} S_{ij}(t)$, $S_{ij}(0) = \frac{A_{ij}}{\rho}$, ρ is a positive real number that is sufficiently large, δ_j represents the same meaning as in Sect. 9.4.1, $\mathbb{R} \ni \zeta_{ij} = \zeta_{ji} \geqslant 0$, and $\zeta_{ij} = 0$ if and only if $A_{ij} \times A_{ji} = 0$.

Then, we obtain

$$\begin{aligned}\frac{\partial w_i(x,t)}{\partial t} &= C\Delta w_i(x,t) - Dw_i(x,t) + Hf(w_i(x,t)) + J\\ &\quad + a\sum_{j=1}^{N} A_{ij}\Gamma w_j(x,t-\tau_1) + a\sum_{j=1}^{N} A_{ij}\Gamma w_j(x,t-\tau_2) + \cdots\\ &\quad + a\sum_{j=1}^{N} A_{ij}\Gamma w_j(x,t-\tau_\eta) + a\sum_{j=1}^{N} S_{ij}(t)\Gamma w_j(x,t), \qquad (9.32)\end{aligned}$$

where $i = 1, 2, \ldots, N$.

Evidently, $(\delta_1, \delta_2, \ldots, \delta_N)S(0) = 0$ and

$$S_{ii}(0) = -\frac{\sum_{\substack{j=1\\ j\neq i}}^{N}\left(\delta_i S_{ij}(0) + \delta_j S_{ji}(0)\right)}{2\delta_i}, \quad i = 1, 2, \ldots, N. \qquad (9.33)$$

Using the adaptive law (9.31), we can acquire

$$\sum_{j=1}^{N} \delta_j \dot{S}_{ji}(t) = \sum_{j=1}^{N} \delta_i \dot{S}_{ij}(t) = 0.$$

Then,

$$\dot{S}_{ii}(t) = -\frac{\sum_{\substack{j=1\\ j\neq i}}^{N}\left(\delta_i \dot{S}_{ij}(t) + \delta_j \dot{S}_{ji}(t)\right)}{2\delta_i}, \qquad (9.34)$$

$$\sum_{j=1}^{N} \delta_j S_{ji}(t) = 0, \qquad (9.35)$$

where $i = 1, 2, \ldots, N$.

Furthermore, from (9.33) and (9.34), we have

$$S_{ii}(t) = -\frac{\sum_{\substack{j=1\\ j\neq i}}^{N}\left(\delta_i S_{ij}(t) + \delta_j S_{ji}(t)\right)}{2\delta_i}, \quad i = 1, 2, \ldots, N.$$

Letting $\tilde{w}(x,t) = \sum_{i=1}^{N} \delta_i w_i(x,t)$ and $\vartheta_i(x,t) = (\vartheta_{i1}(x,t), \vartheta_{i2}(x,t), \ldots, \vartheta_{in}(x,t))^T = w_i(x,t) - \tilde{w}(x,t)$, one gets

$$\begin{cases} \frac{\partial\vartheta_i(x,t)}{\partial t} = C\Delta\vartheta_i(x,t) - D\vartheta_i(x,t) + Hf(w_i(x,t)) \\ \qquad - H\sum_{l=1}^{N}\delta_l f(w_l(x,t)) - \sum_{l=1}^{N}\delta_l v_l(x,t) \\ \qquad + a\sum_{s=1}^{\eta}\sum_{j=1}^{N} A_{ij}\Gamma\vartheta_j(x,t-\tau_s) \\ \qquad + a\sum_{j=1}^{N} S_{ij}(t)\Gamma\vartheta_j(x,t), \\ \dot{S}_{ij}(t) = \zeta_{ij}\delta_j \int_{\Omega} \left(\vartheta_i(x,t) - \vartheta_j(x,t)\right)^T \Gamma\left(\vartheta_i(x,t)\right. \\ \qquad \left. - \vartheta_j(x,t)\right)dx, i \neq j, \quad i = 1, 2, \ldots, N. \end{cases} \tag{9.36}$$

Define

$$\begin{aligned} \vartheta(x,t) &= \left(\vartheta_1^T(x,t), \vartheta_2^T(x,t), \ldots, \vartheta_N^T(x,t)\right)^T, \\ \delta &= \mathrm{diag}(\delta_1, \delta_2, \ldots, \delta_N) \\ \tilde{S}_{ij}(t) &= \begin{cases} \delta_i S_{ij}(t) + \delta_j S_{ji}(t), & \text{if } (i,j) \text{ and } (j,i) \in \mathcal{B}; \\ -\sum_{l=1, l\neq i}^{N} \tilde{S}_{il}(t), & \text{if } i = j; \\ 0, & \text{otherwise}, \end{cases} \\ G_{ij} &= \begin{cases} \delta_i S_{ij}(0), & \text{if } (i,j) \in \mathcal{B} \text{ and } (j,i) \notin \mathcal{B}; \\ \delta_j S_{ji}(0), & \text{if } (j,i) \in \mathcal{B} \text{ and } (i,j) \notin \mathcal{B}; \\ -\sum_{l=1, l\neq i}^{N} G_{il}, & \text{if } i = j; \\ 0, & \text{otherwise}. \end{cases} \end{aligned}$$

Theorem 9.15 *If there is a matrix $\hat{S} = \left(S_{ij}\right)_{N\times N} \in \mathbb{R}^{N\times N}$ such that*

$$\Phi + a\left(\hat{S} + G\right) \otimes \Gamma < 0, \tag{9.37}$$

where $\Phi = \delta \otimes \left(-\sum_{k=1}^{\sigma} \frac{2}{l_k^2}\underline{C} - 2\underline{D} + \varsigma I_n + \Theta\right) + a\eta\left[(\delta A)\otimes\Gamma\right]\left[(A^T\delta)\otimes\Gamma\right] + a\eta I_{nN}$, $G = (G_{ij})_{N\times N}$, $\hat{S}_{ii} = -\sum_{j=1, j\neq i}^{N} \hat{S}_{ij}$, $i = 1, 2, \ldots, N$, $\hat{S}_{ij} = \hat{S}_{ji} (i \neq j)$ are non-negative real numbers, and $\hat{S}_{ij} = 0 (i \neq j)$ is equivalent to $A_{ij} \times A_{ji} = 0$, the network (9.23) is synchronized under the adaptive state feedback controller (9.31).

Proof The Lyapunov functional for the network (9.36) is chosen as follows:

$$V_2(t) = \sum_{i=1}^{N} \sum_{\substack{(i,j)\in\mathcal{B} \\ (j,i)\in\mathcal{B}}} \frac{a}{8\zeta_{ij}\delta_i\delta_j}\left(\delta_i S_{ij}(t) + \delta_j S_{ji}(t) - \hat{S}_{ij}\right)^2$$
$$+a\sum_{s=1}^{\eta} \int_{t-\tau_s}^{t} \int_{\Omega} \vartheta^T(x,h)\vartheta(x,h)dxdh$$
$$+\sum_{i=1}^{N} \delta_i \int_{\Omega} \vartheta_i^T(x,t)\vartheta_i(x,t)dx. \tag{9.38}$$

Subsequently, we obtain

$$\dot{V}_2(t) = 2\sum_{i=1}^{N} \delta_i \int_{\Omega} \vartheta_i^T(x,t)\frac{\partial \vartheta_i(x,t)}{\partial t}dx$$
$$+\sum_{i=1}^{N} \sum_{\substack{(i,j)\in\mathcal{B} \\ (j,i)\in\mathcal{B}}} \frac{a}{4\zeta_{ij}\delta_i\delta_j}\Big(\delta_i S_{ij}(t) + \delta_j S_{ji}(t)$$
$$-\hat{S}_{ij}\Big)\Big(\delta_i \dot{S}_{ij}(t) + \delta_j \dot{S}_{ji}(t)\Big) + a\sum_{s=1}^{\eta} \int_{\Omega} \vartheta^T(x,t)\vartheta(x,t)dx$$
$$-a\sum_{s=1}^{\eta} \int_{\Omega} \vartheta^T(x,t-\tau_s)\vartheta(x,t-\tau_s)dx$$
$$= 2\sum_{i=1}^{N} \delta_i \int_{\Omega} \vartheta_i^T(x,t)\Big[C\Delta\vartheta_i(x,t) - D\vartheta_i(x,t) + Hf(w_i(x,t))$$
$$-Hf(\tilde{w}(x,t)) + Hf(\tilde{w}(x,t)) - H\sum_{l=1}^{N} \delta_l f(w_l(x,t)) - \sum_{l=1}^{N} \delta_l v_l(x,t)$$
$$+a\sum_{s=1}^{\eta}\sum_{j=1}^{N} A_{ij}\Gamma\vartheta_j(x,t-\tau_s) + a\sum_{j=1}^{N} S_{ij}(t)\Gamma\vartheta_j(x,t)\Big]dx$$
$$-a\sum_{i=1}^{N}\sum_{j=1}^{N}\Big(\tilde{S}_{ij}(t) - \hat{S}_{ij}\Big)\int_{\Omega} \vartheta_i^T(x,t)\Gamma\vartheta_j(x,t)dx$$
$$-a\sum_{s=1}^{\eta} \int_{\Omega} \vartheta^T(x,t-\tau_s)\vartheta(x,t-\tau_s)dx + a\eta\int_{\Omega} \vartheta^T(x,t)\vartheta(x,t)dx$$
$$\leqslant 2a\sum_{s=1}^{\eta}\sum_{i=1}^{N}\sum_{j=1}^{N} \delta_i A_{ij} \int_{\Omega} \vartheta_i^T(x,t)\Gamma\vartheta_j(x,t-\tau_s)dx$$

$$
\begin{aligned}
&+\sum_{i=1}^{N}\delta_i\int_{\Omega}\vartheta_i^T(x,t)\Big(-\sum_{k=1}^{\sigma}\frac{2}{l_k^2}C-2D+HH^T+\Theta\Big)\vartheta_i(x,t)dx\\
&+a\eta\int_{\Omega}\vartheta^T(x,t)\vartheta(x,t)dx-a\sum_{i=1}^{N}\sum_{j=1}^{N}\Big(\tilde{S}_{ij}(t)\\
&-\hat{S}_{ij}\Big)\int_{\Omega}\vartheta_i^T(x,t)\Gamma\vartheta_j(x,t)dx\\
&+2a\sum_{i=1}^{N}\sum_{j=1}^{N}\delta_iS_{ij}(t)\int_{\Omega}\vartheta_i^T(x,t)\Gamma\vartheta_j(x,t)\\
&-a\sum_{s=1}^{\eta}\int_{\Omega}\vartheta^T(x,t-\tau_s)\vartheta(x,t-\tau_s)dx\\
\leqslant&\int_{\Omega}\vartheta^T(x,t)\Big[\delta\otimes\Big(-\sum_{k=1}^{\sigma}\frac{2}{l_k^2}\underline{C}-2\underline{D}+\varsigma I_n+\Theta\Big)\Big]\vartheta(x,t)dx\\
&+a\eta\int_{\Omega}\vartheta^T(x,t)\big[(\delta A)\otimes\Gamma\big]\big[(A^T\delta)\otimes\Gamma\big]\vartheta(x,t)dx\\
&+a\eta\int_{\Omega}\vartheta^T(x,t)\vartheta(x,t)dx+a\sum_{i=1}^{N}\sum_{j=1}^{N}(\hat{S}_{ij}\\
&+G_{ij})\int_{\Omega}\vartheta_i^T(x,t)\Gamma\vartheta_j(x,t)dx\\
=&\int_{\Omega}\vartheta^T(x,t)\Big\{\delta\otimes\Big(-\sum_{k=1}^{\sigma}\frac{2}{l_k^2}\underline{C}-2\underline{D}+\varsigma I_n+\Theta\Big)\\
&+a\eta\big[(\delta A)\otimes\Gamma\big]\big[(A^T\delta)\otimes\Gamma\big]+a\eta I_{nN}+a\big(\hat{S}\\
&+G\big)\otimes\Gamma\Big\}\vartheta(x,t)dx\\
\leqslant&\ \phi\|\vartheta(\cdot,t)\|_2^2,
\end{aligned}
$$

where $\phi=\lambda_M(\Phi+a\big(\hat{S}+G\big)\otimes\Gamma)$.

Similarly, it is easy to obtain $\lim_{t\to+\infty}||\vartheta(\cdot,t)||_2=0$. Therefore, the network (9.23) is synchronized under the adaptive state feedback controller (9.31).

Remark 9.16 Some authors have recently studied the synchronization problem for CRDNNs by exploiting the passivity property [29, 40]. However, in these existing studies [29, 40], the CRDNNs models with single time-delay consisting of several identical certain RDNNs were discussed. In this chapter, we discuss the passivity and synchronization for a CRDNNs with multiple time-delays and uncertain parameters, and several passivity (see Theorems 9.5–9.11) and synchronization (see Theorems 9.14 and 9.15) criteria have been established. In addition, we discuss the relationship between output strict passivity and synchronization of the proposed network (see Remark 9.8 and Remark 9.12).

Remark 9.17 The main challenges in investigating the synchronization and passivity of CRDNNS with multiple time-delays and uncertain parameters arise from the coupling weights and uncertain parameters. In this chapter, serval passivity and synchronization criteria for the proposed network model have been derived by exploiting the matrix theory and inequality techniques [42, 43].

9.5 Numerical Examples

Example 9.18 The CRDNNs with multiple time-delays and uncertain parameters is described by

$$\begin{aligned}\frac{\partial w_i(x,t)}{\partial t} &= C\Delta w_i(x,t) - Dw_i(x,t) + Hf(w_i(x,t)) + Bu_i(x,t) + J \\ &\quad + a\sum_{j=1}^{6} A_{ij}\Gamma w_j(x,t-\tau_1) + a\sum_{j=1}^{6} A_{ij}\Gamma w_j(x,t-\tau_2) \\ &\quad + a\sum_{j=1}^{6} A_{ij}\Gamma w_j(x,t-\tau_3),\end{aligned} \tag{9.39}$$

where $i = 1, 2, \ldots, 6$, $a = 0.2$, $f_k(\varepsilon) = (|\varepsilon+1| - |\varepsilon-1|)/4$, $k = 1, 2, 3$, $\tau_1 = 0.1$, $\tau_2 = 0.2$, $\tau_3 = 0.3$, $\Omega = \{x| - 0.5 < x < 0.5\}$, $J = (0.6, 0.4, 0.8)^T$,

$$\begin{cases} C_I := \{C = \text{diag}(c_i) : 0.9 \leqslant c_1 \leqslant 1, 0.7 \leqslant c_2 \leqslant 0.8,\ 0.9 \\ \qquad \leqslant c_3 \leqslant 1.1\}, \\ D_I := \{D = \text{diag}(d_i) :\ 0.6 \leqslant d_1 \leqslant 0.7,\ 0.6 \leqslant d_2 \leqslant 0.8, \\ \qquad 1 \leqslant d_3 \leqslant 1.5\}, \\ H_I := \{H = (h_{ij})_{3\times 3} : 0.4 \leqslant h_{11} \leqslant 0.5, 0.5 \leqslant h_{12} \leqslant 0.6, \\ \qquad 0.6 \leqslant h_{13} \leqslant 0.7,\ 0.5 \leqslant h_{21} \leqslant 0.6,\ 0.3 \leqslant h_{22} \leqslant \\ \qquad 0.4,\ 0.4 \leqslant h_{23} \leqslant 0.5, 0.2 \leqslant h_{31} \leqslant 0.3, 0.1 \leqslant h_{32} \\ \qquad \leqslant 0.2, 0.7 \leqslant h_{33} \leqslant 0.8\}, \end{cases}$$

$$B = \begin{pmatrix} 0.2 & 0.3 \\ 0.1 & 0.4 \\ 0.3 & 0.2 \end{pmatrix}, \quad \Gamma = \begin{pmatrix} 0.5 & 0 & 0 \\ 0 & 0.6 & 0 \\ 0 & 0 & 0.4 \end{pmatrix},$$

$$A = \begin{pmatrix} -0.1 & 0.1 & 0 & 0 & 0 & 0 \\ 0.2 & -0.4 & 0.2 & 0 & 0 & 0 \\ 0 & 0.2 & -0.8 & 0.6 & 0 & 0 \\ 0 & 0 & 0.1 & -0.2 & 0.1 & 0 \\ 0 & 0 & 0 & 0.3 & -0.4 & 0.1 \\ 0 & 0 & 0 & 0 & 0.5 & -0.5 \end{pmatrix}.$$

Obviously, $f_k(\cdot)(k = 1, 2, 3)$ satisfies the Lipschitz condition with $\theta_k = 0.5$. Exploiting the MATLAB function NULL, the positive vector $\kappa = (5, 2.5, 2.5, 15, 5, 1)^T$ that satisfies $\kappa^T A = 0$ can be obtained.

The output vector $y_i(x, t) \in \mathbb{R}^3$ is selected as follows:

$$y_i(x, t) = E\vartheta_i(x, t) + Fu_i(x, t), i = 1, 2, \ldots, 6,$$

where

$$E = \begin{pmatrix} 0.1 & 0 & 0 \\ 0 & 0.2 & 0 \\ 0 & 0 & 0.1 \end{pmatrix}, F = \begin{pmatrix} 0.2 & 0.2 \\ 0.2 & 0.3 \\ 0.3 & 0.2 \end{pmatrix}.$$

Case 1: By taking advantage of the MATLAB YALMIP Toolbox, the following matrices P and $\hat{S}$ that satisfy (9.19) can be obtained:

$$P = I_6 \otimes \begin{pmatrix} 9.8869 & 10.5599 \\ 0.9050 & 1.2271 \\ 19.8072 & 12.9063 \end{pmatrix},$$

$$\hat{S} = \begin{pmatrix} -95.0888 & 95.0888 & 0 & 0 & 0 & 0 \\ 95.0888 & -189.0255 & 93.9367 & 0 & 0 & 0 \\ 0 & 93.9367 & -198.9734 & 105.0366 & 0 & 0 \\ 0 & 0 & 105.0366 & -219.8938 & 114.8571 & 0 \\ 0 & 0 & 0 & 114.8571 & -235.0246 & 120.1674 \\ 0 & 0 & 0 & 0 & 120.1674 & -120.1674 \end{pmatrix}.$$

In accordance with Theorem 9.9, the network (9.17) is passive.

Case 2: By taking advantage of the MATLAB YALMIP Toolbox, the following matrices P, $\hat{S}$ and F_1 that satisfy the condition of Theorem 9.10 can be obtained:

$$P = I_6 \otimes \begin{pmatrix} 7.1901 & 7.7767 \\ 0.1745 & 0.4682 \\ 14.6358 & 9.4744 \end{pmatrix},$$

$$F_1 = I_6 \otimes \begin{pmatrix} 3.6024 & 2.7615 \\ 2.7615 & 2.2429 \end{pmatrix},$$

$$\hat{S} = \begin{pmatrix} -70.3644 & 70.3644 & 0 & 0 & 0 & 0 \\ 70.3644 & -149.5851 & 79.2208 & 0 & 0 & 0 \\ 0 & 79.2208 & -178.7847 & 99.5640 & 0 & 0 \\ 0 & 0 & 99.5640 & -193.7822 & 94.2183 & 0 \\ 0 & 0 & 0 & 94.2183 & -189.9096 & 95.6913 \\ 0 & 0 & 0 & 0 & 95.6913 & -95.6913 \end{pmatrix}.$$

In accordance with Theorem 9.10, the network (9.17) is input strictly passive.

Case 3: By taking advantage of the MATLAB YALMIP Toolbox, the following matrices P, $\hat{S}$ and F_2 that satisfy the condition of Theorem 9.11 can be obtained:

$$P = I_6 \otimes \begin{pmatrix} 8.3063 & 8.7570 \\ 0.2278 & 0.5629 \\ 17.2282 & 11.2000 \end{pmatrix},$$

$$F_2 = I_6 \otimes \begin{pmatrix} 18.6117 & -0.3044 & 0.7080 \\ -0.3044 & 0.4828 & 0.1412 \\ 0.7080 & 0.1412 & 25.0195 \end{pmatrix},$$

$$\hat{S} = \begin{pmatrix} -177.1641 & 177.1641 & 0 & 0 & 0 & 0 \\ 177.1641 & -376.5800 & 199.4159 & 0 & 0 & 0 \\ 0 & 199.4159 & -442.5999 & 243.1840 & 0 & 0 \\ 0 & 0 & 243.1840 & -470.0446 & 226.8606 & 0 \\ 0 & 0 & 0 & 226.8606 & -451.9288 & 225.0682 \\ 0 & 0 & 0 & 0 & 225.0682 & -225.0682 \end{pmatrix}.$$

In the light of Theorem 9.11, the network (9.17) is output strictly passive. Fig. 9.1 presents the evolution processes of the norms of error vectors $\vartheta_i(x,t)$, output vectors $y_i(x,t)$ and input vectors $u_i(x,t)$, $i = 1, 2, 3, 4, 5, 6$. Fig. 9.2 illustrates the evolution tendencies of adaptive feedback gains.

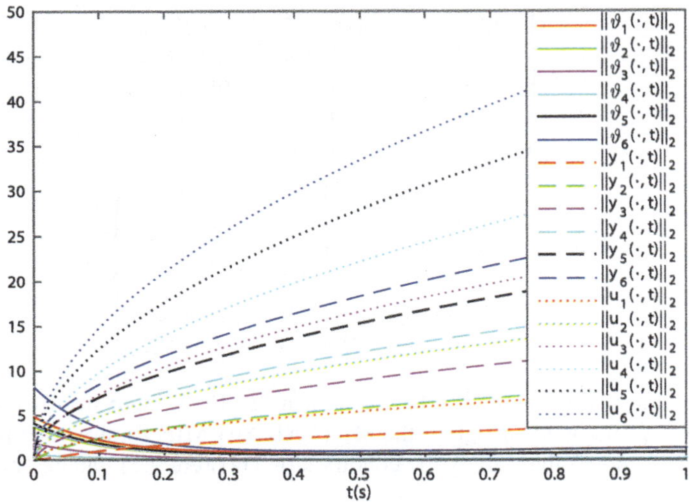

Fig. 9.1 $\|\vartheta_i(\cdot,t)\|_2$, $\|y_i(\cdot,t)\|_2$, $\|u_i(\cdot,t)\|_2$, $i = 1, 2, \ldots, 6$

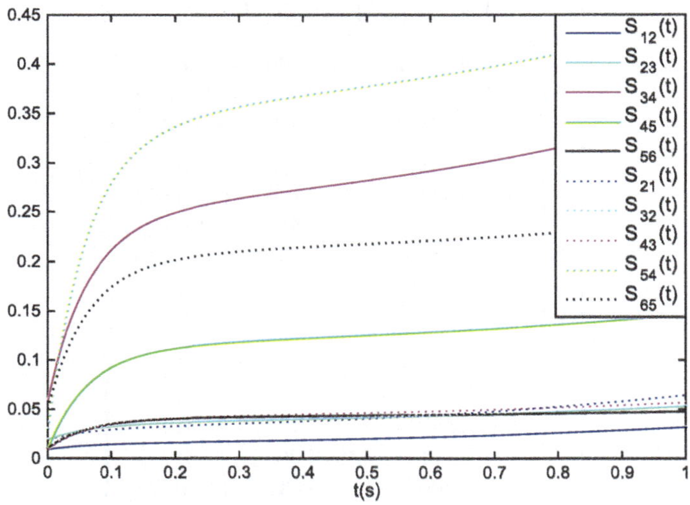

Fig. 9.2 Adaptive feedback gains

Example 9.19 Letting $u_i(x,t)=0$ in the network model (9.39), we have

$$\begin{aligned}\frac{\partial w_i(x,t)}{\partial t} &= C\Delta w_i(x,t) - Dw_i(x,t) + Hf(w_i(x,t)) + J\\ &\quad + a\sum_{j=1}^{6} A_{ij}\Gamma w_j(x,t-\tau_1) + a\sum_{j=1}^{6} A_{ij}\Gamma w_j(x,t-\tau_2)\\ &\quad + a\sum_{j=1}^{6} A_{ij}\Gamma w_j(x,t-\tau_3),\end{aligned}$$

where $i=1,2,\ldots,6$.

By taking advantage of the MATLAB YALMIP Toolbox, the following matrix $\hat{S}$ satisfying the condition in Theorem 9.15 can be obtained:

$$\hat{S} = \begin{pmatrix} -126.1812 & 126.1812 & 0 & 0 & 0 & 0\\ 126.1812 & -249.4128 & 123.2316 & 0 & 0 & 0\\ 0 & 123.2316 & -251.6667 & 128.4351 & 0 & 0\\ 0 & 0 & 128.4351 & -256.0715 & 127.6363 & 0\\ 0 & 0 & 0 & 127.6363 & -255.5727 & 127.9364\\ 0 & 0 & 0 & 0 & 127.9364 & -127.9364 \end{pmatrix}.$$

In accordance with Theorem 9.15, the network (9.23) is synchronized under the adaptive state feedback controller (9.31). Fig. 9.3 shows the evolution processes of the norms of error vectors $\vartheta_i(x,t), i=1,2,3,4,5,6$. Evidently, the error vectors $\vartheta_i(x,t), i=1,2,3,4,5,6$ can converge to zero. Fig. 9.4 displays the evolution tendencies of adaptive feedback gains. From Fig. 9.4, we can clearly observe that the feedback gains can converge to finite positive values.

From the adaptive state feedback controllers (9.12) and (9.33), and the conditions of Theorems 9.9–9.11 and 9.15, we can clearly observe that they do not depend on the parameters c_i, d_i and h_{ij}. Therefore, the adaptive state feedback controllers (9.12) and (9.33) and the conditions of Theorems 9.9–9.11 and 9.15 are robust to the uncertainties of the parameters c_i, d_i and h_{ij}.

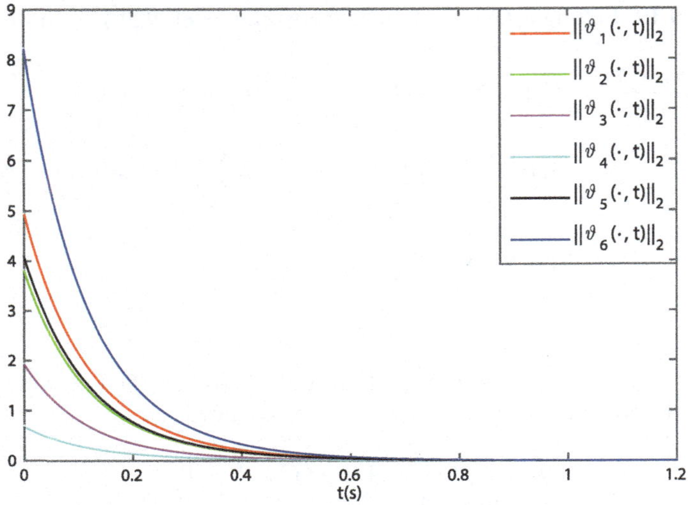

Fig. 9.3 $\|\vartheta_i(\cdot, t)\|_2, i = 1, 2, \ldots, 6$

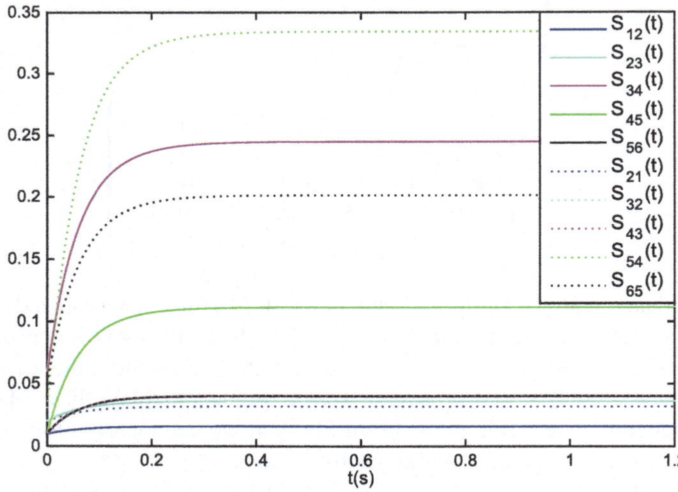

Fig. 9.4 Adaptive feedback gains

9.6 Conclusion

In this chapter, we have proposed a CRDNNs model with multiple time-delays and uncertain parameters, which is completely different from some existing network models. Subsequently, we have acquired several sufficient conditions that ensure the passivity of the proposed network model. Furthermore, an adaptive state feedback controller has also been developed to ensure the passivity of the proposed network model. Similarly, we have also studied the synchronization and adaptive synchronization in such network model. Finally, we have provided two numerical examples to verify the effectiveness of the passivity and synchronization criteria.

References

1. Chen, S., Lim, C. C., Shi, P., & Lu, Z. (2018). Synchronization control for reaction-diffusion FitzHugh-Nagumo systems with spatial sampled-data. *Automatica, 93*, 352–362.
2. Wu, K. N., Sun, H. X., Shi, P., & Lim, C. C. (2018). Finite-time boundary stabilization of reaction-diffusion systems. *International Journal of Robust and Nonlinear Control, 28*(5), 1641–1652.
3. Wei, T., Lin, P., Zhu, Q., Wang, L., & Wang, Y. (2019). Dynamical behavior of nonautonomous stochastic reaction-diffusion neural-network models. *IEEE Transactions on Neural Networks and Learning Systems, 30*(5), 1575–1580.
4. Shen, H., Wang, T., Cao, J., Lu, G., Song, Y., & Huang, T. (2019). Nonfragile dissipative synchronization for Markovian memristive neural networks: A gain-scheduled control scheme. *IEEE Transactions on Neural Networks and Learning Systems, 30*(6), 1841–1853.
5. Xiao, Q., Huang, T., & Zeng, Z. (2019). Global exponential stability and synchronization for discrete-time inertial neural networks with time delays: A timescale approach. *IEEE Transactions on Neural Networks and Learning Systems, 30*(6), 1854–1866.
6. Peng, X., Wu, H., & Cao, J. (2019). Global nonfragile synchronization in finite time for fractional-order discontinuous neural networks with nonlinear growth activations. *IEEE Transactions on Neural Networks and Learning Systems, 30*(7), 2123–2137.
7. Xu, Y., Li, J. Y., Lu, R., Liu, C., & Wu, Y. (2019). Finite-horizon $l_2 - l_\infty$ synchronization for time-varying Markovian jump neural networks under mixed-type attacks: Observer-based case. *IEEE Transactions on Neural Networks and Learning Systems, 30*(6), 1695–1704.
8. Yang, X., Song, Q., Cao, J., & Lu, J. (2019). Synchronization of coupled Markovian reaction-diffusion neural networks with proportional delays via quantized control. *IEEE Transactions on Neural Networks and Learning Systems, 30*(3), 951–958.
9. Lu, J., Ho, D. W. C., Cao, J., & Kurths, J. (2011). Exponential synchronization of linearly coupled neural networks with impulsive disturbances. *IEEE Transactions on Neural Networks, 22*(2), 329–336.
10. Tang, Y., & Wong, W. K. (2013). Distributed synchronization of coupled neural networks via randomly occurring control. *IEEE Transactions on Neural Networks and Learning Systems, 24*(3), 435–447.
11. Zhang, H., Gong, D., Chen, B., & Liu, Z. (2013). Synchronization for coupled neural networks with interval delay: A novel augmented Lyapunov-Krasovskii functional method. *IEEE Transactions on Neural Networks and Learning Systems, 24*(1), 58–70.
12. Yang, X., Cao, J., & Lu, J. (2013). Synchronization of randomly coupled neural networks with Markovian jumping and time-delay. *IEEE Transactions on Circuits and Systems I: Regular Papers, 60*(2), 363–376.
13. Liu, X., Cao, J., Yu, W., & Song, Q. (2016). Nonsmooth finite-time synchronization of switched coupled neural networks. *IEEE Transactions on Cybernetics, 46*(10), 2360–2371.
14. Liu, X., Su, H., & Chen, M. Z. Q. (2016). A switching approach to designing finite-time synchronization controllers of coupled neural networks. *IEEE Transactions on Neural Networks and Learning Systems, 27*(2), 471–482.
15. Wang, J., Zhang, H., Wang, Z., & Shan, Q. (2017). Local synchronization criteria of Markovian nonlinearly coupled neural networks with uncertain and partially unknown transition rates. *IEEE Transactions on Systems, Man, and Cybernetics: Systems, 47*(8), 1953–1964.

16. Chen, H., Shi, P., & Lim, C. C. (2017). Exponential synchronization for Markovian stochastic coupled neural networks of neutral-type via adaptive feedback control. *IEEE Transactions on Neural Networks and Learning Systems*, *28*(7), 1618–1632.
17. Wang, K., Teng, Z., & Jiang, H. (2012). Adaptive synchronization in an array of linearly coupled neural networks with reaction-diffusion terms and time delays. *Communications in Nonlinear Science and Numerical Simulation*, *17*(10), 3866–3875.
18. Wang, J., & Wu, H. (2014). Synchronization and adaptive control of an array of linearly coupled reaction-diffusion neural networks with hybrid coupling. *IEEE Transactions on Cybernetics*, *44*(8), 1350–1361.
19. Wang, J. L., Wu, H. N., & Guo, L. (2014). Novel adaptive strategies for synchronization of linearly coupled neural networks with reaction-diffusion terms. *IEEE Transactions on Neural Networks and Learning Systems*, *25*(2), 429–440.
20. Wang, J. L., Wu, H. N., Huang, T., Ren, S. Y., & Wu, J. (2016). Pinning control for synchronization of coupled reaction-diffusion neural networks with directed topologies. *IEEE Transactions on Systems, Man, and Cybernetics: Systems*, *46*(8), 1109–1120.
21. Wang, J. L., Wu, H. N., Huang, T. W., & Ren, S. Y. (2016). Pinning control strategies for synchronization of linearly coupled neural networks with reaction-diffusion terms. *IEEE Transactions on Neural Networks and Learning Systems*, *27*(4), 749–761.
22. Dharani, S., Rakkiyappan, R., & Park, J. H. (2017). Pinning sampled-data synchronization of coupled inertial neural networks with reaction-diffusion terms and time-varying delays. *Neurocomputing*, *227*, 101–107.
23. Liu, X., Chen, Z., & Zhou, L. (2017). Synchronization of coupled reaction-diffusion neural networks with hybrid coupling via aperiodically intermittent pinning control. *Journal of the Franklin Institute*, *354*(15), 7053–7076.
24. Xu, M., Wang, J. L., & Wei, P. C. (2017). Synchronization for coupled reaction-diffusion neural networks with and without multiple time-varying delays via pinning-control. *Neurocomputing*, *227*, 82–91.
25. Wang, S. X., Huang, Y. L., & Xu, B. B. (2017). Pinning synchronization of spatial diffusion coupled reaction-diffusion neural networks with and without multiple time-varying delays. *Neurocomputing*, *227*, 92–100.
26. Wu, Z. G., Shi, P., Shu, Z., Su, H., & Lu, R. (2017). Passivity-based asynchronous control for Markov jump systems. *IEEE Transactions on Automatic Control*, *62*(4), 2020–2025.
27. Li, N., & Cao, J. (2016). Passivity and robust synchronisation of switched interval coupled neural networks with time delay. *International Journal of Systems Science*, *47*(12), 2827–2836.
28. Ren, S. Y., Wu, J., Wang, S. X., & Huang, Y. (2018). Passivity and pinning control of coupled neural networks with and without time-varying delay. *Transactions of the Institute of Measurement and Control*, *40*(9), 2708–2717.
29. Wang, J. L., Wu, H. N., Huang, T., & Ren, S. Y. (2015). Passivity and synchronization of linearly coupled reaction-diffusion neural networks with adaptive coupling. *IEEE Transactions on Cybernetics*, *45*(9), 1942–1952.
30. Xu, B. B., Huang, Y. L., Wang, J. L., Wei, P. C., & Ren, S. Y. (2016). Passivity of linearly coupled reaction-diffusion neural networks with switching topology and time-varying delay. *Neurocomputing*, *182*, 274–283.
31. Xu, B. B., Huang, Y. L., Wang, J. L., Wei, P. C., & Ren, S. Y. (2016). Passivity of linearly coupled neural networks with reaction-diffusion terms and switching topology. *Journal of the Franklin Institute*, *353*(8), 1882–1898.
32. Chen, W. Z., Huang, Y. L., & Ren, S. Y. (2018). Passivity and synchronization of coupled reaction-diffusion cohen-grossberg neural networks with state coupling and spatial diffusion coupling. *Neurocomputing*, *275*, 1208–1218.
33. Huang, Y. L., Xu, B. B., & Ren, S. Y. (2018). Analysis and pinning control for passivity of coupled reaction-diffusion neural networks with nonlinear coupling. *Neurocomputing*, *272*(10), 334–342.

34. Wang, J., Wu, H., Huang, T., Ren, S., & Wu, J. (2017). Passivity analysis of coupled reaction-diffusion neural networks with Dirichlet boundary conditions. *IEEE Transactions on Systems, Man, and Cybernetics: Systems*, *47*(8), 2148–2159.
35. Ren, S. Y., Wu, J., & Wei, P. C. (2017). Passivity and pinning passivity of coupled delayed reaction–diffusion neural networks with Dirichlet boundary conditions. *Neural Processing Letters*, *45*(3), 869–885.
36. Zhao, Y. P., He, P., Nik, H. S. & Ren, J. (2015). Robust adaptive synchronization of uncertain complex networks with multiple time-varying coupled delays. *Complexity*, *20*(6), 62–73.
37. Wang, J. L., Xu, M., Wu, H. N., & Huang, T. (2018). Finite-time passivity of coupled neural networks with multiple weights. *IEEE Transactions on Network Science and Engineering*, *5*(3), 184–197.
38. Qin, Z., Wang, J. L., Huang, Y. L., & Ren, S. Y. (2018). Analysis and adaptive control for robust synchronization and $\mathcal{H}_\infty$ synchronization of complex dynamical networks with multiple time-delays. *Neurocomputing*, *289*(10), 241–251.
39. Willems, J. C. (1972). Dissipative dynamical systems part I: General theory. *Archive for Rational Mechanics and Analysis*, *45*(5), 321–351.
40. Wang, J. L., Wu, H. N., & Huang, T. (2015). Passivity-based synchronization of a class of complex dynamical networks with time-varying delay. *Automatica*, *56*, 105–112.
41. Wang, J. L., Wu, H. N., Huang, T., Ren, S. Y., & Wu, J. (2017). Passivity of directed and undirected complex dynamical networks with adaptive coupling weights. *IEEE Transactions on Neural Networks and Learning Systems*, *28*(8), 1827–1839.
42. Wu, X., Tang, Y., Cao, J., & Mao, X. (2018). Stability analysis for continuous-time switched systems with stochastic switching signals. *IEEE Transactions on Automatic Control*, *63*(9), 3083–3090.
43. Tang, Y., Gao, H., Zhang, W., & Kurths, J. (2015). Leader-following consensus of a class of stochastic delayed multi-agent systems with partial mixed impulses. *Automatica*, *53*, 346–354.

Correction to: Passivity of Complex Dynamical Networks

Correction to:
J.-L. Wang et al., *Passivity of Complex Dynamical Networks*, https://doi.org/10.1007/978-981-33-4287-3

In the original version of the book, the following belated corrections have been incorporated:

(a) Reference 18 has been changed from "Neurocomputing, 7(9), 1782–1788" to "Neurocomputing, 72(7–9), 1782–1788" of Chap. 1.

(b) Corrections have been incorporated in Sects. 4.3.2 and 4.4.2 of Chap. 4.

(c) Corrections have been incorporated in Sect. 5.3.2 of Chap. 5.

(d) Corrections have been incorporated in Sect. 6.4.2 of Chap. 6.

(e) Corrections have been incorporated in Eqs. 7.4, 7.5, 7.7, 7.9, 7.16, 7.17 and 7.19 of Chap. 7.

(f) In Chap. 8, the sentence "As a natural extension of the existing models, we introduce a new CDN model with time delay which describes a cooperative interaction of N-species that benefits each other in a bounded habitat Θ." has been changed to "As a natural extension of the existing models, we introduce a new CDN model with time-delay which describes a cooperative interaction of N -species that benefits each other in a bounded habitat Θ."

(g) In Chap. 9, (a) the sentence "In [27], the authors presented a switched coupled uncertain NN, and derived several delay-independent and delay-dependent

---

The updated version of these chapters can be found at
https://doi.org/10.1007/978-981-33-4287-3_1
https://doi.org/10.1007/978-981-33-4287-3_4
https://doi.org/10.1007/978-981-33-4287-3_5
https://doi.org/10.1007/978-981-33-4287-3_6
https://doi.org/10.1007/978-981-33-4287-3_7
https://doi.org/10.1007/978-981-33-4287-3_8
https://doi.org/10.1007/978-981-33-4287-3_9
https://doi.org/10.1007/978-981-33-4287-3

J.-L. Wang et al., *Passivity of Complex Dynamical Networks*,
https://doi.org/10.1007/978-981-33-4287-3_10

passivity criteria for such a network model." has been changed to "In [27], the authors presented a switched coupled uncertain "NNs", and derived several delay-independent and delay-dependent passivity criteria for such a network model."; (b) the placement of Equation (9.12) has been corrected; (c) the sentence "Figure 9.1 presents the evolution processes of the norms of error vectors ϑi (x , t), output vectors yi (x , t) and input vectors ui (x , t), i = 1, 2, 3, 4, 5, 6. Figure 9.2 illustrates the evolution tendencies of adaptive feedback gains." has been changed to "Fig. 9.1 presents the evolution processes of the norms of error vectors ϑi (x , t), output vectors yi (x , t) and input vectors ui (x , t), i = 1, 2, 3, 4, 5, 6. Fig. 9.2 illustrates the evolution tendencies of adaptive feedback gains."; and (d) Chap. 9, Page 239, the sentence "Figure 9.3 shows the evolution processes of the norms of error vectors ϑi (x , t), i = 1, 2, 3, 4, 5, 6. Evidently, the error vectors ϑi (x , t), i = 1, 2, 3, 4, 5, 6 can converge to zero. Figure 9.4 displays the evolution tendencies of adaptive feedback gains." has been changed to "Fig. 9.3 shows the evolution processes of the norms of error vectors ϑi (x , t), i = 1, 2, 3, 4, 5, 6. Evidently, the error vectors ϑi (x , t), i = 1, 2, 3, 4, 5, 6 can converge to zero. Fig. 9.4 displays the evolution tendencies of adaptive feedback gains."

The book and the chapters have been updated with the changes.

[illegible]
[illegible]
[illegible]
[illegible]

Printed and bound by CPI Group (UK) Ltd, Croydon, CR0 4YY

[illegible]

The manufacturer's authorised representative in the EU is Springer Nature Customer Service Centre GmbH, Europaplatz 3, 69115 Heidelberg, Germany. If you have any concerns regarding our products, please contact ProductSafety@springernature.com

Printed and bound by CPI Group (UK) Ltd, Croydon, CR0 4YY
30/09/2026
02225288-0003